Commercial Poultry Farming

Commercial Poultry Farming

by

T. Burr Charles

Professor and Head of the Department of Poultry Husbandry
University of New Hampshire
(Formerly Instructor in Poultry Husbandry, Cornell University and Associate Professor of Poultry Husbandry at the Pennsylvania State College)

&

Homer O. Stuart

Professor and Head of the Department of Poultry Husbandry
Extension Director and County Agent Leader, and Supervisor of the Rhode Island Egg Laying Test, Rhode Island State College
(Formerly Instructor in Poultry Husbandry, University of New Hampshire)

BIOTECH

2022

BIOTECH BOOKS®

Charles, Thomas Burr (1890-1967)
Stuart, Homer Ohliger (1903-1982)

2022 Reserved
ISBN 978-81-7622-221-1

Published by
BIOTECH BOOKS
4762-63/23, Ansari Road, Darya Ganj,
New Delhi-110002 (India)
Phone : 011 4300 3222
E-mail: biotechbooks@yahoo.co.in

PRINTED IN INDIA

TABLE OF CONTENTS

CHAPTER I

CHAPTER II

CHAPTER III

CHAPTER IV

CHAPTER V

CHAPTER VI

CHAPTER VII

CHAPTER VIII

CHAPTER IX

CHAPTER X

CHAPTER XI

CHAPTER XII

CHAPTER XVI

CHAPTER XVII

CHAPTER XVIII

CHAPTER XIX

CHAPTER XX

CHAPTER XXI

CHAPTER XXII

ACKNOWLEDGMENTS

The authors wish to express their appreciation for the criticisms, suggestions, and confirmation of subject matter to Dr. R. C. Ringrose and Dr. C. W. Hess, University of New Hampshire, and to Dr. J. P. Delaplane and Thomas C. Higgins, Rhode Island State College.

We are indebted to the American Poultry Association for permission to use certain information from the Standard of Perfection; to the American Poultry Journal, Hatchery Tribune, and The Turkey World, for the use of illustrations and tables.

In addition, we acknowledge use of illustrative material taken from "Practical Poultry Management" by Rice and Botsford and published by John Wiley & Sons. This has been used by special permission from the authors and publishers. We express our gratitude to the many individuals and commercial organizations which have contributed descriptive illustrations and granted permission for their use. Source of material in each individual case is indicated. The authors also appreciate any other aid which has been rendered by individuals not specifically mentioned above.

INTRODUCTION

Poultry production is not a business in which everyone may engage successfully. Too often people start in the business under the illusion that little is to be done other than to see that the birds have feed and water, and then to gather and sell their eggs.

Before engaging in a poultry enterprise it is advisable to investigate the profitableness of the business in the locality concerned, and the markets available. In addition, capital must be correctly proportioned to the estimated size of flock to be handled. Therefore, it is necessary not only to consider the adaptability of the person concerned, but also the location and capital requirement of the poultry business.

In many sections poultry-keeping is maintained to a large extent as a side-line proposition. In others, it has become more extensively commercialized. The problems to the individual commercialized producer are more serious because of the greater financial risk. With commercialization and greater intensification have come disease problems that, although new, cause more acute financial losses due to the numbers of birds involved. Chances of disease-spread are materially enhanced. It is, of course, obvious that a person having initiative, experience and who is well informed has a greater chance for success than those less fortunately equipped.

Educational agencies can help. They offer a source for the solution of problems confronting the poultry keeper. It is to the distinct advantage of poultrymen to avail themselves of such information and to keep abreast of the time by being well informed on the development of new practices. The object of this book, therefore, is an attempt to place at the disposal of poultrymen at large, information relative to established and dependable practices. It is designed to aid the beginner as well as the established poultryman and may serve as a reference for the solution of management problems of others. Naturally, it is impossible to cover all the various phases of poultry keeping in extensive detail. Where some specific additional information is desired it is advisable to consult state agricultural colleges or universities and textbooks dealing extensively with but one phase of the business.

CHAPTER 1

INCUBATION

Subject Matter

Artificial incubation has been practiced for centuries. The strides it has made during that time have been remarkable. It has developed from the Egyptian incubator rooms with their need for diligent and constant attention to the mammoth machines of the present day with their numerous automatic control devices. Naturally improvements will be made from time to time in the future, but the poultry industry now owes much in the way of its commercialization to the adoption and development of artificial incubation.

Selecting Eggs for Hatching Purposes

The selection of eggs for hatching purposes is becoming more widely practiced. This is in line with the lessons of both research and field observation. Certainly too little attention was given in the past to the quality of eggs for producing salable chicks. So long as the eggs would hatch they were considered good enough.

Quality is a keynote of businesses that are trying not only to establish markets for their commodities but to hold those markets against increasing competition. In the poultry industry this necessitates not only the production of good-sized chicks of high livability, but of chicks with the inherited ability to lay a large number of eggs of good size, color, shape and texture. Consequently, to have birds that will lay a large percentage of eggs meeting these qualifications, care is obviously necessary in the selection of the eggs that are to produce such individuals.

Possible results of selecting hatching eggs are:

1. Better size of resultant chicks.
2. More uniformity.
3. Greater potential production and breeding value of stock.
4. More uniformity in color of eggs from pullet offspring.
5. Less variation in shape of pullet eggs.
6. Larger eggs from pullets.
7. Increased hatchability through the elimination of cracked eggs, blind checks, thin-shelled and dirty eggs.
8. As an accumulative effect of the above points, a larger percentage of the eggs from the ensuing generation will be suitable for hatching purposes.

Progress may be made along one or all of these lines by establishing a definite program for standardization of hatching eggs. In many instances rapid advancement has been made by following such a procedure. An additional step may be taken by trapnesting the birds for at least a two or three months' period in order to establish the identity of the females which consistently produce eggs of good size, type and color. Such females would then be used in the breeding unit.

Basis for Egg Selection:

Eggs should be selected for shape, size, color, soundness of shell and cleanliness. It is unwise to set eggs of freakish shape as this procedure seldom gives a good hatch. A good-shaped egg should be the ideal, and extremes of length and width should be avoided.

It has been definitely established that size of egg is an inherited characteristic. Small egg size is incompletely dominant over large egg size and there is a tendency for decreased egg size with the increased productivity of domesticated flocks. The problem of the breeder is to manage these two factors so as to secure a high average production of good-sized eggs. A standard sometimes used is to select from birds in their first laying year, eggs that will weigh 24 ounces or over to the dozen, and on a comparable basis eggs that will weigh 26 ounces or over per dozen from hens in their second or later years of production.

Color is also an inherited characteristic. Primary attention should be given to market requirements, but within the breed selection should aim at reasonably uniform color. In white eggs avoid tinted shells and in brown eggs avoid extremes and wide variations in color.

Setting cracked eggs, blind checks, and eggs having porous or rough shells is inadvisable. Eggs should be reasonably clean or if dirty, should be cleaned to remove possible sources of contamination and substances which interfere with evaporation and respiration.

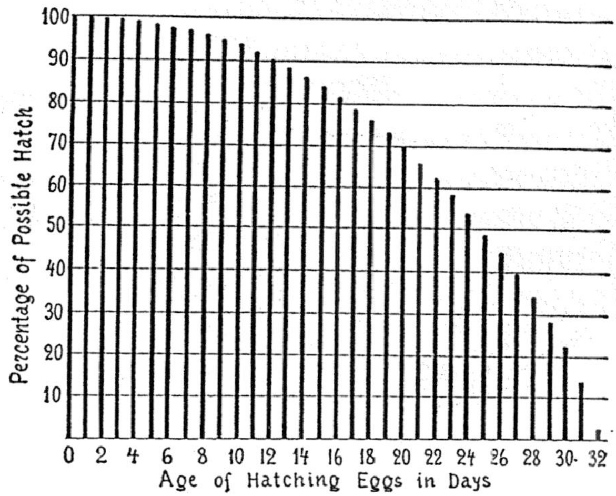

Fig. 1—Effect of age of eggs on their hatchability. Based on the experimental data of Waite.
[Courtesy of New England Poultryman]

Care and Holding of Hatching Eggs:

While the egg is in the oviduct of the hen embryonic development begins; so incubation should not be unnecessarily delayed. When eggs are held for incubation purposes it is advisable to keep them at a temperature of 50 to 60 degrees Fahrenheit to prevent further growth of the embryo during this period.

Eggs are held for incubation purposes for two main reasons: (1) to obtain a sufficient number of proper quality to supply the desired number of chicks, and (2) to follow a definite hatching schedule.

Hatchability of hens' eggs decreases rapidly after 10 days, decreasing slightly previous to this time. The sooner they can be placed in the incubator, the better will be the hatching results. The seven-day holding period is in general usage as it permits efficient use of available hatching trays and the cleaning of the trays between hatches.

Eggs can be held for incubation most easily in 30-dozen cases. They should be placed in the cases with the pointed end down, as when packing for market.

Eggs should be stored in a well-ventilated room that is free from the odors of oils and gases.

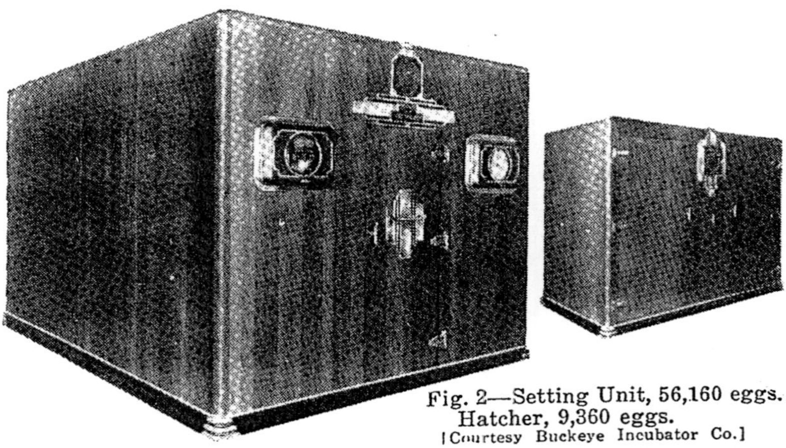

Fig. 2—Setting Unit, 56,160 eggs.
Hatcher, 9,360 eggs.
[Courtesy Buckeye Incubator Co.]

Incubation Principles

Successful incubation depends on a number of conditions, although the principal ones are: (1) correct temperature, (2) proper humidity, (3) sufficient ventilation, (4) turning of eggs, (5) sanitation, (6) adequate temperature of incubation room, (7) quality of eggs, (8) quality and condition of breeding stock, and (9) personal attention and interest.

Proper Temperature:

Irrespective of the type of incubator in use it is wise to follow the directions of the incubator manufacturer, unless experience under your conditions indicates the advisability of a change.

In cabinet-type incubators it is customary to maintain an incubating temperature of 99½ to 99¾ degrees Fahrenheit. Where separate hatchers are to be utilized care must be taken to follow the manufacturer's directions carefully as to temperature and humidity.

Results are indicative of the suitability of the temperature. A premature hatch indicates that the incubator was operated at too high a temperature. A late hatch may point to a temperature maintained at too low a level. Corrective adjustments should be made for ensuing hatches. Uniformity of temperature is highly important; variation should be small since normal embryonic development is dependent on normal incubation temperature.

All thermometers need to be tested by comparison with a clinical thermometer before each hatching season to ascertain if readings are correct. If incorrect, allowances must be made.

Fig. 3—Setting Unit on left; Hatcher on right; Total Capacity, 19,936 hen eggs. [Courtesy Petersime Incubator Co.]

Proper Humidity and Adequate Ventilation:

Embryonic development depends on evaporation of the moisture from the egg contents, a supply of oxygen, and the removal of gases formed by respiration. With the embryos consuming oxygen and giving off carbon dioxide, the quantity of the latter in the incubator increases materially unless sufficient ventilation is provided.

When the amount of carbon dioxide in the air approaches or exceeds two per cent, the hatchability of eggs decreases rapidly. On the 21st day of incubation approximately 140 to 150 times as much air is required as on the first day. This emphasizes the need for good ventilation and particularly the necessity of keeping the ventilation outlets well opened during the later stages of incubation.

Under normal incubating conditions it is best to allow enough ventilation so that the eggs will lose 13 to 16 per cent of their original weight in 18 days.

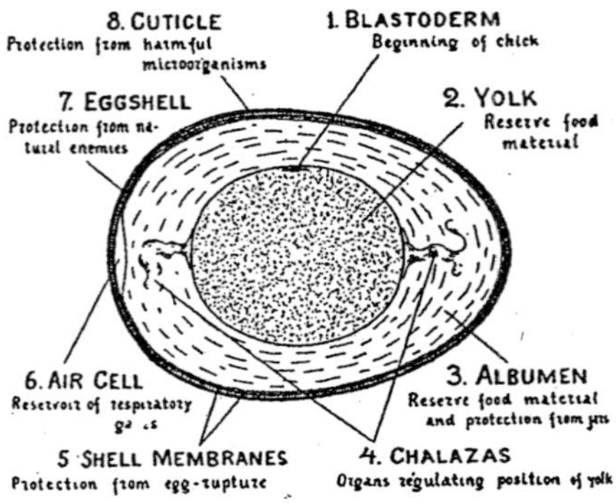

Fig. 4—Diagram of a Fertilized Egg.

1—Blastoderm; 2—yolk; 3—albumen; 4—chalazas; 5—shell membranes; 6—air cells; 7—eggshell; 8—cuticle.

Each part of the egg carries on a definite service in the development of the embryo.

[Courtesy of N. E. Poultryman, from Cornell Bul. 205.]

This indicates reasonable leeway in conditions of ventilation and humidity within which good results may be secured.

Evaporation should occur rapidly enough so that the depth of the air cell measures approximately as follows:

Period	Depth of air cell in inches
Third day	2/16—3/16 inch
Eighth day	3/16—4/16 inch
Fifteenth day	5/16—8/16 inch
Eighteenth day	8/16—or over

When a wet-bulb thermometer is used for recording humidity conditions, a reading of 75 to 85 degrees appears to indicate correct conditions. Relative humidity of 50 to 60 per cent also gives satisfactory results. The humidity of air going into the incubator will largely determine the extra quantity of moisture required within the machine. Ventilation and humidity determine evaporation. It is advisable to follow the incubator manufacturer's recommendations, unless experience under your own conditions justifies a change.

Modern incubators are provided with satisfactory ventilators, except where extremely high room temperatures are encountered. When the cubic air content of the incubating room is relatively small compared to the cubical contents of the incubator, frequent replenishment of the air in the room is an advisable precaution.

Romanoff (1936) stated "that for chicken eggs the relative humidity of 60 per cent during the largest part of incubation and 65 per cent at the end of hatching was found to be most favorable. For the eggs of other species of birds, however, these values have to be slightly modified. Turkey and pheasant eggs require higher humidity at the first part of incubation and somewhat lower at the end, while quail eggs require relatively high humidity throughout or even slightly raised humidity at the end of incubation."

Barott (1937), in reporting results of research at Beltsville, states "The optimum conditions for incubation (hens' eggs) and for growth of the chick embryo, as determined by energy metabolism, so far as the physical factors are concerned, are temperature 100° F., relative humidity 61 per cent, oxygen 21 per cent and carbon dioxide below 0.5 per cent." He further states "the temperature should not vary more than 0.5 per cent from the optimum. The relative humidity may vary 10 per cent with very little detrimental effect. The carbon dioxide should not be allowed to accumulate to a greater extent than 1 per cent or the oxygen to drop below 21 per cent."

Turning of Eggs:

Eggs should be turned daily, starting with the first day they are placed in the incubator. The number of turnings daily is optional, but for best results eggs should be turned four to six times daily.

Eggs may be turned often during the early stages of incubation, but the number of turnings should be decreased as the hatching date approaches. The developing embryo as it grows assumes a new position in the egg with greater difficulty. In commercial hatcheries using large cabinet-type machines it is often impracticable to vary the number of turnings at different stages of incubation. Consequently, if a uniform number of daily turnings is decided upon, it must be suitable to eggs in the latter period of incubation.

It is desirable to give the eggs as complete a turn as the device permits; that is, eggs should not be permitted to remain in the original position. Caution should be exercised in turning trays so as not to jar the eggs out of position, or breakage may occur. Where trays are provided with locks to hold them in place while in a slanted position, care should be taken to be sure that trays are properly latched before the turning device is used.

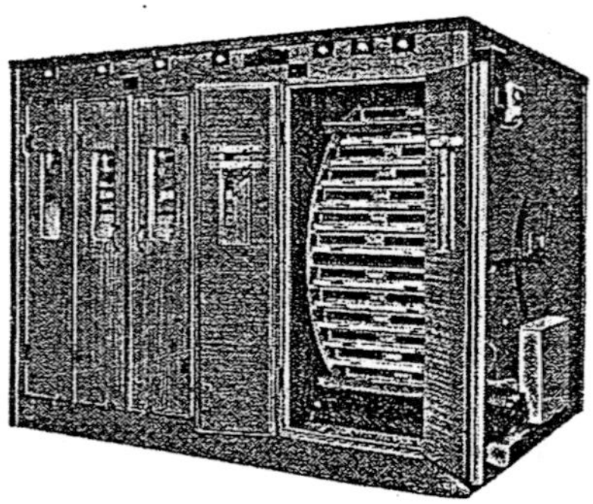

Fig. 5—Hatchibator designed especially for hatching Duck eggs. Drum capacity, setting 12,000; hatching 3,000.

[Courtesy Petersime Incubator Co.]

Sanitation:

Incubation cannot be deemed successful unless the livability of the chicks so produced is satisfactory. Many times good hatches may be secured, yet mortality may be excessive due to faulty incubator management. For instance, the spread of Pullorum disease may be excessive in incubation unless precautionary disinfection is applied to minimize the spread.

Before the incubation season starts all incubators should be cleaned, scrubbed out, and disinfected. Many disinfectant compounds on the market will give satisfactory results. Developments indicate that formaldeyde-gas treatment is valuable in checking the spread of disease.

As far as practical, the amount of fluff or down in the machine should be reduced by artificial means. In this respect vacuum cleaners have been resorted to in many instances and down or "fluff" catchers are on the market. Increasing the humidity is also helpful. Down or "fluff" catchers, while they may be ideal for reducing the spread of this material through the incubator, may, however, serve as an additional source of trouble. As they become clogged with down, it is necessary to clean them thoroughly and at regular intervals; also in those types where porous cloth is used to catch the down, this cloth should be renewed when necessary, to prevent these "down eliminators" from shutting off air circulation in the machine, and thus interfering with air movement and renewal of oxygen for the embryos.

Unless incubators are cleaned and disinfected between hatches when Pullorum and other diseases are present, the accumulative effect as the hatching season progresses is likely to result in excessive chick mortality. Chicks ought to be removed from the hatching compartment as soon as they are dried off sufficiently, so that they will not be subjected to disease spread of this nature longer than necessary. They should be placed either in shipping boxes, storage brooders or brooder houses.

Fig. 6—Bundy Separate Hatcher and Incubating Unit. Trays pulled out to show hatching end of unit.

[Courtesy Bundy Incubator Co.]

1. *Formaldehyde Gas Treatment*—Formaldehyde is a strong, non-poisonous, disinfectant. The gas is liberated by interaction between potassium permanganate and formalin. The correct dosage for each 20 cubic feet of incubator content has been found to be approximately one-fourth ounce of a 40-per-cent formaldehyde solution mixed with one-eighth ounce of potassium permanganate. This is based on .35 cubic centimeters of a 40-per-cent formaldehyde solution and .175 grams of potassium permanganate per cubic foot of space.

Thus by determining the cubical contents of the compartment or machine, the total amount of disinfectant needed can be learned. Due to the quantity of heat generated, the permanganate should be put in an earthenware vessel, partially submerged in a pan of water, before the formalin is poured in. This treatment should be used principally between hatches.

2. *Other Disinfectants*—Various products are on the market and are satisfactory for general disinfection. The disinfecting power of these products varies greatly. They should be purchased with the relative germicidal value and price in mind.

Temperature of Incubation Room:

Under average incubating conditions a room temperature of 65 to 70 degrees is maintained so that the air entering the incubator is preheated.

Auxiliary heat in the room furnishes insurance against excessive drop in incubator temperature when incubator heating equipment fails to function, and also conserves the fuel cost.

Fig. 7—Humidaire Separate Hatching Unit. Heating elements, moisture pan, and circulating fans in top of unit.

[Courtesy New Madison Incubators, Inc.]

Better hatches are normally secured where the incubator room is heated as it is easier to maintain and regulate the incubator temperature. Working conditions are also more satisfactory for the attendant, in setting and candling eggs and in the routine incubation management. In addition, chicks being removed from the machine are less likely to be chilled.

In hot weather provision should be made for holding the temperature of the room at a satisfactory level. This can be accomplished by insulation of the building, increased room ventilation and, where necessary, the provision of cooling equipment.

Quality of Eggs

No matter how accurate the management of the incubator, maximum results depend on hatching eggs of high quality. Eggs weakened through improper handling or of poor fertility cannot be expected to produce large numbers of strong, livable chicks. High-quality eggs properly incubated will produce high-quality chicks. Quality is the basis for success. All the factors of egg selection should be adhered to in choosing eggs for incubation purposes.

Quality and Condition of the Breeding Stock

For quality hatching eggs, close supervision should be given the breeding flocks. The flocks should be carefully selected for health and vigor and mated to males possessing desired characteristics. Roughly, with dual-purpose breeds about seven or eight males should be mated to 100 females. With the light breeds fewer males may be used.

The health of the birds and the sanitation measures applied are also important factors. Birds weakened by disease and abnormal management conditions are certainly not in the best condition for the production of hatchable eggs. Regardless of good management, housing, and feeding of the birds during the breeding season, however, poor results may be forthcoming if the

Fig. 8—Hatching compartment at left; center and right setting compartments. Total capacity 34,720 Hen eggs.

[Courtesy Robbins Incubator Co.]

health of the flock has previously been impaired through unsatisfactory conditions. Many poultrymen have attempted without success to get suitable hatches from birds whose health had been undermined by coccidiosis, worm infestation and other ailments.

Rate of production is an indication of the condition of a flock and to a certain extent may be used as an index in the selection of hatchable eggs. In flocks where this rate is low or where laying has just commenced, fertility is liable to be inferior with resultant poor hatchability.

The feeding of breeding birds should meet their needs for maintenance and production. This necessitates a ration providing the necessary nutrients.

Other factors in the handling of breeding stock are: (1) provision for comfortable housing quarters and (2) careful attention to management details.

The quality and condition of the breeding stock is the result of a number of factors, the omission of any or all of which may mean inferior hatchability. The breeding back of the flock itself is by no means of minor importance.

Personal Attention and Interest

Success in any business or any phase of it is dependent to a particularly large extent upon the personal factor. Two people given identical equipment, education, training and practical experience, and with hatching eggs from a common source will produce hatches varying slightly or greatly, depending on the ability and interest exhibited by each individual. In many cases this determines success or failure. It also explains why some individuals are successful in but one or a few phases of the poultry business.

Embryonic Development

A study of chick embryology deals with the development of the individual from the origin of the germ cells to the time of hatching. Fertilization of the ovum (female germ cell) by the spermatazoon (male cell) takes place at the upper end of the oviduct, probably before albumen is deposited around the yolk. As soon as the male and female cells unite embryonic development begins and continues as long as the incubating temperature is supplied, until the embryo dies or the chick hatches. Thus the evidence is that embryonic development progresses while the egg is held within the body of the female, and that it continues until the egg is placed in a temperature which holds development in abeyance.

A thorough knowledge of embryonic development of the chick is of great value to those who can apply it in the operation of incubators. It is an aid in securing satisfactory hatches and can be used to check up mistakes in the incubation practice and to rectify them in the future. From the laymen's viewpoint important observable developments are given in the following chart:

Changes Day by Day: (Chick Embryo)

First Day—The germinal disc becomes larger; part of the yolk around the disc gets darker.

Second Day—Embryo takes definite shape. The tubular heart is formed and its curvature begins. At about 44 hours of incubation it beats rhythmically. The embryo lies directly across the top of the yolk.

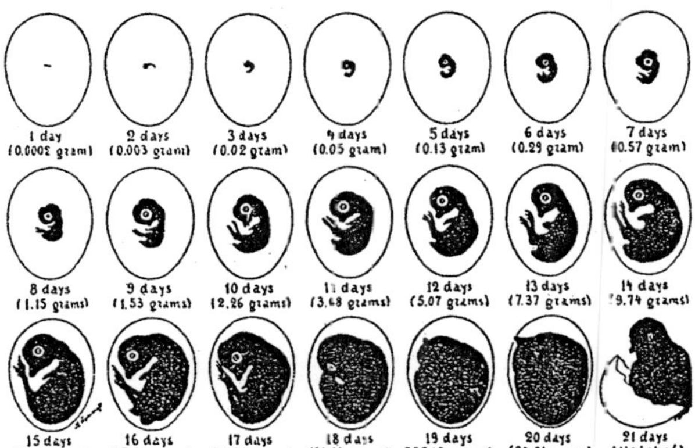

Fig. 9—Daily changes in the weight and the form of the developing chick embryo (White Leghorn).

[Courtesy of N. E. Poultryman, Cornell Bul. 205.]

Third Day—Embryo starts to turn over onto left side with top of head toward air cell. Development of the eyes has begun.

Fourth Day—By this time the body is completely turned onto the left side. Ramifying blood vessels should be clearly evident by this time. Two fetal membranes, namely the amnion and allantois, are clearly visible. The allantois serves as temporary lungs.

Fifth Day—The vascular area extends over about two-thirds of the yolk. Limb stocks begin to bend for elbows of wings and the knees of legs. Heart takes on definite shape.

Sixth and Seventh Days—Distinct avian characters are observable, such as the crop, gizzard, caeca, beak and the feather tracts. Obvious movements begin in limbs and amnion (sac in which developing embryo is rocked in the amniotic fluid).

Eighth, Ninth, and Tenth Days—Feather tracts are more distinct and appear along back, neck and thighs. Toes have appeared on the feet and there is a chalky deposit about the mouth which forms the horny beak.

Eleventh Day—Abdominal walls are definitely established. Embryo is nearing completion.

Twelfth and Thirteenth Days—Ossification of the bones well distributed over the body. The cartilaginous skeleton is completed, the nails are visible and the beak is horny but soft. From this time on embryonic development is merely a matter of growth and the complete functioning of all organs.

Fourteenth and Fifteenth Days—The embryo accommodates itself to the long axis of the egg, head toward the large end.

Sixteenth Day—The white of the egg is disappearing.

Seventeenth, Eighteenth and Nineteenth Days—The intestine is being drawn into the body cavity. Fluid decreased in the amnion. The neck acquires a double bend so that the head is turned forward, with the beak toward the part of the shell membrane next to the air cell. The chick is now in a hatching position. Yolk sac is beginning to enter the body cavity towards the end of this period.

Twentieth and Twenty-first Days—The yolk sac is completely included within the body cavity and practically all of the amniotic fluid has disappeared. The chick now occupies all of the egg outside the air cell and the blood circulation becomes more rapid in the lung tissue. Now the chick raises its head and pierces the shell membrane and slowly breathes the enclosed air. It may then be heard to peep. When the chick pips the shell it begins to breathe normally. The function of the allantois is completed; so it rapidly dries up. The chick is now ready to emerge from the shell and complete the hatching process.

Incubation of Chick Embryo by Days:

(Technical Development of Essential Features Only.)

First Day—(1) the primitive streak is formed and along its axis the primitive groove develops. (2) With the development of the head fold comes the first definite appearance of the head. (3) One or more pairs of somites make their appearance. (4) The first trace of the amnion appears in front of the head-fold. (5) The vascular area begins to be distinguishable.

Second Day—(1) Elevation of the head occurs. (2) The tubular heart and blood vessels form. (3) Eyes appear as hollow buds. (4) Auditory pits converted into pockets. (5) Head-fold moves backward. (6) Curvature of the heart increases. (7) Soon after 33 hours, heart begins to twitch at regular intervals, by 44 hours it beats regularly.

Third Day—(1) Embryo commences to turn over. (2) Circulation of the yolk sac is completed and the curvature of the heart is increased. (3) The eye-lens, nerves, etc., are formed. (4) The hind-brain separates into cerebellum and medulla oblongata. (5) Nasal pits form. (6) The cranial and spinal nerves are established. (7) Formation of fore- and hind-guts, division of fore-gut into esophagus, stomach (proventriculus) and gizzard, duodenum, and later, the large intestine.

Fourth Day—(1) Great increase in blood circulation in vascular area. (2) Amnion becomes more conspicuous. (3) Tail fold increases. (4) Limbs form. (5) Olfactory grooves are shaped. (6) Absorption of the partition between mouth and throat occurs. (7) The ureter forms. (8) Embryo is entirely on left side. (9) Allantois appears as small pear-shaped vesicle.

Fifth Day—(1) Vascular area extends over two-thirds of the yolk. (2) Allantois is very vascular and serves as the chief organ of respiration. (3) Limb stocks begin to bend for elbow and knee; cartilage is forming which precedes bone in limbs. (4) Heart takes on definite shape. (5) Primitive cartilaginous cranium takes shape. (6) Parts of the face develop and nasal processes close in. (7) Large development of gray matter is noticeable in spinal cord.

Sixth and Seventh Days—(1) Distinct avian characters are present. (2) Crop, caeca, and gizzard are visible. (3) Nose develops into a beak. (4) Obvious movements begin in limbs and amnion. Seventh day amnion pulsates slowly and rhythmically,

TRANSFORMATION OF LIVING ENERGY
FROM THE EGG TO THE CHICK

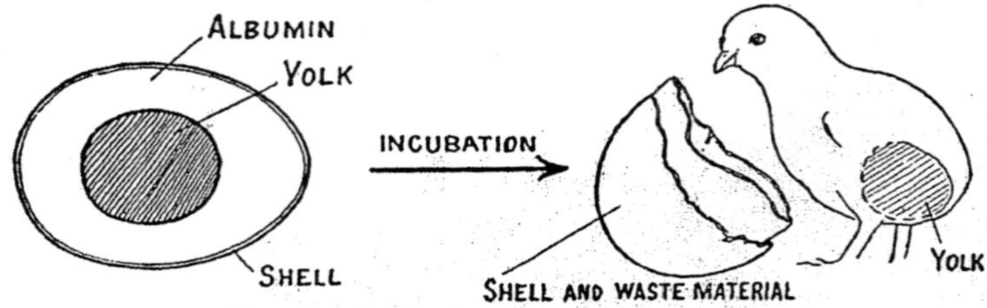

BEFORE INCUBATION

	WEIGHT (GRAMS)	HEAT VALUE (CALORIES)
YOLK	20	60
ALBUMEN	30	30
SHELL	10	TRACE
TOTAL	60	90

AFTER INCUBATION

	WEIGHT (GRAMS)	HEAT VALUE (CALORIES)
BODY OF CHICK	30	40
UNUSED YOLK	10	30
SHELL, ETC.	10	TRACE
TOTAL	50	70

ANALYSES:

EGG CONTENTS (50 GR.)	GRAMS	PER CENT	CHICK (30 GR.)	GRAMS	PER CENT
TOTAL SOLIDS	12.5	100.0	TOTAL SOLIDS	6.0	100.0
ORGANIC MATTER	12.1	96.8	ORGANIC MATTER	5.5	91.7
PROTEIN	6.8	54.4	PROTEIN	3.5	58.4
FAT	5.0	40.0	FAT	2.0	33.3
CARBOHYDRATE	.3	2.4	CARBOHYDRATE	TRACE	TRACE
INORGANIC MATTER	.4	3.2	INORGANIC MATTER	.5	8.3
CALCIUM	.05	.4	CALCIUM	.2	3.3
PHOSPHORUS	.2	1.6	PHOSPHORUS	.15	2.5
OTHERS	.15	1.2	OTHERS	.15	2.5

A. ROMANOFF

"Cornell Countryman," vol. 26, p. 210, April, 1920

Fig. 10—Transformation of Living Energy.

[Courtesy of New England Poultryman]

rocking embryo in the amniotic fluid and permitting normal development without adhesion. (5) Allantois much larger. (6) Body begins to grow more rapidly in comparison with the head. (7) The egg tooth begins to develop on the sixth day. Its function is for a single critical event; the breaking of the shell.

Eighth, Ninth and Tenth Days—(1) About the 10th day little sacks containing the feathers begin to appear, along back, neck and thighs. (2) Chalky deposit shows about the mouth. It will change into the horny beak.

Eleventh Day—(1) Abdominal walls are definitely established. Loops of the intestine which have been hanging in the somatic stock are confined within the cavity of the abdomen. (2) Body of the embryo is therefore established.

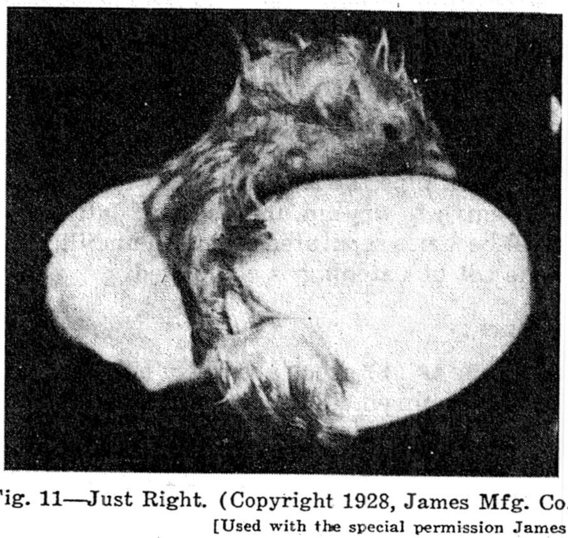

Fig. 11—Just Right. (Copyright 1928, James Mfg. Co.)
[Used with the special permission James Mfg. Co.]

Twelfth and Thirteenth Days—(1) Beak is horny but soft. (2) Nails are visible on 13th day. (3) The cartilaginous skeleton is completed on the 13th day, and the various muscles of the body are fairly distinct. (4) Ossification of the bones is well distributed over the body.

Fourteenth and Fifteenth Days—(1) Embryo accommodates itself to the long axis of the eggs, head towards broad end.

Sixteenth to Twenty-first Days—(1) By the 16th day white of egg has practically disappeared. (2) After 16th day the urates are abundant in the allantoic fluid. (3) Seventeenth to Nineteenth Days:

a. Fluid decreases in the amnion.
b. The neck acquires a double bend so that the head is turned forward with the beak towards the part of the shell membrane next to the air cell.
c. The intestine is retracted completely within the body cavity.
d. On the 19th day the yolk sac begins to enter the body cavity.
e. On the 20th day the yolk sac is completely included and practically all of the amniotic fluid has disappeared.

(4) The chick now occupies all of the egg outside of the air-chamber. The umbilicus is closing over. (5) The blood circulation in the allantois is beginning to contract and more blood is going through the lungs. (6) Now the chick raises its head and pierces the shell membrane and slowly breathes the inclosed air. It may then be heard to peep. (7) Circulation in the allantois is still maintained and it still preserves its respiratory function.

When the chick pips the shell it begins to breathe normally. The allantois begins to dry up and the circulation in it rapidly ceases. It then becomes separated from the umbilicus and the remainder of the act of hatching is completed.

Fetal Membranes:

Amnion—Acts as (1) shock absorber, (2) protection against dessication of the embryo, (3) guard against adhesions.

Chorion—Not important in chick development.

Allantois—Serves as embryonic organ of respiration, receives excretion of the kidneys, and absorbs the albumen.

Yolk sac—Absorbs yolk material.

Malpositions of Embryos

Anyone who makes a careful study and examination of unhatched eggs will readily note that there are a number of abnormalities which are so severe as to result in the failure of many embryos to hatch. Normally, a chick ready to hatch lies with its head in the large end of the egg, its legs toward the small end, the head turned to the right and lying under the right wing, with the beak directed towards the air cell.

According to W. F. Dove of the Maine Experiment Station, ". . the fundamental causes of the defective condition in an embryo, whether they be from the effects of inbreeding, of faulty nutrition, or of careless incubation methods, will interfere with the normal sequence of body shiftings and as a consequence, will leave the embryo in a retarded or strained position when death occurs. Malposition I of Sanctuary, together with the extreme form, malposition V of Smith, may now be used to indicate three possible causes of death: (1) Air hunger when the air cell is misplaced, (2) gravity when the egg is incubated upside down, and (3) factors delaying development previous to the fifteenth day of incubation."

TABLE I—MALPOSITIONS

Position No.	Described by	Description
I	Sanctuary—1925	Head between thighs
II	Sanctuary—1925 Reaumur—1751	Head in small end
III	Sanctuary—1925	Head to left instead of under right wing
IV	Hutt—1929	Head normal, but rotated away from air cell
V	Smith—1930	Feet over head
VI	Hutt & Cavers—1931	Beak over right wing or over and under wing

F. B. Hutt of Cornell University states in this connection, "It has been shown that the frequency of position IV is reduced, when the eggs are incubated with the large end up. Unfortunately, however, the large-end-up position causes an increase in the number of malpositions I and II. If some way could be found to combine the desirable features of both positions, the number of malpositions would be materially reduced and it has been estimated that hatches would be increased from six to eight per cent. Further experiments may show that more frequent turning or turning in more than one plane might reduce the number of abnormalities. So far, there has been found no adequate proof that there is a hereditary tendency to produce any of them, or that the diet of the breeding stock has any effect on their frequency. It has been shown, however, that position III is markedly more frequent in the larger eggs."

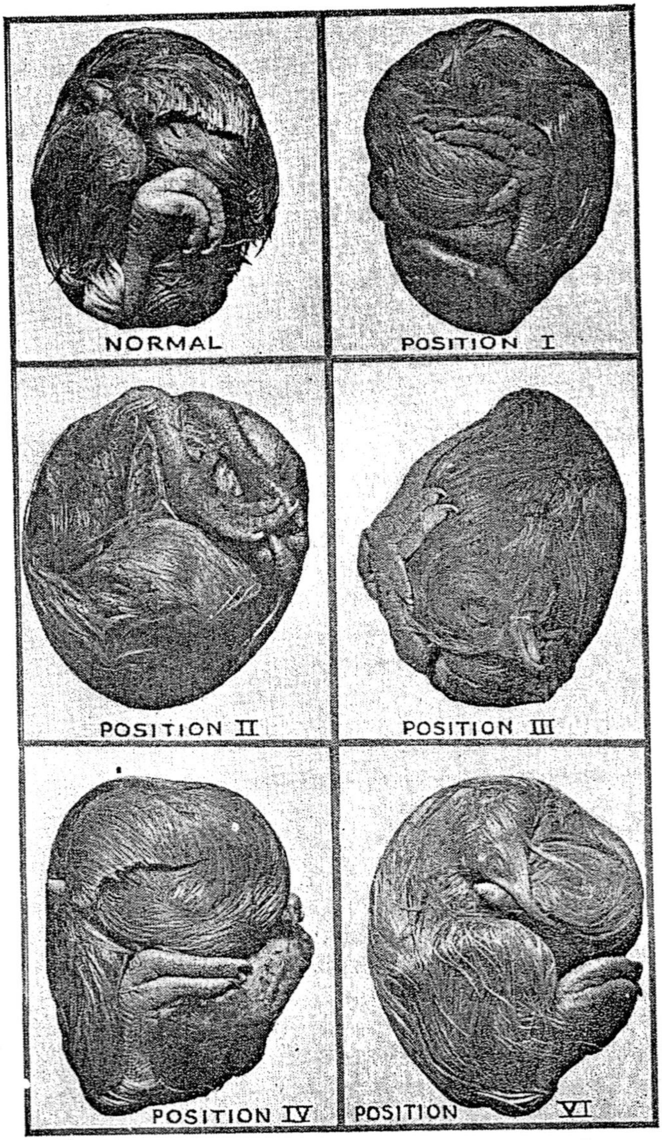

Fig. 12—This shows the normal position of the chick embryo (with a portion of the right wing cut away to show the head more clearly) and five common malpositions. These are described in the text. Position V, referred to in the text, is not shown because it is not common.

[Courtesy American Poultry Journal.]

Dove also mentions a few unclassified conditions that may interfere with hatching. These are listed under abnormal feet positions and include the following:

1. Agrippa, feet first.
2. Legs crossed.
3. Feet tied by membrane or by the yolk stalk.
4. Membranes tied about the body or neck.
5. Hernia due to kicking.
6. Broken yolk sac due to kicking.

Hatching Periods

In order that one may be successful in securing satisfactory hatchability for the various types and species of domestic poultry it is highly desirable that the relative length of hatching period be known.

The following should be an aid in incubating eggs from the various species listed: Hen, 21 days; duck, 28 days; Muscovy duck, 33 to 35 days; geese, 28 to 32 days; turkey, 28 days; guinea, 26 to 28 days; pheasant, 21 to 24 days; peafowl, 28 days; ostrich, 42 days, and pigeon, 18 to 20 days.

Questions on Chapter I

1. What are the possible results that might be secured through careful selection of hatching eggs?
2. What are the points to look for in selecting hatching eggs?
3. Successful incubation depends on a number of factors. What are these?
4. Does excessive carbon dioxide in the incubator have any effect on the hatchability of the eggs?
5. How does one calculate the formaldehyde and potassium permanganate requirements for incubator fumigation?
6. What is the normal position of a chick just previous to hatching?
7. How many days are required for eggs to hatch of the following:
   a. Hens
   b. Ducks
   c. Geese
   d. Turkeys
   e. Guineas
   f. Pheasants
8. Crippled chicks can be traced to what factors of incubation?

CHAPTER II

HATCHERY PRACTICES

Subject Matter

Management practices of different hatcheries vary considerably, depending on local condition; equipment, labor and capital available; size and demands of the business; and other factors. Management practices, which from the light of experience and observation, are vital to the success of hatchery operation are here discussed. It is evident that sound business and management practices must be applied if the venture is to be placed on an efficient basis.

Type and Location of Incubator Building

Forethought should be given to location of the incubator building. Features to consider are accessibility, centralization, advertising value, possibilities of expansion and territory to be served. A building centrally located in the town or community will be easily reached by the regular trade, prospective customers and hatching-egg producers. A good location has the added value of advertising the hatchery and keeping it before the interested

Fig. 13—Inside a 52,000-Egg Smith Incubator.
[Courtesy Meade Summers, Purina Mills.]

buying public. An obscure hatchery is at a serious disadvantage with more-favorably-located competitive interests. Location, and the overhead it will incur, should be carefully weighed.

In planning hatchery buildings the past custom was to partially submerge them in the ground, but the growing tendency is to erect buildings entirely above ground level. This facilitates receiving and shipping, makes excavation unnecessary, and simplifies ventilation.

Floor space should be planned to meet present needs and to allow for economical expansion of the building as the demands of the business grow. Office, display, storage, receiving and shipping space may need to be included.

A labor-efficiency study is needed if each hatchery operation is to be handled with minimum labor requirements.

With the advent of above-ground rooms, insulation became increasingly important as a guard against abrupt climatic changes that might seriously affect results. The proper application of insulation, heat, and ventilation controls room conditions.

Sufficient ventilation will furnish a constant supply of fresh air for respiration and will remove carbon dioxide. Carbon-dioxide concentration is closely linked with impaired hatchability. Most important is the relative proportion between the cubic content of the room and the total egg capacity of its incubators. In some cases no special ventilation system is necessary. In others the advice of ventilation engineers may be needed. Electric current, water supply and drainage will also prove helpful.

Often old buildings can be remodeled into suitable hatchery quarters with a minimum of expense. Many commercial incubator manufacturers will furnish plans of various types to meet individual needs.

Purchase of Incubator

In the selection of the incubator too much importance cannot be attached to the reliability and stability of the manufacturing concern. The trade-in value of the machine is another point to be ascertained. The enormous strides made in perfecting incubators during the past decade leaves many machines obsolete with resultant high depreciation.

There is a considerable variation in the cost per egg capacity. This is not necessarily the most important criterion of a good machine, but needs to be weighed carefully. Not only capacity cost but also the expense of operating and of housing should be borne in mind. Certain types may be more economical in these respects. Compactness, from the viewpoint of operating and housing costs, is a more important factor in some cases than in others.

Another element sometimes overlooked in making the selection is that of fuel supply. Whatever its nature, whether coal, oil, gas or electricity, it should be dependable and consistent with conditions existing in your locality. Many hatcheries have installed a generator which can be used for operations in emergencies when commercial service is interrupted.

The use of modern devices, such as humidifiers, hygrometers, recording thermometers and alarms, will be an advantage. These make possible more accurate control of the incubator, providing optimum conditions for the development of livable chicks.

It is wise when purchasing an incubator to look forward to the possible development of the business and provide sufficient capacity for at least the needs of the immediate future. Similarly the building should be planned with later expansion in mind. In anticipating this, avoid planning for over-expansion of the business, requiring a heavy investment in permanent equipment with the possible depletion of working capital. Resources and size of enterprise should be well balanced.

Installment Buying:

The purchase of commodities on the installment basis has permeated many phases of the manufacturing field, offering one an opportunity of securing his needs at an earlier date than would otherwise be possible. This applies to incubator equipment. Whether one is to purchase on the installment plan will depend on his financial condition. Whatever the method of purchasing, it is obvious that the machine must be paid for sooner or later before it becomes the property of the purchaser.

Where possible, paying cash is undoubtedly most satisfactory. Usually this is more economical, due to cash discounts and the absence of the interest charges connected with installment purchasing.

Under the finance plan do not overestimate ability to meet heavy payments. Regular overhead and current expenses may limit the amount of cash on hand. One must be reasonably sure of the source and amount of income so as to allow a safe margin over expenses to meet the periodic installments.

Do not depend entirely on business expectations to meet these notes, as considerable time is usually required to establish a business and a clientele. Not all of one's capital should be tied up in the incubators; enough should be available to purchase hatching eggs, chick boxes, advertising and other items incidental to hatchery operation. The hatchery business may be primarily a seasonal enterprise and finance plans should be adapted to this condition.

Normally installment buying of incubators is based on a carrying charge of six per cent. Finance corporations exact a higher interest rate, depending on the article and the risk assumed. If the rate appears excessive your problems of financing should be discussed with the local banker. It is well at any rate to establish a credit rating in order to meet emergency requirements for cash. Naturally the banker's attitude on loans is likely to be influenced by his previous experiences in the same field. If failure has marked a few similar ventures, he is likely to be conservative. A frank, honest statement of all the details of your business will establish confidence and assist materially in arranging loans to your satisfaction.

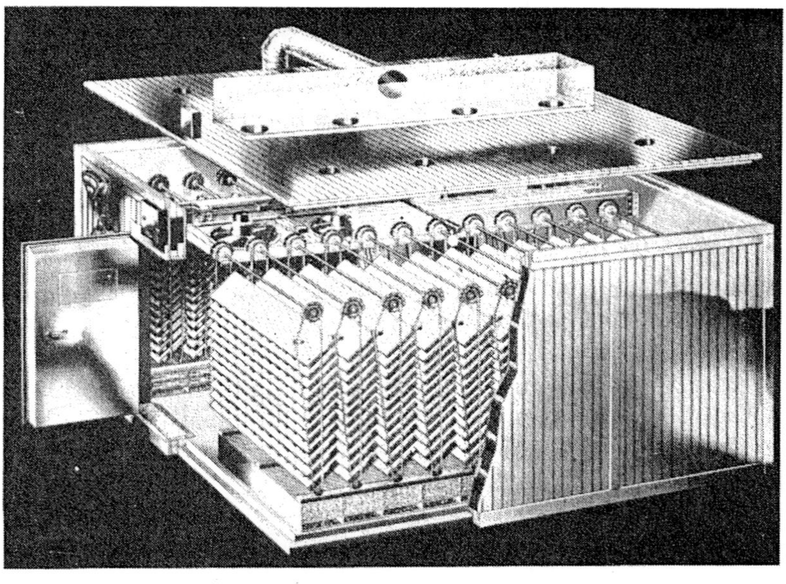

Fig. 14—Phantom view of Smith Model 52. Total capacity 52,416 Hen Eggs.

[Courtesy Smith Incubator Co.]

UNIFORM SCORE CARD FOR JUDGING EGGS*

Exterior Quality	
Size (total weight of exhibit)	12
Uniformity of weight	12
Uniformity of color	12
Uniformity of shape	6
Shell texture	12
Condition	10
Interior Quality	36
Total	100

Instructions For Scoring Eggs

Size (total weight of exhibit)

A perfect score will be given to the dozen of eggs weighing between 24 and 27 ounces. If the weight of the dozen is above or below these weights the entire 12 points shall be cut. Weights and similar cuts shall prevail for other classes.

Fig. 15—The Smith Separate Hatcher. Capacity from 5,616 to 12,168 Hen Eggs.

[Courtesy Smith Incubator Co.]

Uniformity of weight

The individual weight of each egg shall be determined in ounces per dozen. Each egg weighing at the same rate as the total weight of the entry shall receive a perfect score of 1. A cut of ½ point per egg shall be made for ½ ounce variation from the total weight of the entry with a maximum cut of 1 point per egg.

Uniformity of color

All eggs in the entry shall be of the same shade of color. If white, all eggs should be chalk white and free from creamy or tinted coloring.

* Northeastern Poultry Producers Council.

If brown, all eggs should be uniformly the same shade of color. Cuts of ¼ to ¾ points per egg may be made according to the actual shade of each egg and how well it conforms in uniformity of shade with the other eggs in the dozen.

Uniformity of shape

Each egg to conform as nearly as possible in shape. Extremes in shape, such as short, round, and long eggs should be avoided. Cut ¼ to ½ point for each egg off in shape.

Shell texture

The egg shell should be strong and of uniform thickness. Cuts of ¼ to ½ point per egg may be made for porous, rough or ridged shells. A cut of ¼ point is to be made for a blind check, ½ point for a check, and 1 point for a leaker or smashed egg.

Condition

Eggs should be clean and unwashed.

Interior quality	Perfect score per egg	Cuts per egg
Fancy	3	0
Extra		¾
Standard		1 and ½
Trade		2 and ¼

Inedible eggs

The following are classed as inedible eggs: Blood spots, bloody whites, meat spots, molds, and rots and should disqualify the dozen. A cut of three points per egg will be made for each egg in the entry showing meat or blood spot.

Number of Eggs to Set

The total twice weekly or weekly setting should be based on the number of chicks booked, plus a reasonable allowance for orders which may be received between the setting and hatching date. This will also aid in filling orders if hatchablity of the setting is not up to expectancy.

This expectancy will vary widely under different conditions. Some hatcherymen book orders based on 70 per cent hatchability, while others find a different level exists under their conditions and they make adjustments on the weekly settings in accordance. A good plan is to summarize the hatchability results of past seasons in order to arrive at a dependable figure, and then produce a slight surplus so that all customers who have booked orders may be accommodated on the date desired. Overestimating hatchability will result in dissatisfied customers. Good will is a valuable asset in the continuance of a business.

The more or less standard number of hens required to furnish an adequate number of hatching eggs for a hatchery is between 100 and 120 hens for each 1,000-egg capacity. Thus for a hatchery with a total capacity of 10,000 eggs, it is generally assumed that between 1,000 and 1,200 hens will be required. Also that an estimated total of 107 hens was available for the production of hatching eggs for each 1,000-egg incubator capacity. It is probable that this estimate of 107 hens is lower than 15 years ago because of the increased egg production per bird, as well as the increased hatchability of eggs.

Testing Thermometer

Thermometers need to be tested previous to each hatching season, and during the season if indications point to an incorrect reading. They may be checked with a doctor's clinical thermometer, with all under identical conditions and placed side by side to facilitate reading. Many commercial organizations will test thermometers for a small charge.

Correcting Separations in Mercury in Thermometers:

The thermometer should be heated slowly until the mercury enters the slight enlargement at the top of the tube and the separation disappears. Avoid entirely filling this enlargement or the thermometer will break. After separation disappears, remove thermometer from the heat, holding it in an upright position until the mercury recedes towards the bulb. Slight tapping may also help in joining the separated parts of mercury.

Candling

In commercial hatching establishments candling of eggs is customary the eighteenth day of incubation. At that time the live germs are transferred from turning to hatching trays. Sometimes, due to lack of capacity, candling of white eggs is desirable the fifth to seventh day, and brown eggs the eighth or ninth day, in order to remove dead germs and infertile eggs and secure additional capacity by doubling up. This also makes possible an early check on the fertility of various flocks, giving the hatcheryman a basis for hatching expectations. Where slanted trays, other than stay-wire trays, are in use, however, the removal of dead germs and infertiles before the eighteenth day is often unwise, as this necessitates the additional labor of retraying or packing. In large hatcheries the efficiency of candling is increased by using a portable candling device. It is an advantage to candle a few eggs at frequent intervals to check up on the progress of embryo development and also on evaporation, the latter to check the adequacy of ventilation and humidity.

Common defects of fresh eggs which might interfere with hatchability are blood clots, bloody albumen, double yolks, blind checks, cracks, air bubbles, and ruptured air cell. External defects, such as cracks and blind checks, may be readily detected by handling; the few internal defects which affect hatchability do not occur frequently enough under ordinary conditions to justify candling before the eggs are put into the incubator. Eggs in the process of incubation are candled principally for the removal of infertiles and dead germs. An infertile egg is one which has not been fertilized, but many germs die at such an early stage that distinguishing them from infertile eggs is impossible by candling. Dead germs may be classified into several groups, but for practical purposes three groups are sufficient:

1. Those which start to develop and die during the first week of incubation. Practically all of these may be classified as blood rings.
2. Those which die between the 7th and 18th days of incubation.
3. Those which die between the 18th and 21st days, including not only those dead in the shell but also those which pip but fail to get out of the shell.

A good live germ at 7 to 10 days is one in which there is obvious movement in the limbs and amnion—(the sac in which the embryo develops) ; in which radiating blood vessels show distinctly before the candle.

Fertility Determination

Research work has demonstrated that it is possible to determine fertility in eggs, after about 12 to 18 hours of incubation. According to the findings of Olsen and Knox (U.S.D.A.), "The method of determining fertile and infertile eggs is relatively simple and may be done by the average hatcheryman after a few hours' practice. All the equipment needed is an ordinary egg candler with a 75-watt bulb. However, a daylight bulb or a blue-green bulb or blue eye glasses aid in seeing the embryo and make identification somewhat more accurate."

"The embryos that can be seen at these ages contain no blood, cover an area on the yolk about the size of a dime, and appear before the candle as a small spot. This effect is caused by a slight bulge of the embryo on the surface of the yolk. Such eggs are known as 'heated eggs' in commercial candling. In the infertile eggs no development can be seen."

"Some embryos develop faster than others and some are farther developed than others when they are placed in the incubator. Some embryos are developed enough in fresh eggs to tell fertility but the investigators have found, after candling and then break-

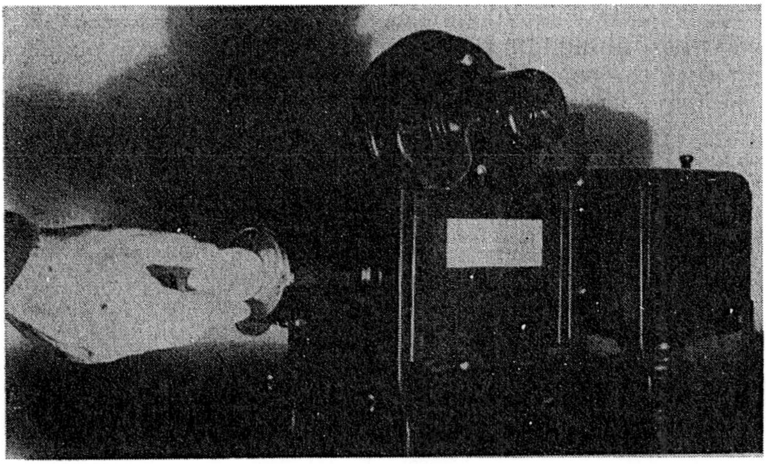

Fig. 16—Candling device for detecting fertility in eggs, at about 15 hours of incubation.

[Courtesy Prof. D. F. King, Inventor.]

ing over 4,000 eggs to test the accuracy of their identification, that most eggs take 12 hours or more of incubation before the two classes can be identified with fair accuracy. They were able to identify accurately 88.2 per cent of the embryos at 12 hours, 92.1 per cent at 14 hours, 97.6 per cent at 16 hours, and 97.1 per cent at 18 hours."

At least two commercial concerns have manufactured special lights, provided with proper filters, and cooled by electric fans, for this phase of hatchery operations. These machines may be leased to the hatcheryman at a small yearly fee. It is usually conceded that about 10 per cent of all eggs are infertile; therefore, removing the infertile eggs at 14 to 16 hours of incubation would increase the capacity of an incubator by that amount.

One disadvantage would be that all eggs would have to be retrayed, which would involve a lot of labor, and tend to offset some of the advantage gained by the use of such lights. Some hatcherymen have sought to overcome this by having a special preheating incubator or room, where the eggs are held during the first 14 to 16 hours. The eggs are then candled, as above indicated, and only fertile eggs put in the regular incubators.

Traying Eggs

To facilitate traying in cabinet-type incubators, slant the trays when filling. In this type the eggs should be placed upright with the pointed end downward, starting at the lower end of the tray. Then the eggs will rest against each other and not fall out of place. Vacant tray space may be packed with paper to hold the eggs in their original position when placed in the incu-

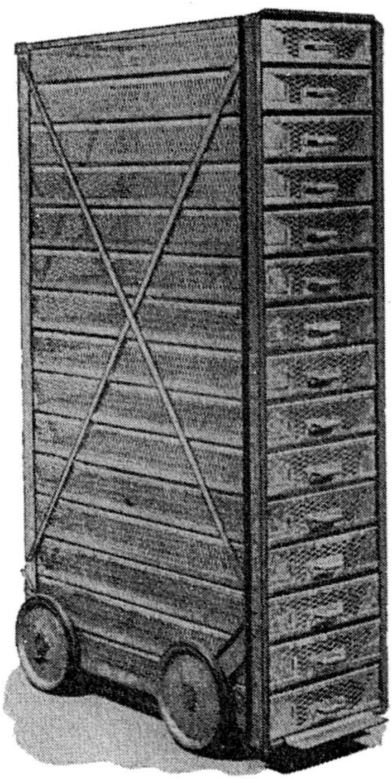

Fig. 17—Tray Buggy and trays. Smith Separate Hatcher.
[Courtesy Smith Incubator Co.]

Fig. 18—Showing Fan, Ducts and Cooling Coil. Smith Separate Hatcher.
[Courtesy Smith Incubator Co.]

bator and while being turned. This does not apply to trays equipped with stay wires. When stay-wire trays are used do not force abnormally wide eggs in position, for the wire may cut the shell.

To avoid unnecessary fluctuations in temperature in the incubator put the trays in the machine as they are filled. If they are all put in at the same time after filling, the drop in temperature is greater and a much longer period will be required to raise the temperature to the normal incubation level.

Where custom hatching is done and trays are only partially filled, it is good practice to add enough eggs of a different color so that definite incubator records may be kept. In incubators having trays of variable sizes, fill according to the traying schedule suggested by the manufacturer.

Custom Hatching

Custom hatching is the practice of hatching a customer's eggs for a stipulated compensation. This consideration may be based on the number of eggs set or chicks hatched.

Custom hatching is normally a side line, with the operator often open to criticism if the eggs do not hatch well, even through no fault of his own. This discourages many.

Many machines are not constructed to lend themselves efficiently to custom hatching. The job further involves careful separation of eggs previous to hatching, and careful attention to chick separation. A special hazard is that of introducing Pullorum disease into the incubator through hatching from unclean flocks. Because of the probabilities of the spread of the infection, the mortality in your own chicks may be greatly increased and more than offset any cash advantage gained through custom hatching.

Hatching Pedigreed Stock

In hatching eggs from pedigreed stock it is necessary to group together the eggs from individual hens so that the parentage of the chicks may be recorded. This segregation is necessary at the 18th day of incubation, either in a hatching compartment in the incubator, in wire baskets, or in mosquito netting bags. To designate the parentage of the chicks during the growing season, and later, the chicks are wing-banded and the band numbers recorded. This may be done when the chicks are removed from the incubator. The operation involves piercing the web of the wing, inserting a wing band through the opening and clinching it.

Overheating and Underheating

For normal embryonic development, a normal incubating temperature should be maintained. Deviation from this in either direction will affect maximum results.

Its importance can be readily seen from the following table:

Effects of Temperature Levels of Embryo on Rate of Embryo Growth:*

Temperature— Fahrenheit Degrees	93.2	95.0	100.4	102.2	104	105.8
Per cent of Normal Development	65	80	100	106	125	151

Low temperatures retard development, produce subnormal embryos, cause excess mortality in incubation and result in lower hatches. On the other hand temperatures above normal cause excessively rapid growth of embryos, the abnormal condition resulting in high incubation mortality and lower hatchability.

Excessive heat early in the hatch kills many embryos during this period. Similar temperature conditions during the later stages will result in chicks coming out bloody. The raising of temperatures at hatching time does not give the chick embryo sufficient time to transfer from the temporary to the permanent respiratory system, with the result that the blood circulation has not been entirely retracted and bloody hatches result. Often the yolk sac protrudes from the abdomen and many navels are imperfectly healed.

Auxiliary Generator

In many instances hatcherymen have installed auxiliary generators as a precautionary measure against power failure. This is particularly true where substantial hatchery capacity is being operated and where past experience dictates the desirability of such protection. Where such equipment is installed it should be maintained in operational condition ready for emergency use.

*Adapted from "The Development of the Chick," Lillie.

Removal of Hatch

Chicks should be removed from the incubator as soon as the hatch is completed and they are thoroughly dried off. If they are to remain on the premises they may have access to feed at this time. When the hatching trays are crowded, a preliminary removal of the early hatched chicks will give additional space for those still to hatch. Removing the egg shells will aid in the prevention of the telescoping of eggs still to hatch or in the process of hatching.

Generally it is not considered a wise policy to help chicks out of shells. Chicks not vigorous enough to kick out of the shell are usually inferior and undesirable.

Chick boxes should be prepared in advance with litter of such a nature that it will provide firm footing for the chicks and prevent them from catching their feet under partitions and piling

Fig. 19—Mammoth Brood-Unit for broiler and pullet production.
[Courtesy Petersime Incubator Co.]

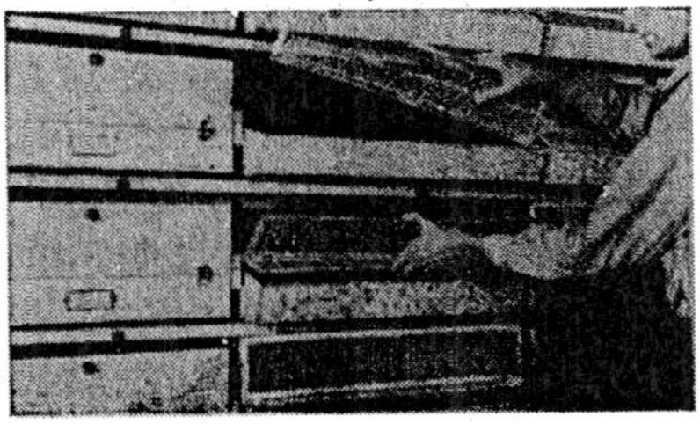

Fig. 20—Brood-Unit, showing removable self-closing gates.
[Courtesy Petersime Incubator Co.]

up in the corners. This material should be fine enough so that the chicks are not likely to become entangled in it. The chicks should be removed directly from incubator to chick boxes and the loaded boxes placed in a reasonably warm room so that chilling or overheating may be avoided.

Proper ventilation of chick boxes is essential if chicks are to arrive at destination in good condition. Common sense and experience are necessary in determining this. Weather reports will serve as a basis for estimating the possible fluctuations of conditions during transit and the number of ventilation openings needed. Manufacturers of chick boxes furnish estimation charts of ventilation requirements for varying climatic conditions.

Ventilating Chick Boxes

The following schedule* refers to the number of holes to be opened in the cover and body of a 25-chick compartment in a standard hundred capacity box.

0 to 20 degrees— 2 holes in both the side walls and cover.
20 to 32 degrees— 4 holes in both the side walls and cover.
32 to 50 degrees— 6 holes in both the side walls and cover.
50 to 70 degrees—8 to 10 holes in both the side walls and cover.
70 or over—about 20 holes in both the side walls and cover.

Unless provisions are made in the construction of the box to prevent the closing of vent openings when boxes are piled one on the other, cleats should be securely nailed on the cover. This provides ventilation between boxes and prevents overheating and suffocation.

In gauging temperature, a report of the Massachusetts State College covering five years' records for March and April may be of help. Their records show that 9:14 A. M. was the time to ascertain the average temperature for the 24-hour period. About 3:00 P. M., is the hottest time of day and 5:00 A. M. the coldest.

Many types of boxes have been developed, some for warm- and the others for cold-weather shipments. Some hatcherymen prefer to use the summer box throughout the entire season providing additional litter during the colder weather.

The hatch should be set so that it will come off at a suitable day and time of day. A careful study of train schedules and transfer points in transit will help in getting shipments to customers with a minimum of delay and exposure of chicks.

Crippled and deformed chicks should be sorted out, killed and burned. Infertiles, dead germs, shells and other incubator waste should be disposed of, or destroyed by burning. Infertiles may be denatured and sold to tanneries. Off-color individuals should be sold as mixed-grade chicks. This procedure will result in uniformity in chicks, adding materially to the quality of the output.

Storage Brooders

Battery brooders have come into extensive use for both brooding and storage purposes. Storage is really the brooding

*Courtesy of Norb Schaefer, Mgr. Anderson Box Company, Indianapolis, Indiana.

of chicks which are kept until one to four weeks of age. Batteries are also used for the additional period up to broiler age.

The development of this type of equipment has opened a field for the production of started chicks. Such chickens are purchased when past a critical period in their development. An additional charge is made depending on age of chicks. Local conditions should be considered in establishing a price for started chicks.

These brooders are well adapted to caring for the hatcheryman's surplus chicks. They are particularly convenient for handling hatches that are better than anticipated, for maintaining a supply of chicks to fill late orders, and for caring for chicks on cancellations. They also present a way for the proprietor to check on the livability of his chicks and on the livability of chicks from various sources of hatching eggs. An accurate check will enable him to discontinue buying from flock owners supplying eggs of low chick livability. The livability of his future sellings will thus be increased.

The storage of chicks in batteries releases the hatcheryman from the necessity of "dumping". Previously, surplus chicks had to be disposed of immediately and in many cases represented a material loss due to unfavorable marketing. To move them, prices were often slashed and the chicks "dumped" on a buyers' market. Through the use of the storage brooder the chicks can now be carried until a more favorable outlet is obtained. However, storage brooders should not be located in incubator rooms and preferably not in the incubator building because of the disease hazard.

Battery brooders add to the cost of hatching equipment but provide a splendid opportunity for display of chicks.

The optimum time for carrying chicks in storage brooders is probably not exceeding three weeks. This allows use of the batteries under a weekly hatching schedule, and without excess requirements for battery space and equipment.

The use of such storage implies that the hatcheryman must also be a brooderyman. No system is 100-per-cent foolproof. In any operation involving livestock the personal factor is a very important one.

Business Records

One of the first records to maintain is one on the egg production of the hatchery flocks. This will be necessary if a sufficient number of eggs are to be secured during the hatching season. In addition it will form an index of the egg production your customers may expect.

A hatching record should also be kept on each flock. This will assist in correcting serious leakage in the business, and the accumulation of figures from year to year will enable you to make more accurate predictions on expected hatchability.

To trace eggs through the hatchery a lot number should be given to each shipment received from every individual shipper. And to facilitate payment for hatching eggs, an egg-voucher system will be valuable, whereby a separate voucher is issued upon receipt of each shipment of eggs. This voucher, when presented at the business office of the hatchery, calls for payment which should be made by check in order to keep a permanent record of transactions.

Tray cards must be provided for definitely marking and locating each lot of eggs so that at the completion of the hatch the results may be tabulated and entered in permanent forms. Separate forms for seasonal results on individual flocks are also needed for determining the satisfactory ones.

A chick production chart will also be helpful to indicate the estimated production of chicks at each hatch. This should constitute the basis for the booking of orders, but the estimations ought to be conservative and allow sufficient leeway. It is wise to set only a limited percentage over bookings to take care of the normal number of orders received after the eggs are placed in the machine. Experience over a period of years will make estimates of the amount of such business quite accurate.

In conjunction with a chick-production chart a record of orders received for various hatching dates should be carried. All orders should warrant a reply when recorded. To facilitate record keeping it would be desirable to classify orders under (1) full cash payment, (2) deposit, (3) C.O.D. shipment, and (4) replacement orders.

Information regarding shipments should be filed relative to dates orders are received and filled. Similarly data should be recorded as to the cost of production of each hatch. Advertising

Fig. 21—Type of Incubator with separate hatching unit in the base. Trays in top (incubating unit) are tilted to proper position for the incubating period.

[Courtesy Cugley Incubator Co.]

cost must be considered as a charge against the entire hatching season.

Monthly business summaries as well as a complete statement for the season should be tabulated and this information carefully analyzed in order to take advantage of profitable practices.

The following table represents the operating statement of 18 accredited hatcheries in Ohio in 1927. It lists the important factors entering into the cost of producing chicks.

TABLE 2
STATEMENT OF EXPENSES PER 100 CHICKS
(Average of 18 Ohio Accredited Hatcheries, and Two Individual Reports—Year 1927)
Bureau of Business Research—Ohio State University—Monograph No. 13.

	Average	Hatchery A	Hatchery B
Direct Costs:			
Egg costs	4.60	3.58	6.62
Labor costs	1.79	1.75	2.64
	6.39	5.33	9.26
Indirect Costs:			
Heat, light, power and water	.11	.10	.22
Repairs, fixtures and equipment	.04	.01	.24
Supplies, sundry hatchery expense	.03	.11	.04
Taxes, personal-property	.07	.02	.07
Insurance, fixtures, equipment	.09	.04	.09
Depreciation, fixtures, equipment	.66	.58	.94
Rent, or equivalent building expense	.14	.07	.23
	1.14	.93	1.83
Total Production Costs	7.53	6.26	11.09
Selling and Administrative Expenses:			
Advertising	.51	.86	.22
Telephone, telegraph	.05	.04	.13
Boxes, twine	.17	.22	.17
Postage, express	.25	.46	.08
Automobile expense	.04	.05	.02
Office supplies and expense	.03	.04	.14
Replacements	.01	.26	.03
Accrediting expense	.10	.05	.24
	1.16	1.98	1.03
Total Cost per 100 Chicks	8.69	8.24	12.12
Hatchings as a Percentage of Eggs Set	65.0%	69.0%	49.8%

The chief factors in chick-production costs are egg cost—reflected by hatchability, labor, depreciation, and selling and administrative expense.

TABLE 3

STATEMENT OF AVERAGE REVENUES AND EXPENSES
(19 Ohio Accredited Hatcheries)

Sales:			
Baby chicks		$33,021.39	
Custom hatching		1,235.47	
			$34,256.86
Cost of Sales:			
Direct Cost:			
Egg cost	$12,159.63		
Labor	4,575.05		
	$16,734.68		
Indirect Cost:			
Heat, light, power & water	400.59		
Repairs, fixtures & equipment	245.30		
Supplies, and sundry hatching expense	226.95		
Taxes—personal property	124.92		
Insurance—fixtures and equipment	224.98		
Depreciation—fixtures and equipment	1,822.17		
Rent, or equivalent building expense	402.80		
		$ 3,447.71	
Chick Purchases		2,261.82	
			$22,444.21
Gross Margin			$11,812.65

Expenses—Selling and Administrative:			
Advertising	$ 2,036.00		
Telephone & Telegraph	134.29		
Boxes and twine	432.51		
Postage and express	789.05		
Automobile expense	243.86		
Office supplies and expense $	223.85		
Replacements	213.01		
Accredited expense	310.58		
Bad debts	183.50		
			4,566.65
Net Profit from Hatchery			$ 7,246.00
Other Operations:			
Interest paid on loans		$ 254.60	
Less other income:			
Profit from sale of feed and supplies	33.80		
Miscellaneous income	34.35		
		68.15	
Net expense from other operations			186.45
Net Profit (before Federal Income Taxes)			$ 7,059.55

Incubation Costs

In addition to the figures given for hatchery operation of incubators, the following distribution of costs for poultry farms is of interest. Misner and Lee in their study of New York state farms report as follows:

For 448 upstate farms, they report, human labor 12.3 per cent, use of automobile 0.8 per cent, use of buildings 4.0 per cent, use of equipment 10.8 per cent, miscellaneous 8.9 per cent and hatching eggs 63.2 per cent.

Service

In many lines of commercial endeavor, organizations are placing service men in the field at the disposal of customers and others as a means of creating good will and business contact. To those producing and selling baby chicks, this should be an object lesson as to the importance of good will. Good will is essential for the continuance and expansion of business.

In view of the fact that a sale is never completed until the customer is satisfied, service must be rendered until the customer's poultry project is on a sound basis. This may be given through direct contact or through circulars dealing with the various phases of poultry management, including brooding, feeding, rearing, and management of adult birds. Personal contact will avoid many losses which usually occur through lack of knowledge and poor management, but for which the hatcheryman is often blamed. Help your chicks to help themselves and your business will increase.

Service normally is of two types: service to hatchery flock owners and to chick customers. It should be expanded to include all poultrymen in the community, thus establishing contacts and building up the good will essential to any business enterprise.

Trouble calls should be taken care of immediately. This will facilitate the application of corrective measures. Where the call involves disease, the state laboratories can be resorted to for diagnostic purposes if such laboratories are available.

Inquiries on poultry matters should also be answered promptly.

A service quite necessary is that of culling hatchery flocks and furnishing satisfactory high-grade male birds for breeding purposes. Only through rigid selection can progress be made in the production of baby chicks that have inherent high egg production and other desirable characters.

Many hatcheries maintain side-line enterprises in addition to their baby chick business. These include chiefly the sale of feed and equipment. They may form an integral portion of the total year's business. Care to avoid stocking up excessively will prevent a heavy carrying charge. Money tied up in dead stock or seasonal equipment may be difficult to liquidate. With either feed or equipment sales, service that satisfies the customer is essential. A hatchery operator must determine the amount of servicing that is feasible under his conditions.

TABLE 4

ELEMENTS OF COST OF DOING COMMERCIAL HATCHING IN 683 HATCHERIES*
(By Capacity Classes)

CAPACITY GROUP—	Egg Cost		Labor Cost		Relatively Nonvariable Costs		All Other Costs		Total Costs	
	Per 100 chicks hatched	Total	Per 100 chicks hatched	% of Total	Per 100 chicks hatched	% of Total	Per 100 chicks hatched	% of Total	Per 100 chicks hatched	%
Under 10,000	$3.47	51.6	$1.42	21.1	$ .84	12.5	$ .99	14.8	$6.72	100
10,000 to 24,999	3.54	52.1	1.48	21.8	.77	11.3	1.01	14.8	6.80	100
25,000 to 39,999	3.50	52.7	1.41	21.2	.74	11.1	.99	15.0	6.64	100
40,000 to 59,999	3.39	49.9	1.40	20.6	.69	10.2	1.31	19.3	6.70	100
60,000 to 99,999	3.47	50.8	1.28	18.7	.71	10.4	1.37	20.1	6.83	100
100,000 to 199,999	3.36	50.6	1.28	19.3	.59	8.9	1.41	21.2	6.64	100
200,000 to 499,999	2.99	43.8	1.03	15.1	.48	7.0	2.32	34.1	6.82	100
500,000 and over	4.83	56.3	1.31	15.3	.63	7.3	1.81	21.1	8.58	100
All Groups	$3.55	50.6	$1.29	18.4	$ .65	9.3	$1.53	21.7	$7.02	100

*U.S.D.A. Bulletin G-33.

Quality Chicks

A reliable hatchery is dependent on the quality of chicks produced, for a major portion of the business volume. This is primarily repeat orders. Only under the influence of satisfactory results does the customer have any desire to continue business relations.

TABLE 5

SCALE OF POINTS FOR JUDGING CHICKS*

(25 Chicks to the Entry)

	Perfect Score	
Weight	10	Two pounds and one ounce is considered standard for an entry of 25 chicks. One point is deducted for every ounce under the above weight.
Vigor	30	A cut of ⅛ to 2 points is made for each chick not showing maximum alertness when judged. Roundness of eye, appearance of shank, and condition of down are considered in determining vigor.
Condition	30	A cut of ⅛ to 2 points is made for each chick showing a deformity, an imperfectly healed navel, pasty vent, signs of sticky hatch or short, thin down.
Color	20	A cut of ⅛ to ¼ point is made for each chick whose down or leg color varies from the average. Furthermore a cut of ⅛ to ¼ point will be made for each chick not conforming to the standard requirements for down color or for color of shanks or toes.
Size	10	¼ point is deducted for each chick varying from the average in size.
Dead Chicks		Four points will be deducted from score for each dead chick or for each chick less than the number required for full entry.

Conditions Eliminating Chicks From Competition

An entry will be eliminated from competition if generally inferior and lacking breed characteristics, if showing evidence of disease or evidence of being fed.

Serious Defects

A cut of four points will be made for each chick showing the following:

1. Type of comb not characteristic of the breed.
2. Sprig on comb of single-combed breeds and varieties.
3. Feathers on shanks or down between toes of clean shanked varieties.
4. Color of down on shank foreign to the breed.
5. More or less than the ordinary number of toes or web-footed condition in chicks.

* Northeastern Poultry Producers Council.

Quality is based on condition of the hatch; uniformity, freedom from disease, size, and inherent production qualities of the chicks; and the sanitary conditions under which they were hatched.

Appearance will produce a favorable or unfavorable impression in the mind of the buyer. All crippled or deformed chicks, as well as those undersized or not conforming to the breed characteristics, should be removed as this will add materially to the uniformity of the shipment.

TABLE 6

COST OF HATCHING 100 CHICKS AND EGG COST PER 100 CHICKS*

Region 5—Maine, Massachusetts, Vermont, New Hampshire, Rhode Island, Connecticut, and New York.

CAPACITY GROUP—	Cost per 100 chicks	Egg cost per 100 chicks	Percent Egg cost is of total cost	Cost per 100 chicks not including egg cost
Under 10,000	$ 8.55	$4.69	54.9	$3.86
10,000 to 24,999	8.35	4.95	59.3	3.40
25,000 to 39,999	10.75	7.45	69.3	3.30
40,000 to 59,999	9.34	4.33	46.4	5.01
60,000 to 99,999	7.43	3.77	50.7	3.66
All Groups	$ 8.45	$4.70	55.6	$3.75

More attention is being directed to the production of disease-free chicks than ever before. The disease of special significance in this regard is pullorum disease, which is transmittable through the eggs to the offspring. The reason for this attention is the bearing it has upon the livability of chicks, particularly during the first two critical weeks. If the mortality is high during this time, the effect on next year's business is likely to be noticeable. Effort ought to be directed, therefore, to correcting or at least controlling this problem.

A standard of quality which can be maintained and approved should be established so that customers from year to year may be assured of receiving chicks of a uniform quality.

Quality chicks should be shipped in attractive containers because of the impression they are likely to create. A good package

*U.S.D.A. Bulletin G-33.

Fig. 22—One of the longest rows of Incubators in the World. Capacity 2 Million eggs.
[Courtesy Harold M. Morris, Morris Poultry Farm.]

is a valuable asset, serving as an advertising medium not only to the customer but others that come in contact with it in transit.

Quality and honesty are paramount for continued success.

Advertising

Advertising or publicity is necessary to the development and expansion of a hatchery business. Advertising may be direct, such as through the medium of newspapers and magazines, or indirect, such as that secured through satisfied customers. A satisfied customer is the best advertisement.

Large hatcheries spending considerable money in advertising should consider securing the services of a competent agency which specializes in poultry advertising. This service can usually be secured at very little if any additional charge provided the agency is given the entire advertising contract.

The small hatcheryman should consult the editors of his local papers and the local magazines in which he advertises. The best way to build up a business is first to develop local distribution through local advertising; then if business develops other mediums of publicity may be used in expanding into outlying districts. The latter can be done much more easily after a hatcheryman has developed a local reputation for honesty, service and quality of product.

Your advertising should be truthful, stating just what you have for sale. It should never mislead. Avoid over-advertising. It might act as a boomerang, depriving the business of some of its profits and even creating a demand you cannot supply. A budget of publicity expenditures should be determined for the season and should be spent in a continuous advertising program, rather than a spasmodic or irregular one. Probably a good plan is to set aside a definite sum for each thousand capacity.

The next step in the program is to decide the medium in which advertisements are to be placed. This will depend on whether the business is to be developed as county-wide, statewide or sectional in scope. In this respect information will be needed as to the circulation and area covered by the medium, the rate charged for advertising, the type of advertising—whether classified or display—and the closing date for receiving copy.

Fig. 23—7-2700 Egg, Single Stage Hatchers.

[Courtesy James Mfg. Co.]

Fig. 24—Circulars that catch the eye. [Courtesy Baby Chick Magazine.]

Before definitely deciding on a particular publication it is often wise to test it out by the insertion of a small ad. The results from this can often be used as a guide. The publication to be used should reach the class of people that are interested in your product.

There are two general types of advertisements: (1) classified and (2) display. They may be of the nature to attract inquiries or to draw immediate orders. Where orders are expected directly, prices must be quoted. It is not always possible to completely describe the product; so in many cases ads may be aimed to attract inquiries for catalog or circular distribution.

Following each inquiry it is desirable to present detailed information to the prospective customer in regard to the quality of the product or products for sale. In some cases this can be accomplished by the use of a simple circular with which is enclosed a detailed price list. In other cases a more elaborate catalog may be required. A definite follow-up program ought to be established to consummate sales.

The amount of money to be spent and the size and elaborateness of the catalog will depend on the size of the business, the class of people you wish to reach and the territory to be covered.

A catalog should present the story of the product to be marketed in an attractive and interesting manner. It should be readable and well illustrated, inspire confidence, tell the exact truth concerning the product and give information as to the organization back of the business and its foundation. Usually the price list and order blank should be printed separately from the catalog and included for the convenience of the customer. The order blank should state in detail the terms of payment and call for sufficient information to enable the hatcheryman to meet all the requirements of the purchaser.

Letterheads and slogans may be utilized with a view to capitalizing on their publicity value. Registering of these will protect the hatcheryman from infringement.

Inquiries may be used in developing a suitable mailing list. Purchasing of ready-made mailing lists is a questionable practice. The most satisfactory way to make additions to your list is to secure the names of prospects from satisfied customers. Revision to bring the mailing list up to date will be necessary at regular intervals to avoid unnecessary expense in mailing material to people who are no longer interested.

Records on the sales increase through advertising must be maintained if the maximum value of the advertising allotment is to be realized. The medium through which you secure your greatest returns should be considered when contemplating expansion in advertising.

Factors for Success in the Hatchery Business

Many factors bear on the success of the hatchery, and for continued prosperity recognition of these points is necessary. These are not necessarily listed in order of importance, as they may be affected more or less by local conditions and the personality and initiative of the hatcheryman. They are as follow:

1. Honesty.
2. Sales and productive ability.
3. Modern equipment.
4. Aggressive business methods in face of competition.
5. Quality eggs and chicks.
6. Efficient advertising.
7. Efficient marketing.
8. High per cent of repeat orders.
9. Efficient and economical production.
10. Service rendered.
11. Accurate record keeping.
12. Guarantee back of product.

13. Personal ability and initiative.
14. Sufficient working capital.
15. Location in a good sales field.
16. Foresight.
17. "Stick-to-it-ive-ness."

TABLE 7

EGG COST PER 100 CHICKS

Price Paid Per Dozen Eggs	Percentage Hatch							
	50	55	60	65	70	75	80	85
40	$ 6.66	6.05	5.55	5.12	4.76	4.44	4.16	3.92
50	8.33	7.57	6.94	6.40	5.95	5.55	5.20	4.90
60	9.99	9.08	8.33	7.68	7.14	6.66	6.24	5.88
70	11.66	10.60	9.71	8.97	8.33	7.77	7.28	6.86
80	13.32	12.11	11.10	10.25	9.52	8.88	8.33	7.84
90	14.99	13.63	12.49	11.53	10.71	9.99	9.37	8.82
100	16.66	15.14	13.88	12.81	11.90	11.10	10.41	9.80
110	18.32	16.66	15.27	14.09	13.09	12.21	11.45	10.78
120	19.99	18.17	16.66	15.37	14.28	13.32	12.49	11.76

TABLE 8

FIXED CHARGE COST PER 100 CHICKS*

Total investment in hatchery, inspections, land, equipment, etc.	Number of chicks produced (000 omitted).												
	10	25	50	75	100	125	150	200	250	300	400	500	600
$ 1,000	$ 1.40	$ .56	$ .28										
2,500	3.50	1.40	.70	$.47									
4,000	5.60	2.24	1.12	.74	$.56								
5,000	7.00	2.80	1.40	.95	.70	$.56							
7,500	10.50	4.20	2.10	1.42	1.05	.84	$.70						
10,000	14.00	5.60	2.80	1.90	1.40	1.12	.93	$.70					
12,500	17.50	7.00	3.50	2.37	1.75	1.40	1.65	.90	$.70				
15,000	21.00	8.40	4.20	2.85	2.10	1.68	1.40	1.00	.84	$.70			
20,000	28.00	11.20	5.60	3.80	2.80	2.24	1.90	1.40	1.12	.93	$.70		
25,000	35.00	14.00	7.00	4.75	3.50	2.80	2.30	1.80	1.40	1.14	.87	$.70	
50,000	70.00	28.00	14.00	9.50	7.00	5.60	4.60	3.60	2.80	2.21	1.74	1.40	$1.16
etc.													

VOUCHER

					Fill this out when you get returns from Hatchery.			
Date	Lot No.	No. Cases	No. Eggs	No. Doz.	No. Crax	No. Culls	Amount Received	REMARKS

*Reprinted by permission—The Baby Chick Magazine.

EGG TRAYING RECORD

Machine Nos.____________________

Hatch No.____________ Date Set____________ Date Off____________

Shipr.	Lot	No. Cs.	No. Eggs	Culls	Crax	Tryr	Date	REMARKS

EGG SHIPMENTS RECEIVED

Date Rec'd in Hatchery	Shipper	No. Cases	How Shipped	Received by	REMARKS

HATCHERY RECORD

Hatch No.____________________

Machine Nos.____________ Date Set____________ Date Hatch____________

Shipper	Lot No.	No. Eggs Rec'd	Cull Eggs	No. Eggs Set	Test Out	S. K.	Brd.	Chicks Counted	No. Chicks Hatched	%

TESTING RECORD

Hatch No.____________________ Date Tested____________________

Shipper	NO. EGGS TESTED OUT	Total Out

HATCHING RECORD

Season____________________ Breed____________________

Hatch No.	Date Test	Hour	Date Set	Date Hatch	Machines	No. Eggs Set

SUMMARY OF ORDERS FOR SHIPMENT

Date Hatch____________________

Date Feed____________________ Hatch No.____________________

Order No.	Name	Address	No. Chix	Flock No.	Breed	Ship By	Time

(Forms used by courtesy of R. E. Waldo—Skookum Chix Hatcheries.)

Questions on Chapter II

1. What factors should be considered in the location of the hatchery building?
2. What is the best basis on which to plan purchase of your needed incubator capacity?
3. How should thermometers be tested?
4. How should one determine the number of eggs to set?
5. How can fertility be determined?
6. What is the hazard in custom hatching?
7. What is the effect of (a) underheating, (b) overheating?
8. How should chick boxes be ventilated?
9. What business records should be kept by the hatcheryman?
10. What services should a hatcheryman endeavor to render his clients?
11. To what extent does advertising enter into the plans for selling chicks?

CHAPTER III

BROODING PRACTICES

Subject Matter

- Brooder Houses
  - Portable Colony Brooding System
  - Permanent Brooding System
- Brooder Appliances
  - Heating Units
  - Feeding Units
  - Water and Milk Devices
  - Appliances of Convenience
- Principles of Brooding
  - Disease-Free Chicks, Healthy Breeding Stock
  - Maintenance of Proper Temperature
  - Fresh Air, Ventilation Control
  - Keep the Chicks Busy
  - Safety from Fire
  - Dryness
  - Sanitary Conditions
  - Proper Feed and Management
  - Litter
  - Proper Housing and Ease of Attendance
  - Protection from Weasels, Rats, etc.
  - Skillful Operation
- Chick Mortality
- Rearing Expectancy
- Purchasing Chicks
- Confinement Brooding
- General Practice
- Brooder Houses

Artificial brooding is the natural accompaniment of the commercialization of the poultry industry. It involves the application of artificial means to the rearing of chicks in large numbers. With this intensification newer methods have been evolved, some of which are fast becoming antiquated with still greater and more advanced developments.

Even though many of the factors of brooding are under man's control, simplicity of operations is not found. Many of the problems facing the brooderyman are intricate and call for good judgment.

The brooding and rearing of a crop of chickens for the annual replacement of all or part of the poultry flock is probably the most important operation on a poultry farm. Unless a man can successfully rear a large percentage of chicks he had better shun at least the production end of the poultry business. If brooding results are not satisfactory, and a crop of poor pullets is raised, certainly a poultryman cannot expect maximum egg production.

A successful poultryman, therefore, is normally a good brooderyman. The results in brooding and rearing tend to prophesy what may be expected.

Brooder Systems

To raise a large crop of quality pullets, an efficient brooding system is necessary. For a matter of convenience the types may be grouped as portable or permanent.

Portable Colony Houses:

The term, portable brooding system, usually refers to the colony houses that are moved by team or tractor. These vary in size from 6x8 to 12x16 feet, and may be constructed with a shed,

Fig. 25—New type brooder or laying house.
[Courtesy R. W. Duck.]

two-third-span or gable roof. They are usually designed to accommodate from 200 to 400 chicks, depending on the floor space provided by the various sizes. In regard to construction see Chapter XVII.

For heating purposes it is customary to provide an individual brooder unit, which may depend on coal, oil, gas, wood, or electricity for supplying heat. The best type of brooder unit to use depends on local conditions, climatic variations and the season of the year that brooding is to be practiced.

The advantages of a portable colony-brooder system may be outlined as follows:

1. Movability—The colony brooder may also be used for a range house, as it can be moved to new land each year or as occasion demands.
2. Convenience of size—It is a desirable size, and as the business develops additional units may be built as necessary.
3. Adaptability—It can be used for both brooding and range-house purposes and if necessary will shelter adult laying birds.
4. Reduction of fire danger—Where separate houses are scattered over the range there is less danger of fire destroying the entire brooding equipment.
5. Ease of construction—Due to its simplicity a great deal of technical skill is not required to construct a satisfactory brooding unit. Such houses may also be purchased ready-made in sections.

The disadvantages of such a system are:

1. Rigidity of structure impaired by moving—Continuous moving of the house from year to year causes the building to get out of shape and numerous cracks to open in side walls and floors. The doors and windows also soon fail to operate properly.
2. Repairs frequent—Due to the rapid deterioration of these buildings an excessive amount of repairing is necessary, entailing considerable expense for material and labor.
3. Annual moving costs large—Where the houses are moved to new range annually the labor cost for moving represents a considerable item.
4. Cleaning difficult—With a wood floor, cracks and crevices develop and accumulate filth, making thorough cleaning difficult if not impossible. When the floor is scrubbed the boards absorb water and tighten up, gripping the dirt in the cracks. After the brooder stove is started and the boards again dry out, the chicks constantly scratch out the dirt and mix it with the fresh litter. This old filth is a potent source of trouble from coccidiosis and possibly other disorders.

5. Depreciation and replacement costs high—With more individual heating units operating, proportionately more parts will need replacement and consequently depreciation will be higher. Yearly up-keep can be reduced, however, if such equipment is properly stored and serviced. Likewise due to the repairs necessary on buildings, as stated under (2), depreciation on these is also high.
6. Excessive chore route labor.

Permanent Brooding System:

The term, permanent brooding system, normally refers to the long, continuous brooder house, but may be used in reference to colony houses which are on permanent foundations. The long type has been gaining in popularity since improvements have been made in brooding equipment adapted to this type of building. In general this type of house can be heated by coal or oil-burning brooder stoves, by electric units, or by a piped hot-water system. In the case of barns or two-story buildings, piped hot-air furnaces can be used.

A growing tendency is for poultrymen to draw their colony houses together in a straight line to approximate the long-type

Fig. 26—Permanent Brooder House, Showing Concrete Platform and Sun-Porch.

[Courtesy N. H. Agr. Exp. Sta.]

brooder house from the standpoint of ease of attendance. Laying houses can be adapted for brooding purposes without additional expenditures when the decrease in the laying flock by the advent of the brooding season is sufficient to permit the vacating of one or more pens. A person starting in the business could construct a laying house and use it entirely for brooding purposes the first year, thus temporarily saving expense.

The advantages of the permanent long-type brooder system are as follows:

1. Ease of attendance—With adjacent pens the convenience is obvious. One person can brood larger numbers of chicks.
2. Permanency—Permanent construction means a life-time investment. It permits the use of concrete floors which are rat- and vermin-proof.
3. Reduced labor requirements—Considerable labor is saved in taking care of the routine work of feeding, watering, cleaning and other phases of general management.
4. Better supervision—Because of the relative simplicity of operation a person will observe the chicks oftener during the day. The closer the operator comes in association with the chicks the better the supervision and the more satisfactory the results secured.
5. Easier to heat and more draft proof—Due to less exposed area and the type of construction it is easier and more economical to heat. A permanent building usually has tighter walls and floor and is, therefore, less drafty.
6. More sanitary—A permanent house makes possible the use of a concrete floor. which is more easily and thoroughly cleaned.
7. Suitable for broiler production—Due to the efficiency of operations, including such items as labor and fuel consumption, the long-type house lends itself readily to the economical production of broilers.
8. Lower depreciation and equipment-replacement costs—While the original costs may be slightly higher. the replacement requirements on equipment are likely to be lower than where individual brooding units are used. Similarly the depreciation on building and equipment is lower.

9. Suited to the use of labor-saving equipment—Because of the continuity of pens, litter and feed carriers, automatic waterers and automatic feeders may be employed. In addition to being great conveniences they are also labor savers.
10. Adaptable to electric lights and appliances—Adjacent pens can easily be wired to permit the use of pilot lights which are of low wattage yet provide sufficient light so

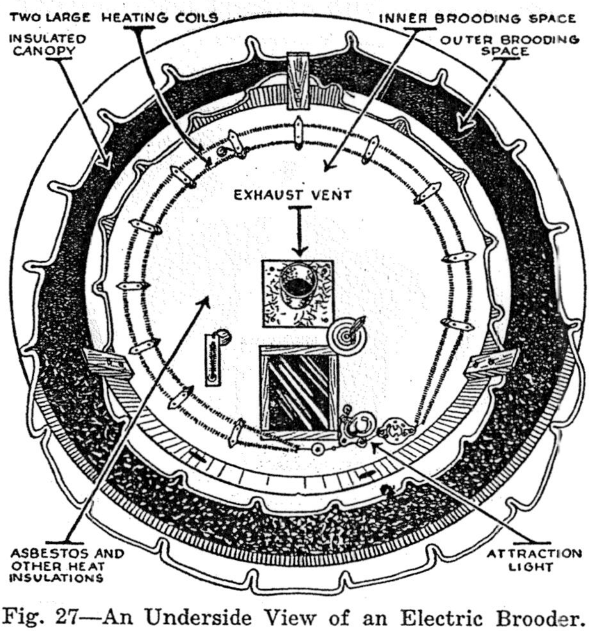

Fig. 27—An Underside View of an Electric Brooder.
[Courtesy Lyons Electric Co.]

that the chicks may drink at night. With such light the chicks also tend to spread around the canopy in a more satisfactory manner. The light should be placed directly over the hover.
11. Adapted to the installation of a ventilation system.
12. Permits efficient use of a central feed storage—With a feed room placed in the center of the building, it is readily accessible to either wing for feed supplies.

The disadvantages of such a system include the following:

1. Soil contamination—With continued use of the land directly in front or rear of house, soil contamination is in-

evitable. It is a decided advantage to employ summer shelters for ranging the chicks where a permanent brooder house is in operation. It is also distinctly helpful to use wire platforms or sun porches with a view to preventing contamination.

2. Fire hazard—Where individual stoves are in use in a long-type house, danger of fire destroying the entire brooding facilities is increased. Where a central heating system is used this danger is materially decreased.
3. Disease danger—With adjacent pens an outbreak of infectious and contagious diseases may spread more rapidly and be much more disastrous.

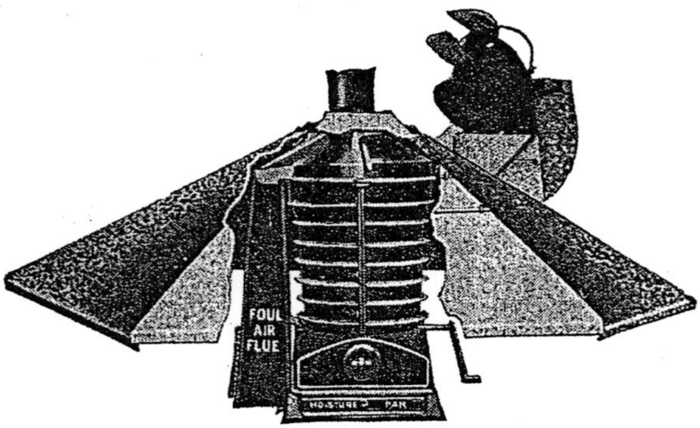

Fig. 28—Coal-burning Brooder Stove with Moisture Pan.
[Courtesy James Mfg. Co.]

Brooder Appliances

The extensive and intensive development of the poultry industry, particularly along the line of artificial incubation and brooding, has brought about the necessary development of brooding appliances. Probably one of the most important of these appliances has been the development of brooder stoves or units. Different types of heating units have been evolved employing coal, oil, gas, wood or electricity as a heat source specifically developed to supply sufficient heat for normal brooding temperatures.

Heating Units:

Coal-Burning Brooders—Generally speaking, coal-burning brooders have been entirely satisfactory in brooding operations in cold climates during the bulk of the brooding season. When used during warm weather, however, difficulty has been experienced

rienced in dampering down the fire sufficiently to prevent overheating.

Care should be taken to select a stove with a grate that is large enough and fits smoothly in position.

The coal stove will usually perform best if chestnut-size anthracite coal is used, as it requires less attention. If only soft coal is obtainable, a mixture of this and coke of suitable size may be used.

Fig. 29—Electric brooder with supplementary oil heater for emergency use.
[Courtesy Shenandoah Manufacturing Co.]

Oil Burning Brooders—These brooders usually do better under milder climatic conditions or where late-season brooding is practiced, as they can easily be regulated to take care of brooding during warm days and cool nights.

Some are designed to use fuel oil and others kerosene or coal oil. Some are wickless; others equipped with a wick. Many require regular cleanings, which necessitates turning out the fire. This constitutes an important consideration for cold-weather brooding when the house is likely to cool off rapidly.

In cases where practical, many stoves can be connected in series to utilize a single storage tank which may be placed inside or outside the building. Some burners are designed for use in the coal stoves. In either case, the fuel line must be leveled carefully, according to directions.

Gas-Burning Brooders—In some sections gas brooders command considerable popularity. Their use offers advantages of convenience coupled with satisfactory brooding results. Operators look upon them as dependable brooding units. Because of possible variation in the burners, they should be operated by the

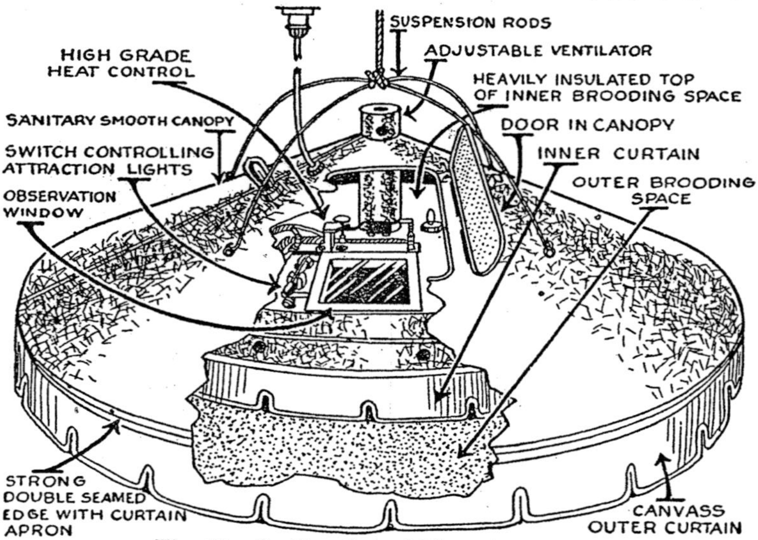

Fig. 30—Section Cut of Electric Brooder.

[Courtesy Lyons Electric Co.]

type of gas recommended by the brooder manufacturer. Where a dependable commercial gas supply is available various types of brooders may be outfitted with burners which will utilize gas.

Electric Brooders—Brooders designed to use electricity for heat are undergoing rapid improvement and are becoming more popular. In colder climates an auxiliary heating system may be necessary to supplement the electric-unit and maintain a reasonable room temperature.

Research work at the New Hampshire Agricultural Experiment Station has, however, demonstrated that it is possible to successfully brood chicks with electric hovers, in sub-zero weather, in uninsulated portable brooder houses without auxiliary heat. These experiments covered a period of four years. The brooding period each year extended over a period of twenty weeks, starting about the middle of November, and thus cover some of the most severe winter conditions as well as the normal spring brooding season. Quoting from New Hampshire Station Circular No. 46: "Comparisons between chicks in insulated and non-insulated houses as to feed consumption, weight gains, condition of chicks, condition of litter and kilowatt-hours consumed,

Fig. 31—Stand for Waterer Used to Avoid Wet Spot Around Fountain.

show practically no difference. The growth of chicks was very favorable, uniformity was well in evidence, and feathering was somewhat more rapid than in similar lots under hot-water brooding."

Current-consumption for the entire group of brooding periods, which averaged six weeks each, varied from 1.36 to 1.87 kilowatt-hours per chick.

Factors to consider in selecting electric brooders are: (1) cost of current, (2) electrical service without interruptions, and (3) seasonal climatic conditions when brooding is to be practiced.

Hot-Water Brooders—This system consists of a central heating plant with hot water pipes running through a number of adjacent pens. The heating plant and radiating pipes may be placed in various places in the house. In recent years the imbedding of the pipes in the concrete floor has attracted considerable favorable attention. Hot water is circulated through copper tubing or wrought iron piping. Some variation exists in the size of pipes and in the distance they are placed apart. Generally ¾ inch copper tubing or 1-inch wrought iron pipe is used. Poultrymen have placed the radiating pipes about a foot apart. The utilization of radiating heating of this type has resulted in a saving of labor.

Wood-Burning Brooders—During the past few years, research work by commercial concerns has resulted in placing on the market dependable wood-burning brooder stoves. Where wood is available on the farm or can be purchased at a reasonable price, wood-burning brooder stoves can be used to advantage.

Battery Brooders—Refer to Chapter IV.

Feeding Units:

Feeding troughs or hoppers should be so constructed that they waste little feed and prevent the chicks from getting into and contaminating the feed. A variety of sizes is necessary to take care of requirements from baby chick to mature bird. Sufficient hopper space is also needed so that a majority of chicks may eat at the same time. Suitable wooden feeders may be made, or metal feeders purchased. Some types are more adapted to small groups of chickens and others to larger broods.

Water-and-Milk Devices:

Ample drinking troughs or fountains should be supplied to meet the requirements of the brood. These ought to be large enough to hold a reasonably large quantity of water. If the waterers are placed at the proper height, contamination from litter will be avoided. The height is dependent on the size of the chicks.

Convenient and more sanitary is the plan of setting the water fountain on a wire platform under which a drip pan is placed. Where possible a large amount of labor may be saved by the use of automatic water fountains. In any case water should be available at all times, and a fresh supply at regular intervals will be appreciated by the chicks.

It is probably a wise policy when feeding liquid milk consistently, to provide it sour or sweet to avoid possible digestive disorders.

Appliances of Convenience:

Convenient appliances for the caretaker are: (1) training guard, (2) wire floors, (3) wire frames under feeders and waterers, (4) wire sun porch, (5) wire rack for green food, (6) electric lights, (7) automatic waterers, (8) ventilating system, (9) vat for disinfection, (10) drain for washing out the pen, (11) thermometers, and (12) cleaning utensils. Many of these are usable mainly in the permanent brooder house.

Fig. 32—Coal-burning Brooder Stove with Tilting Canopy.
[Courtesy Buckeye Incubator Co.]

Principles of Brooding

Successful brooding depends on at least the following 12 fundamentals:

1. Disease-free chicks and healthy breeding stock.
2. Maintenance of proper temperature.
3. Fresh air, ventilation control.
4. Keeping the chicks busy.
5. Safety from fire.
6. Dryness.
7. Sanitary conditions.
8. Proper feeding and management.
9. Litter.
10. Proper housing and ease of attendance.
11. Protection from weasels, rats, etc.
12. Skillful operation.

Disease-Free Chicks and Healthy Breeding Stock:

For maximum brooding results the operator should endeavor to get chicks free from Pullorum Disease, as this disease exacts a heavy annual mortality from the poultry industry. If chicks are affected heavy losses are likely even under the best of brooding conditions. Consequently, the poultryman in attempting to insure his success must give due consideration to this factor.

In so far as quality of chicks is a partial determiner of results, attention should be paid to the health and condition of the breeding stock. The fact that like tends to beget like furnishes the progressive poultryman with a working tool for improvement.

Maintenance of Proper Temperature:

General recommendations for brooder temperature are 90 to 95 degrees Fahrenheit for the first week, with a drop of 5 degrees a week until a reading of 80 degrees is reached. Further reductions will depend on weather conditions and the judgment of the operator.

The action of the chicks should be used as a guide in judging proper temperature. The chicks ought to be comfortable at all times. An intelligent operator soon learns how to judge when the chicks are too warm or too cold and makes stove adjustment accordingly. It is possibly better to run the stove a little too warm, provided the chicks can get away to a cooler area, than to have the stove temperature too low and endanger chilling the chicks.

The effects of chilling are: general enfeeblement, smothering from crowding, "denginess" of chicks in which the wing feathers develop abnormally large in proportion to the size of the chick's body, a loose diarrheal condition, and a higher mortality.

On the other hand the effects of overheating may be evidenced by the shriveling and wasting away of the chick, particularly noticeable in the shanks; crowding in the brooder corners and smothering, in attempting to get away from the intense heat; and a sapping of the vitality of the chicks, leading to increased mortality and uneven growth.

To train chicks to the source of heat a training guard should be used. It may consist of one-foot-inch-mesh wire around the canopy at a distance of a foot or two from the outer edge. This circle may gradually be expanded in size, and can be entirely removed at the end of a week or ten days. Boards, corrugated paper or metal may be used for the same purpose.

Pilot lights or lights of low wattage are used at night, preferably over the brooder stove or under the canopy to dimly illuminate the brooder pen and help chicks that have strayed from the stove to find their way back. This will prevent a considerable amount of chilling and crowding in corners. It will also lead many of the chicks to drink and feed at night.

Fresh Air and Ventilation Control:

An abundance of fresh, pure air without draft ought to be provided, whether by a crude make-shift arrangement or by the installation of a special ventilating system. During the first two

Fig. 33—Method of starting chicks using a guard to keep them near the source of heat. "The ring" should be large enough so that chicks can get away from the heat, so as not to become overheated. [Courtesy N. H. Agr. Exp. Station.]

or three weeks, ventilation requirements are small and no particular provision for this is necessary. In fact, heavy mortality may result from excess ventilation where the air in the room is changed rapidly. As the chicks grow and feather, however, frequent air replacement becomes necessary. A person must use "chicken sense" in ventilation and management as well as in other phases of poultry production.

Controlled ventilation in poultry houses has been increasing in popularity during recent years. Colony brooder houses are not usually supplied with special ventilation equipment but depend on the use of (1) muslin curtain, (2) cow-barn type windows which are hinged at the bottom and tip back from top, and (3) narrow ventilating doors near plates in front and rear of house. For the long permanent-type brooder house special ventilating systems are being manufactured by commercial concerns and can be purchased for installation. Regardless of the system employed, provide sufficient ventilation *without drafts.*

Keeping the Chicks Busy:

Hopper and drinking space is sufficient when the majority of the chicks can eat at the same time and a large number drink together. This will help in keeping the chicks busy and aid in preventing pernicious vices such as cannibalism. Ample hopper space will also materially decrease the percentage of culls as the timid chicks will have greater opportunity to eat.

Supplementary feedings of green foods will help to keep the chicks busy.

Safety from Fire:

Even with the most modern brooding equipment the fire hazard must be guarded against, and certainly antiquated and inefficient equipment should be discarded because of this. The overheating of a single stove may result in the loss of the entire house its equipment and chicks, as well as the labor, feed and overhead expense already involved.

Elevation of the stove above the floor by means of legs or bricks is advisable. The base of stoves setting directly on the floor must be properly insulated with asbestos. Keep litter away from the base of the stove as a preventative measure, and in removing ashes and live coals take care not to drop any into the litter.

A check draft in the smoke pipe is desirable to prevent overheating during windy weather. Similarly the thermostat must be in working condition to regulate the stove and maintain proper temperature. A practice only too common with broodery men is to leave the ash door of the stove open for greater draft. This is to be discouraged as hot coals might escape into the litter and cause fire.

A metal collar or roof saddle is another good investment, as it will prevent the stove pipe from coming in contact with the roof boards and roofing material.

Dryness:

For best results the brooder house needs to be dry. Dampness aids parasitic intestinal diseases to gain a foothold. A moist condition with the heat that is naturally present in a brooder house provides ideal conditions for the development of these troubles. A clean, fairly dry, environment is a good precaution.

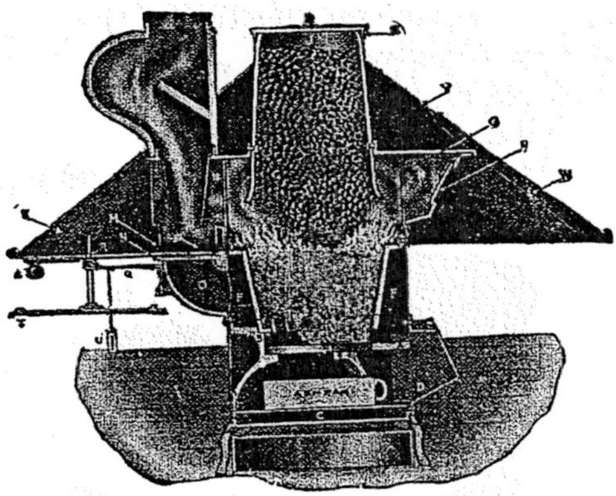

Fig. 34—Newton Coal Burning Brooder Stove. Note large coal capacity. Metal thermostat control at left.
[Courtesy Buckeye Incubator Co.]

Sanitary Conditions:

Many states have inaugurated extensive programs relative to successfully rearing chickens in which the maintenance of sanitation has been an outstanding feature, thus indicating its importance. Follow sanitation measures not only during the brooding season but throughout the entire cycle of management practices.

Proper Feed and Management:

Feeding and management of chicks lend themselves to a rather wide range within which successful results can be obtained. Only clean wholesome foods should be fed, and these only in waste-proof hoppers into which the chicks cannot gain access or void their droppings. The ration must be one that satisfies the requirements of the chicks for maintenance and growth. For further information on feeding see Chapter VIII.

In most instances, particularly during cold weather, it will be found necessary to tend the stove four times daily, preferably the first thing in the morning, at noon, at supper time and at night just before retiring. Normally stoves will require fueling twice daily, morning and night. The night fueling should be done before the chicks "go to bed" so as not to unnecessarily disturb them later.

When they are first allowed out of doors they ought to be permitted to remain outside for only short periods of time. They should be watched carefully to see that they do not become chilled. As they acclimate themselves, lengthen their outdoor periods until eventually they are given complete freedom.

A small yard temporarily wired will be found advantageous in teaching the chicks the location of the brooder house. With contaminated ground and disease conditions present, wire porches

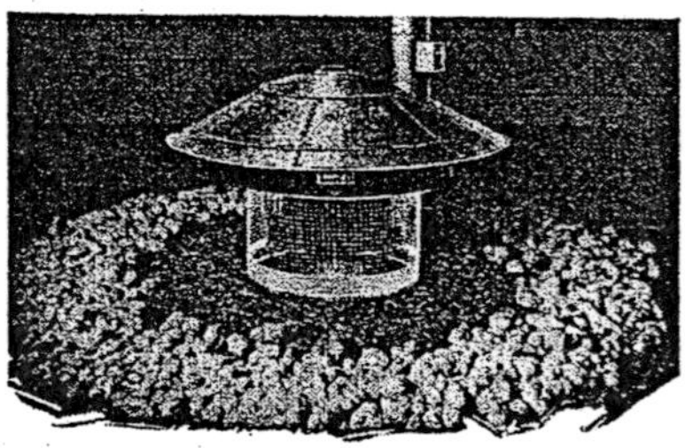

Fig. 35—Oilburning Brooder, Showing Location of Chicks at Night.
[Courtesy Simplex Brooder Stove Co.]

or sun parlors are advisable to confine the chicks to the house until 8 or 10 weeks of age, when they are transferred to shelter houses on clean range. Where yards and wire porches are used the waterers and feeders may be placed outdoors as an aid in getting the chickens out of the house, at least until they become accustomed to being out.

Yards or porches tend to relieve congestion in the house by giving the chicks a greater area over which to distribute their activities.

Litter:

Many kinds of litter are available for brooding chicks. Whatever is used ought to be of fine quality, and a type which will not tangle. It must be dry and absorptive and reasonably free from dust. Avoid moldy litter.

Provide enough bedding so that the chicks will not come in direct contact with the cold floor and so that the droppings will be taken up readily.

The use of cut alfalfa or clover or similar material which a chick will readily pick at and eat is to be discouraged, because of the possibility of digestive disorders from the consumption of fibrous material and droppings.

Economical use of litter must be considered; however, it should be replaced as often as necessary to maintain dryness.

Proper Housing and Ease of Attendance:

A satisfactory house may be economically constructed without undue expense. Nothing elaborate is necessary. However, the brooder house must be tightly boarded, afford ample protection from the elements and have a leak-proof roof.

Locate the brooder house so that it will be convenient for the operator in attending to the routine work accompanying brooding. An unexposed or protected position will prove helpful.

Consideration of the ease of disinfection and cleaning, and of the use of labor-saving devices where possible, is desirable.

The house must be large enough to meet the requirements of the number of chicks to be placed in it. Crowding, one of the most common mistakes, is an unnecessary evil resulting in stunted birds and unsatisfactory growth.

Do not overestimate the capacity of either the brooder stove or house. Most 60-inch hovers are rated at a thousand-chick capacity but results are better if only 350 or 400 are allotted. About one-half square foot of floor space per chick is essential for the first few weeks and this should be increased to one square foot at approximately six weeks of age when the sexes ought to be separated by removing the cockerels. In some of the heavy slow-maturing breeds it will not be possible to distinguish sex in all the individuals this early.

Protection From Weasels, Rats, and Vermin:

Predatory animals are likely to cause heavy loss to the poultryman if he is not adequately protected against these enemies. Take every precaution in planning and constructing brooder houses to make them rat and vermin proof. In one night a single rat or weasel may kill the majority of the chicks in a brooder house.

Such losses cannot be consistently sustained. In many cases they can be charged to carelessness on the part of the attendant.

Skillful Operation:

No one thing is so important in the successful brooding of chickens as the skill of the operator. In any phase of poultry keeping the human factor determines failure or success to a large extent. Practical experience must be gained by each individual, either under the supervision of an experienced poultryman or through the school of hard knocks.

Given the same opportunity for education and experience, and possessing identical facilities in stock and equipment, two persons will seldom attain equal success in their chosen profession. To succeed in the poultry business, one must possess an interest in it, combined with natural initiative, good judgment and ambition. The urge to succeed must be backed by careful attention to details. In other words an arm-chair philosopher must apply his ideals.

A successful poultryman has a sympathetic understanding for his birds, ability to observe, and the physical as well as the mental ability to put his observations into operation.

Chick Mortality:

In studying the causes of chick mortality we find that it varies in different sections of the country, but in general the major disease conditions that prevail in the adult flocks are very liable to be transmitted to the growing stock. There is also a variation from year to year as to the chief causes of losses in the growing stock. New diseases appear and old ones may disappear or become less troublesome, or be brought under control by vaccines, sanitation, or other measures.

A summary of autopsy records at the Agricultural Experiment Stations indicates that there are from 35 to 50 or more disease conditions that cause loss to poultry growers, during the normal brooding and rearing season.

It is wise to establish a policy of checking up on any mortality that occurs to see whether it is due to management, or disease. In each case, where doubt exists, it is well to check up with the established Poultry Pathology Laboratory of your own State College or Department of Agriculture. A prompt diagnosis may immediately locate the cause of trouble, and save you monetary loss, as well as many regrets and unnecessary mortality.

For further discussion of the principal diseases affecting growing chickens, see Chapter XIX.

Rearing Expectancy:

Number of eggs required to produce one chick.........................1.5
Number of eggs required to produce one mature pullet............3.3
Number of chicks required to produce one mature pullet..........2.5
Per cent of mortality first three weeks.................................2-15%
Per cent of mortality after first three weeks to maturity....2-10%
Number of cockerels necessary to retain at broiler age to produce one mature selected breeding male of good quality.. 3

Purchasing Chicks:

Several factors must be borne in mind in securing stock of a desirable quality from the viewpoint of livability of the chicks, growth of the chickens, and the egg production of the pullets. As a rule few poultrymen investigate thoroughly enough the source of stock for the replacement of their individual flocks.

Prevention of excessive disease and mortality during the brooding season demands great caution in selecting stock most likely to show the best livability. The greatest assurance of this perhaps, is the purchasing of chicks from stock known to be free from Pullorum Disease. It is not at all uncommon for chicks to have a very high mortality when produced from infected flocks. This, however, is but one point to consider. Vigor must not be forgotten. Baby chicks are similar to pullets or adult fowls. The inherent production qualities are likewise another factor to be considered.

It is a wise policy to purchase chicks close to home. The buyer is more likely to receive satisfaction and more certain to know the man with whom he is dealing. He also has an opportunity to check on the vigor, the production of the flock in question, and the livability of the chicks the man is producing.

One of the guides used extensively in the past has been the breeder's reputation for producing good chicks and his relations with his clientele.

If one contemplates buying chicks indirectly by the purchase of eggs for hatching, an additional check on their prospective vitality is available in the size of the eggs. It is logical to expect good-sized chicks from good-sized eggs. It is advisable to set only eggs weighing in the neighborhood of 24 ounces to the dozen if they came from pullets, and 26 ounces to the dozen if they were produced by old hens.

It is usually advantageous to purchase chicks on the basis of the purpose for which they are intended. If they are purchased for production of broilers entirely, for instance, it would be well to consider the market adaptability of the various strains and breeds before making a final choice.

Fig. 36—Layout for Utilizing Brooder Houses Where Chicks Must Be Raised in Confinement Under Fly-Screening.

Heavy breeds are marketable at a greater price per pound than light breeds. Where broiler production is the main object a poultryman can take advantage of a more favorable market price by considering this point in advance. The same consideration holds true where a man intends to market a large percentage of the cockerels as capons. The problem of purchasing baby chicks is, consequently, a serious one calling for a good deal of forethought. The solution will link itself quite definitely with profitable production or failure.

Confinement Brooding

Confinement brooding is an interesting development in poultry management and has been evolved primarily to assist in controlling parasitic diseases. The newer knowledge of feed requirements has made it possible to rear chicks in confinement without encountering some of the troubles previously experienced.

Confinement falls within two distinct classes of usage: (1) chicks confined for a few weeks during the intense brooding stage and then given access to range; (2) chicks confined throughout the entire brooding and rearing season. This is usually followed by confinement as mature birds.

In many instances chicks are confined for periods varying from 6 to 10 weeks before having access to range due to unfavorable weather conditions encountered by early season brooding, or due to contaminated soil conditions. In restricting chicks from ranging, sun porches may be helpful as they materially increase the available floor space and eliminate considerable crowding of the chicks. These porches may have a solid or wire floor. Wire will permit the passage of droppings and thus keep the chicks out of contact with them.

Fig. 37—Wire Floors Are Used Inside These Houses and the Outdoor Yards. All Brooder Houses Are Also Fly-Screened.

[Courtesy Allan Buckley.]

If the motive of confinement is the control of tapeworm infestations, the porch and house should be thoroughly screened against flies. In some cases where under ordinary brooder management it has been impossible to successfully raise pullets, brooding in confinement to maturity has had to be resorted to. In these cases the poultryman has had no option as to his rearing methods if he intended to remain in the business.

This type of rearing is not advised for all to follow; only where conditions make it necessary should such a step be taken. Where chickens are confined to maturity, additional brooding houses and other equipment are necessary to separate the birds and give them better opportunity for development. Additional space is also essential because houses cannot be used for successive broods of chicks as is feasible where summer range shelters are in use.

Advantages of Confinement Rearing

Some of the advantages claimed for confinement rearing to maturity may be listed as follows:

1. Greater supervision is possible, permitting constant check on all the factors of rearing.
2. Because of the elimination of the use of range, less land is necessary.
3. Losses from predatory animals tend to be prevented.
4. Regulation of feeding conditions for different ages of chickens is possible.
5. Accurate feed cost records are obtainable.
6. Rigid sanitation assists in the control of parasitic diseases.
7. Chicks mature earlier than where allowed free range.
8. Houses may be permanently and conveniently located.

Fig. 38—Portable 10′ x 20′ Brooder House, with wire sun-porch. This house is connected to the power line, so that electric brooders can be used.

[Courtesy N. H. Agr. Exp. Sta.]

As the other side of the picture we cite the following disadvantages:

1. Additional housing space is necessary to properly "thin out" the developing birds.
2. Limited space per bird tends to cause trouble from (a) cannibalism, and (b) feather picking.
3. Where houses are close together the fire hazard is increased.
4. Labor costs may be increased.
5. Unless sanitary measures are properly applied severe epidemics of parasitic diseases may occur.
6. The construction of sun porches is practically necessary.
7. Litter cost is increased.

Fig. 39—Brooder Houses Placed for Convenience of Operator. Sun-Porches in Use.
[Courtesy N. E. Poultryman.]

General Practice:

The popular practice of rearing chicks is to confine them to the houses, with or without sun porches, for a limited period from 6 to 10 or 12 weeks, dependent upon weather conditions. After this they are placed in summer shelters or remain in colony houses, with access to clean range. The summer shelter permits the efficient and economic housing of the developing birds and the use of the brooder house for an additional brood.

This method allows the best utilization of brooder houses and equipment, and the combination of short confineme period with the use of clean range has proved extremely efficient in the production of quality pullets.

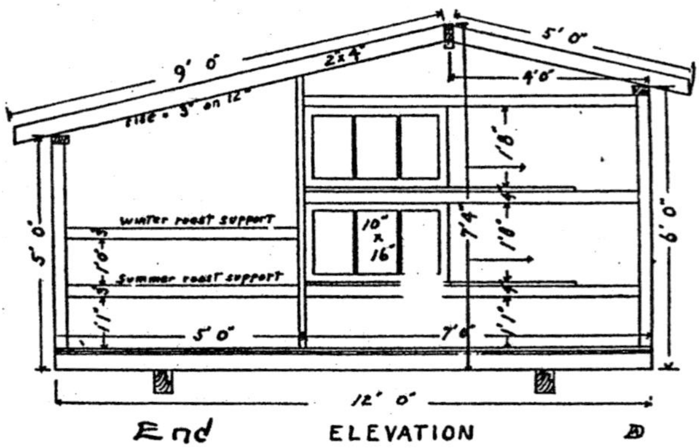

Fig. 40—End View Framing for Connecticut Brooder House.

Connecticut 12′x12′ Portable Brooder House or Colony Laying House

BILL OF MATERIAL

Lumber List:

Skids, 2—4″x6″x14′
Sills and Plates, 11—2″x4″x12′
Rafters, 7—2″x4″x14′
Center support, 1—2″x6″x12′
Studding, 12—2″x3″x12′
Roosts, 9—2″x3″x12′
Matched Boards—900 board feet
Window Sills, Slides, Frames and Finish: 9—1″x4″x12′

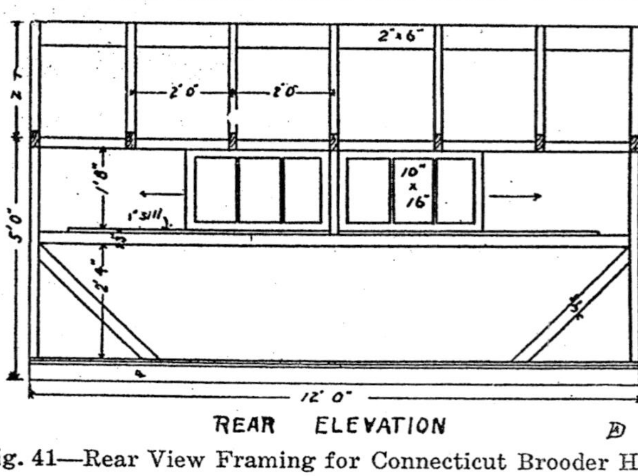

Fig. 41—Rear View Framing for Connecticut Brooder House.

Roof Trim, 4—1"x2"x14'
Pine for door, 2—12' boards 5" wide;
1—12' board 8" wide;
2—16' boards 6" wide.

Fittings:

Roofing paper, 4 rolls
Windows, 4 cellar sash, 3 lights 10"x16"
Unbleached sheeting, 2 yards, 2' 8" wide
Wire netting, 25 ft., 1" mesh 3' wide
Glass substitute, 7½ ft. x 3 ft. wide
Nails, 15 lbs. 8d—5 lbs. 10d—4 lbs. 20d—2 lbs. 8d finish
Staples, 2 lbs. wire
Hinges, 2 pr. 3" loose pin butts.
1 pr. 6" strap
1 pr. 4" strap
Hook, 1—6" wire
Building paper, 150 sq. ft.

Cornell 10'x12' Brooder House

MATERIAL REQUIRED

Lumber

Size	Use
4—3" x 6"-14' or 8—2" x 6"-14'	runners (chestnut or hemlock)
2—2" x 4"-16' 6—2" x 4"-12'	braces, joists, plates, studs (chestnut, hemlock, or yellow pine)
6—2" x 4"-10'	rafters (chestnut, hemlock, or yellow pine)
1—2" x 3"-16' 1—2" x 2"-14' 1—2" x 2"-10' 7—2" x 2"-12'	roost supports, roosts, girts, door, and window framing (chestnut, hemlock, yellow pine)
1—⅞" x 12"-4½'	window guards (spruce, yellow pine, or fir)
2—⅞" x 8"-12'	front and rear ventilator (spruce, yellow pine, or fir)
2—⅞" x 3"-14' 2—⅞" x 4"-14'	corner casing (spruce, yellow pine, or fir)
3—⅞" x 4"-12'	window casing (spruce, yellow pine, or fir)
2—⅞" x 4"-12' 2—⅞" x 4"-10'	frieze
1—⅞" x 4"-16' 1—⅞" x 1"- 6'	door and exit casing (spruce, yellow pine, or fir)
2—⅞" x 4"-10' 2—⅞" x 5"-14'	cornice (spruce, yellow pine, or fir)
3—⅞" x ⅞"-12'	window and exit stops (spruce, yellow pine, or fir)
1—⅞" x 6"-14'	guard between runners (spruce, yellow pine, or fir)
1—⅞" x 6"-12' 1—⅞" x 3"- 6'	window and exit sills (white pine)

120 feet 16' matched flooring (hemlock, spruce, pine, or fir)
160 feet 12' matched material for vertical siding, front and rear (chestnut, hemlock, or yellow pine)
140 feet 14' matched material for vertical siding, ends (hemlock, chestnut, or yellow pine)
175 feet 14' matched roofers (chestnut, hemlock, or yellow pine)

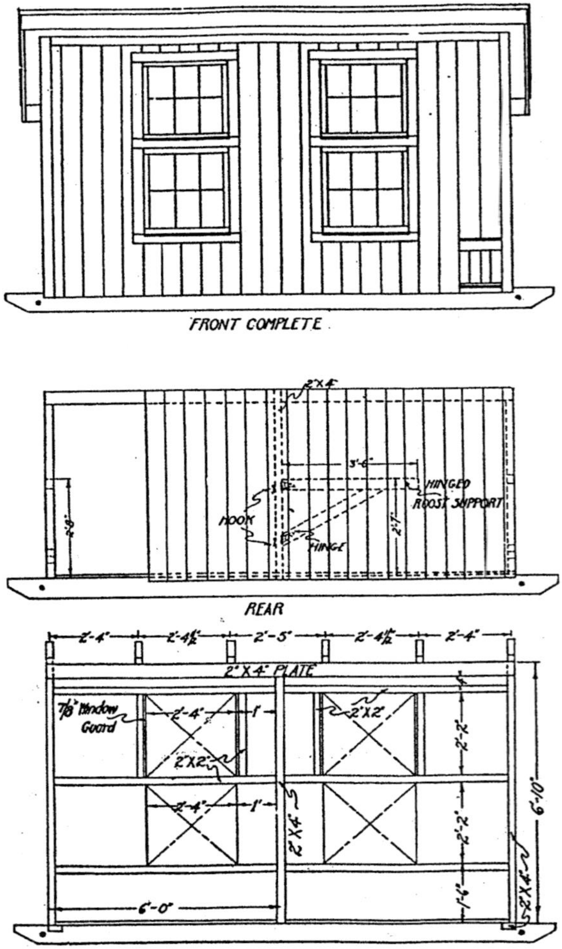

Fig. 42—Plan of Cornell 10′ x 12′ Brooder House.
[Cornell Ext. Cir.

Hardware

4—6-light 8" x 10" sash
1 pair 4" strap hinges (roost support)
1 pair 6" T-hinges (door)
5 pair 3" strap-hinges (front and rear ventilators)
10—2" iron buttons (front and rear ventilators)
1—No. 3 thumb latch
14', ¾" mesh, 3' poultry wire (window openings and guards)
2 rolls 3-ply roofing paper
1 pound 20-penny nails
2 pounds 10-penny nails
15 pounds 6-penny nails
½ pound poultry staples
1 pair 3" strap-hinges (upper exit door)
1—2" iron button (upper exit door)

List of Material for 12'x16' Brooder House

Use—	No. or Amt. Required	Size	Length
Runners	4	2" x 6"	18'
Rafters	9	2" x 6"	14'
Floor joists	7	2" x 6"	12'
Plates	3	2" x 4"	16'
Studding and framing	4	2" x 4"	10'
Studding and framing	3	2" x 4"	12'
Studding and framing	4	2" x 4"	14'
Roofers, door (T & G)	300 sq. ft.	6" or 8"	
Siding	360 sq. ft.	Novelty	
Flooring (T & G)	235 sq. ft.		
Sub-flooring	210 sq. ft.		
Trim	170 linear ft.	⅞" x 4"	

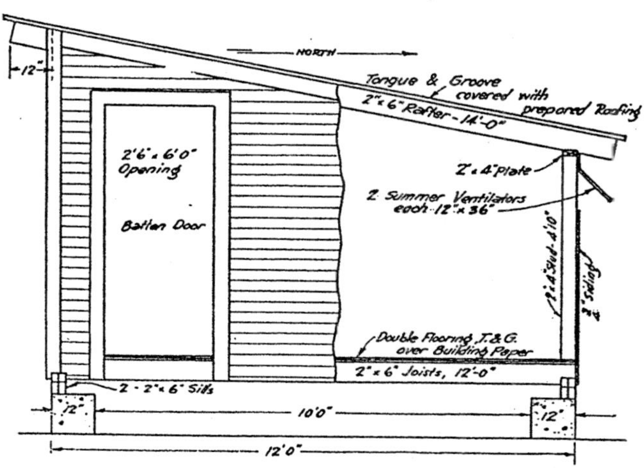

Fig. 43—End View of 12' x 16' Brooder House.
[Penn State College Circ. 115.]

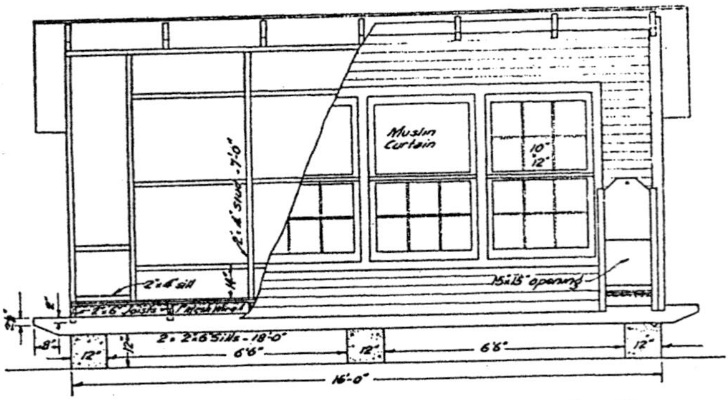

Fig. 44—Front Elevation of 12′ x 16′ Brooder House.
[Penn State College Circ. 115.]

Hardware

Roofing (3-ply)	252 sq. ft. or 3 rolls
Windows	6–6 light @ 10″x12″
Wire netting	224 sq. ft. 1″ mesh (under floor)
Wire netting	20 linear ft. ¾″ mesh x 3′ (windows)
Wire netting	16 linear ft. ¾″ mesh x 1′ (rear ventilator)
Loose-pin hinge	10 pair 3″
Strap hinge	1 pair 6″
Galvanized chimney for attachment to roof	1
Nails, muslin, lock, paint, etc.	

Questions on Chapter III

1. Give the advantages and disadvantages of the portable colony brooder system.
2. Give the advantages and disadvantages of the permanent brooder system.
3. List the types of brooders that are available on the market.
4. What are the principles of successful brooding?
5. Is the action of the chicks a good guide to the correctness of brooding temperature?
6. What factors should be borne in mind in purchasing chicks?
7. In some cases chicks are confined throughout the brooding and rearing season; list the advantages and disadvantages of such a system.

CHAPTER IV

BATTERY BROODING AND LAYING CAGES

Subject Matter

There has been a tendency in the past several years to rear and manage birds under more artificial conditions in order to place the operations under more direct control. This has developed as a result of the poultryman's endeavor to curtail disease, reduce losses from preditory animals, and to secure greater efficiency in production. Improved production efficiency places a given poultryman in a better competitive position. The use of battery brooders and laying cages represents but one of many methods that poultrymen have adopted in their quest for greater efficiency. There are many proponents and opponents of this method of production.

Battery-Brooder Rooms:

Quite often it is possible to remodel existing buildings into suitable battery-brooder rooms at considerable saving. This should be considered before a decision is made to build a new structure.

New buildings are more adaptable as they can be made to embody the principles underlying the successful operation of battery brooders. Before drawing up plans, the capacity to be brooded and the length of this brooding period need to be decided. If chicks are to be carried to broiler age, for instance, a relatively larger capacity must be furnished per chick than if they are to be retained in the battery for only a short time.

The accompanying discussion can be used as a basis for designing a layout which will meet the specific requirements of your own brooding program. Among other considerations special attention should be given to the following factors which are tied up directly with efficiency and optimum results.

Fig. 45—Universal Battery Brooder, with Contact Heat.
[Courtesy James Mfg. Co.]

1. Size of Rooms—Rooms need to be sufficiently large to allow space around the batteries for the attendant to fill feed-and-water troughs and make necessary adjustments for head room for the chicks, and to permit the dropping pans and wire bottoms to be easily withdrawn for cleaning. This means careful measurement, location and arrangements of batteries to most economically utilize available floor space. If feed is to be kept in the battery room, plan a special allotment of space to accommodate it, or a special room conveniently located.

2. Ceiling Height—Most batteries average about six feet in height. Since at least three feet of head room is desirable, a nine-

foot ceiling is preferable. A low-posted ceiling increases the danger of overheating and adds to the difficulty of supplying a sufficient volume of air in warm weather without forced ventilation. Provide sufficient ceiling height to allow approximately 2,500 cubic feet of air space for each 1,000 pounds of birds, live weight. Where forced ventilation is employed ceiling height can be reduced in proportion to the increased air movement.

3. Insulation—Decided advantage is to be found in thoroughly insulating the walls and ceiling of the battery room. If the house is to be built above the surface of the ground the floor may be insulated likewise. The main advantage of insulation is that it prevents the loss of excess heat and thus reduces fuel cost. Fluctuating temperatures brought about by seasonal and other climatic variations are also reduced. Insulation further aids in maintaining the even temperature which is so desirable in battery rooms.

Competent local building authorities should be consulted in regard to details of construction and the type of insulating material to use. Information can be secured from the United States Bureau of Standards as to the insulation value of different types of insulating material.

4. Heat and Ventilation—Make provisions for adequate heating and ventilating equipment. Heat may be supplied through steam or hot-water radiators, electrical heating units or coal stoves. Many devices are on the market for controlling heat from these sources. The more dependable the heat control the better the results that can be secured with the chickens.

Any one of a number of ventilating systems may be used, but a definite one should be incorporated in the original plans of the battery-brooder room. Either a gravity system or forced-air draft can be installed depending on the capacity to be brooded.

5. Light—Few battery rooms are dependent on natural light through windows to illuminate all sides of it. Where artificial lights are to be added it is well to place them with regard to the position of the batteries. They should be high enough so that shafts of light in the battery decks are avoided, and situated so that the feed and water pans are illuminated but the battery decks remain dark as a preventive measure against cannibalism.

Although windows are usually discriminated against they do furnish a supplementary means of ventilation during warm-weather brooding.

6. Water Supply and Drainage—The water supply should be convenient for each battery room. It is a means of increasing the humidity of the room when desired. Adequate water and drainage systems aid in reducing the labor of cleaning and filling water pans, cleansing the wire floors of the battery, and scrubbing and hosing the brooder-room floor.

7. Disinfecting Vats—Metal vats consistent with the size of the battery equipment will help materially in the disinfecting operations. Provision should be made for draining the vats, whether they are movable or non-movable. The vat may be located in the battery rooms, if it is large enough, or in an adjacent room. Parts of the battery equipment which are readily movable can be placed in this vat and allowed to soak, preparatory to cleaning and disinfecting.

The Operation of Battery Rooms

In the operation of battery-brooder rooms the main considerations probably are temperature, ventilation and humidity. The proper control of these will determine success to a large extent.

Temperature:

The temperature should be one in which the chicks are comfortable, showing no signs of overheating or chilling. Experience has demonstrated that it is best to reduce the level of temperature as rapidly as possible as the chicks grow older. This will tend to promote feathering and hardening off of the birds. The temperature reading ought to be taken consistently at one location.

Ventilation:

Ventilation is important in batteries for three reasons: (1) to furnish the chicks with oxygen for respiration, (2) to remove the carbon dioxide and moisture produced by the respiration processes, and (3) to remove the moisture from the evaporation of water from the water pans, additional amounts supplied to increase the relative humidity, and that and odors produced from the droppings.

The air requirements of young birds are comparatively small, but increase rapidly with size and activity; so the ventilation system should be large enough to be satisfactory when the battery room is operated at capacity and the birds are at the maximum age at which it is planned to carry them in the batteries. A change of air once in every five or six minutes is advisable. This may be excessive for early season brooding when the room is stocked with young birds, but inlet and outlet flues may be controlled with dampers so that any number of air changes per hour may be provided, up to 10 or 20 per hour when necessary. In warm weather when the difference between outside and inside temperatures narrows and air exchange is restricted, the flues

Fig. 46—Warm and Cool Room Brooding Layout, Showing Ventilation.

[Courtesy Shenandoah Mfg. Co.]

may be adjusted to full capacity. If this does not supply sufficient air during hot weather, the installation of exhaust fans may be necessary. The exhaust capacity of these fans is rated so that they can be purchased in accordance with the size of the battery room and the rapidity of air movement desired.

Where the construction of the battery rooms permit, the intake flue should be arranged in many cases to heat the air as it enters the room. This is particularly essential where forced air draft is used, and may be accomplished by locating the air-intake flues on the side walls directly back of the heater or heating units.

For size and capacity of flues see Chapter XVII.

Automatic electrical devices can be used advantageously for controlling either temperature or ventilation or both.

Humidity:

Humidity has been considered by some to play an important part in the results of battery brooding. Since experimental data is lacking to indicate the desirable range, it is wise for the present to attempt to maintain humidity at the normal level of climatic conditions. In semi-arid regions, however, it is well to raise it above this level.

Fig. 47—"Radiant heat" brooding system under construction.
[Courtesy of A. M. Byers Co., Pittsburgh, Pa.]

The Operation of Batteries

Next in importance to familiarity with the battery room itself is a thorough knowledge of the operation of the batteries. Management practices applied to one specific type or make of battery do not necessarily apply to others, but in general the principles of operation are applicable to all types.

The decks of storage brooders may be varied in depth from 8 to 14 inches to meet the increasing size of the chickens. In addition the sides should be adjusted frequently to give the chickens sufficient room to get their heads through in order to eat and drink. It is essential that the spacing of these adjustable sides be checked often. Never leave the space so wide, however, as to allow chicks to get out of the decks into the feed and water pans and eventually onto the floor.

Cleaning the battery and its equipment calls for considerable effort. All means to speed up and increase the efficiency of this operation should be utilized. Establish a definite routine, cleaning dropping pans and waterers daily, and feeders and wire floors at regular intervals. Obviously it is most convenient to clean the wire floors when the chickens have been removed from the decks. This is also a good time to thoroughly clean the batteries.

To facilitate the removal of manure from the dropping pans various materials may be used, such as papers, sand or litters. Conveyor belts in large installations facilitate cleaning and substantially reduce labor expenditure.

Keep the batteries in good repair. At the end of the season, or occasionally between broods, clean all equipment and replace all defective parts. Keep the castors greased to assist in moving the brooders during operations. Thermostats or electrical contacts of any type need regular inspection and servicing. Where electrically heated batteries are used it is often advisable to rewire them, when decks are in series, so that various decks may be used independently.

It is well to check the accuracy of all important thermometers. This can be done with a doctor's clinical thermometer by comparing readings.

Management of Chicks Brooded in Batteries

Chicks to be brooded in the battery may be removed from the incubator as soon as they are thoroughly dry and placed in the brooder with water and feed before them. Observation will indicate whether the chickens can and do reach the water and feed. Occasionally it is necessary to place a feed hopper inside the deck in order to get them to eating, else they may starve to death through inability to locate the feed and water or to gain access to it.

In starting chicks in a deck approximately three feet square, usually 100 are accommodated. This number is apparently sufficient at the time. As the chicks grow they naturally require more floor and feeding space; so thinning soon becomes imperative. The advisability of distributing them into two decks at the end of the first week is seldom criticized. Regular thinning needs to be practiced, and the more space allotted per chick the better. Manufacturers in many cases have overrated the carrying capacity of their batteries.

As to feeding chicks in storage brooders, the same feeding practices are followed as outlined in the chapter on chick feeding.

Keep the feed and water pans filled so that the chicks by maintaining a maximum intake of feed will reach desired weight at an early age. This will keep the chicks busy and tend to obviate trouble from cannibalism.

It is considered good practice to inspect the chicks at least once daily in order to detect and remove any crippled, injured, weak or dead ones.

Fig. 48—Contact heat starting brooder and growing unit. 100 broilers monthly.

[Courtesy Hawkins Million $ Hen]

In the use of lights in the battery room the operator has the option of a reasonable length of day, an all-night light or a combination of the two. Under average poultry-farm conditions and the usual routine of work, a day of 12 to 14 hours is probably long enough.

As in floor brooding there is a distinct advantage in separating sexes as early as they can be identified. This gives the pullets in particular a much better chance to get their share of feed and avoids the inevitable competition between sexes due both to sex instinct and to greater size of the males. Males may quite often be identified earlier in battery than in floor brooding.

Hardening off the chicks is fundamentally the object of brooding; so they may be taken from the relatively high temperature of the incubator clothed only in down, and in a comparatively short time be completely feathered and ready for the range. In batteries, hardening off is accomplished by placing the chickens in a cooler environment as they grow older, either by reducing the temperature level in the deck or by transferring them to cooler rooms specifically provided for such purposes.

If the chicks have not been hardened off as they developed, rapidity of feather growth is likely to be retarded. Poor feathering is not to be entirely construed as indicative of poor brooding, however, because the rate of feathering is a genetical factor. Many strains of birds feather very poorly. The selection of well-feathered birds at broiler age for use as breeders the following hatching season will help to eliminate poor feathering from the flock.

In handling chickens in the brooders avoid frightening them. Light breeds, possibly because of their temperament, are not so successfully confined to batteries for a long period as the heavier breeds. They become rough in appearance and have a tendency to go off their feed if retained beyond normal limits.

Quality of Stock

The quality of the stock produced under battery brooding is likely to be distinctly different from that of stock produced with normal brooding systems. In the first place the skin is likely to be more tender and more likely to tear and become abraised. The quality of flesh similarly is softer, with the result that shrinkage of live weight in transit is likely to be excessive.

In view of this it will be found advantageous, in cases where the volume is sufficiently large, to kill, dress and pack the birds on the premises where produced and then ship to market. In many cases it may be possible to develop a satisfactory local market, either for live or dressed broilers. Thus shrinkage during transportation may be avoided.

Comb growth is apparently faster in cockerels raised in batteries, the color is usually pale, and with single comb breeds and varieties a percentage of the combs will flop to one side. It is possible to distinguish the sex of battery chickens at an earlier age than under floor-brooding conditions. Cockerels raised in batteries until marketed will in some cases show blisters and calluses in the region of the anterior portion of the keel bone. These will detract from the market value of the carcasses.

Work conducted indicates that where chickens are brooded under normal floor conditions for eight to ten weeks, that if they are then transferred to batteries for finishing as broilers that they will develop breast blisters if left much over two weeks on wire.

The Use of Batteries for Storage

Batteries lend themselves exceptionally well to the display of chicks in the hatchery sales room. Many people wish to see before buying. An attractive display may be responsible for many direct sales. Because of respiratory problems, storage brooders should not be located in the incubator rooms. Furthermore, they should not be located in a room directly connected with incubator rooms.

Storage brooders enable the proprietor to hold for later sales the surplus chicks resulting from an "over hatch," excess settings over bookings, and cancellations of orders. Such facilities eliminate the necessity of dumping chicks on the market. Also with storage chicks on hand the necessity is eliminated of turning away or delaying delivery of the late orders that accompany all hatchery enterprises. If this type of business is encouraged, however, it may place the hatcheryman in the difficult position of not having orders booked previous to setting date and consequently add considerable gamble to his work.

Maintaining storage brooders also gives the proprietor a check on the livability of the chicks he is producing. If the chicks are identified this gives him an idea of what hatchery flocks are producing the best quality chicks.

The sale of started chicks has been an addition to the hatchery business. In a general way poultrymen are able to purchase chicks one to three weeks old, or any designated age, at the rate of the baby-chick price plus three to five cents a week, or on the basis of a sliding scale such as three cents the first week, four the second and five the third depending on the current economic situation.

Exact cost records should be obtained on some lots, to serve as a guide in establishing the weekly charge. Many buyers, particularly those purchasing small lots, are not equipped to handle baby chicks properly. The purchase of started chicks eliminates this drawback, as well as the early mortality that the purchaser would likely sustain.

The Cost Factor

The cost of battery brooding, including batteries, ventilation and insulation, must be given careful consideration by the average poultryman before investing in it. Where he already has a conventional brooder plant the purchase of additional equipment might be unwise.

If batteries are bought they should be operated 15 to 20 weeks during the season to be a justifiable investment. Otherwise the overhead expense will add an excessive charge per unit produced.

Because changes in equipment are rapid, models are soon obsolete, adding a high depreciation charge to the investment.

Management practices have shown the necessity of spreading out the chicks as they grow; so that if chicks are to be kept in batteries much beyond three weeks of age, considerable brooding equipment is necessary and this adds materially to the cost factor.

Many feel there is a saving of labor in brooding chicks in batteries and that this factor may offset to some extent the cost of purchasing the equipment. It is doubtful, however, whether any material saving in labor is afforded when chicks are grown in these brooders to broiler age.

TABLE 9

CHART FOR COMPUTING BATTERY CAPACITY

	Sq. In. Per Chick
1 day to 1 week	12
1 week to 3 weeks	20
3 weeks to 5 weeks	25
5 weeks to 7 weeks	40
7 weeks to 9 weeks	45
9 weeks to 10 weeks	50
10 weeks to 12 weeks	65
12 weeks up	80+

Considerations in the Purchase of Batteries

In considering the purchase of batteries one must determine his need for additional brooding equipment. Where sufficient conventional equipment is available it is doubtful if a poultryman is justified in purchasing battery brooders.

Installation of this type of equipment means considerable expense. The prospective purchaser should, therefore, analyze his finances to make certain that such purchase will not impair his working capital. He should also consult others regarding their experience with similar equipment.

Brooding needs are another factor determining how extensively batteries can be used to increase the general efficiency of management. As mentioned before battery brooders are adaptable for storage of chicks for hatchery operators and for specialized broiler production, but they are less suitable to a commercial egg enterprise.

A comparison of the various types of batteries and their full cost of operation under local conditions is likewise advisable. The reputation and reliability of the battery manufacturer are important considerations. Another feature to consider is whether the type of building necessary to properly house batteries is already on the premises or will need to be built.

In deciding if the construction of the battery meets one's needs, emphasis ought to be placed on its design, mechanical construction, sanitary provisions and ease of cleaning, durability and the convenience for handling chickens. Careful study of these points will make it possible to more intelligently select equipment adapted to one's brooding requirements.

Knowledge and Skill of the Operator:

With intensive confinement methods placing more of the details of management directly under the operator's control, it is imperative that more attention be given to details. The success of brooding operations depends to a considerable degree on the knowledge, skill and personal aptitude of the attendant.

Most successful businesses have been built from small-scale operations without plunging into costly layouts. From a small start, with basic information on battery management, an enterprise can be built that is consistent with the growth and financial status of the business.

Troubles Encountered in Battery Brooding:

Purchase of batteries may often result from a desire to control certain conspicuously damaging diseases. Very often a poultryman is misled into believing that the use of batteries will control disease outbreaks. The truth is they really provide only the means of controlling certain specific diseases through sanitary conditions. Coccidiosis, for example, is supposed to be a self-limiting disease—one that leaves a bird unless reinfestation occurs through the ingestion of fecal material containing the causative organism. Although batteries are provided with wire bottoms to permit the passage of droppings out of reach of the birds, outbreaks of coccidiosis have been known where regular cleaning was not practiced.

Popular articles have been known to stress the assumption that batteries also control mortality from pullorum disease. No brooding equipment can be expected to adequately control disease troubles originating in the breeding stock or spreading during incubation.

Where considerable numbers of chickens are confined to one room in close association, outbreaks of contagious and infectious diseases are likely to be disastrous. Fowl pox and respiratory troubles, such as bronchitis, newcastle, and tracheitis, spread rapidly under these conditions; prompt control measures are essential if serious loss is to be avoided.

Cannibalism because of the close confinement of the birds also presents a serious problem. As precautionary measures keep

Fig. 49—Laying Cages Showing Belt Cleaners, Waterers and Large Type Feeders.

[Courtesy Davis-Watson Mfg. Co.]

the feed and water pans filled; darken the decks somewhat by not permitting direct shafts of light to enter; maintain a reasonably cool yet comfortable temperature; do not overload the capacity of the decks; avoid frightening the birds, as this might result in injury and lead to cannibalistic outbreaks; and remove all dead chicks from the decks.

Water blisters and tender skin are commonly found in birds raised to broiler age in batteries, and are very likely traceable to this method of brooding. Trouble has been experienced with very poor feathering where a high temperature has been maintained.

The transitional change, when birds are removed from batteries to other systems of brooding or to the range, may cause temporary trouble in management. Birds brooded on the floor are more hardened than those reared in batteries.

A minor difficulty encountered in the use of battery brooding equipment is caused by "floorwalkers." These are chicks which have escaped from the batteries through improper adjustments or construction of the sides and may be a source of considerable annoyance.

Increased knowledge of the various management phases of battery brooding may increase the desirability of the utilization of such equipment. Through proper management many of the troubles can be avoided.

Laying Cages

Individual cages for laying hens are used to a limited extent. Any number of birds up to several thousand may be individually caged in such batteries. The main advantages stated in favor of this system are as follows:

1. Accurate individual records of egg production are possible without trapnests, the labor of trapnesting, or the losses of hens from suffocation in the trapnests in hot weather.

2. No litter is necessary with the use of the wire-bottomed batteries. This means considerable saving in a year's time on commercial propositions, and over a period of years is a decided advantage. Obviously this eliminates crop-bound conditions due to litter eating.

3. Capacity of buildings is increased, because batteries are obtainable several decks high. This makes laying batteries ex-

tremely adaptable for the factory system of production at minimum housing costs.

4. Adult mortality is lessened. Mature mortality is one of the pertinent problems confronting poultrymen. With hens individually isolated the danger from cannibalism should be practically eliminated.

5. Definite accurate culling can be done on the basis of the records secured. No guess work as to whether the bird is laying or not, enters into the culling operation. Birds going out of production can be detected immediately and disposed of.

Fig. 50—Partial View of a 4-500-Bird Installation of Laying Cages.

6. All birds may feed at will. Under average housing conditions the timid and weaker birds are forced away from the hoppers by more aggressive individuals and usually secure an inadequate feed supply. The laying batteries permit all birds to feed at will without interference.

Naturally all systems of managing laying birds also have some definite disadvantages which should be considered before changing one's scheme of management. Laying cages may be difficult to clean, especially the wire floors. They are not adaptable for breeding purposes; they make supplementary feeds, such as green or succulent foods, pellets and wet mash, difficult to use; and do not readily accommodate both grain and mash. Additional unattractive points include the cost of installation, difficulty in controlling spread of infectious diseases, the possibility of heavier egg breakage, and the need for artificial heat in the pen to prevent the water in individual drinking pans and the droppings from freezing during cold weather, and breakage of bones in handling and removing birds.

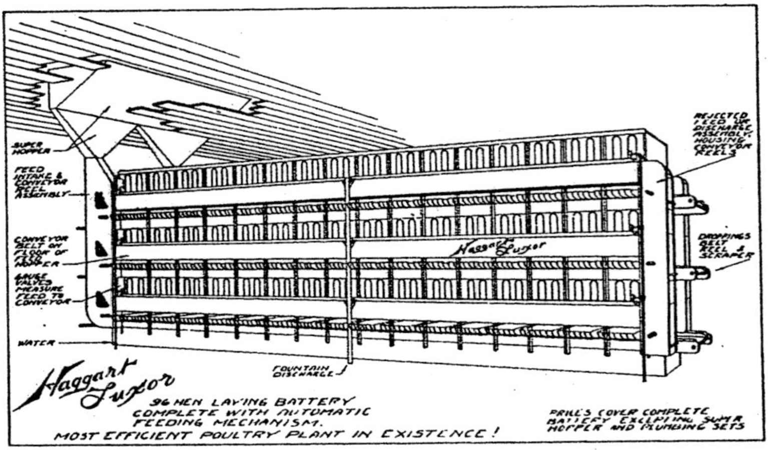

Fig. 51—Laying house with automatic feeding device.
[Courtesy Haggard Mfg. Co.]

Management of Laying Hens in Cages

The laying cage has passed through its "mystic" period. Secret methods, feeds, formulas and practices no longer exist to confuse the issue and confound the beginner as well as the experienced. Management of laying hens in cages is now recognized as essentially the same as for laying hens kept under normal floor conditions.

Seven Essentials in Cage Management

It is the purpose of this brief discussion to point out some of the essentials in securing results with birds in cages, without any attempt to discuss the inherent advantages or disadvantages of laying cages. In the first place in selecting a cage, attention should be paid to cost per bird, possible upkeep and repairs, ease of taking apart to clean and disinfect, simplicity in cleaning and disinfecting, ease and completeness in removal of droppings, cost for renewal of belt or paper used for droppings, ease of cleaning waterers and feeders, ease of getting birds in and out of cages, possibility of producing eggs that keep clean after rolling to egg tray, construction of cages to reduce chance for birds to get legs or wings caught and thus become injured. These are at least a few considerations when planning to invest money in cages in which birds will be restricted to single "cells" for life.

After selecting the cages, there are at least seven essentials for handling the laying hens that are to be placed in them. These include good housing, good birds, good feeds, sanitation, replacement stock, disease control, and the personal skill of the manager.

Good Housing:

Good housing does not necessarily mean expensive buildings. There are extant on many farms substantial buildings that with a reasonable outlay can be remodeled into satisfactory laying-cage rooms. In our own experimental work we have utilized a 30 x 30 laying pen which happened to be insulated and provided with a ventilating system. We had to add only heat and water to take care of 480 birds in cages, where we had previously housed about 250 birds on the floor.

Obviously where running water is provided the room temperature must be kept above freezing. The best room temperature is not yet known, but during the winter probably 40 to 55 degrees is satisfactory; in summer as cool as possible, providing against too high a temperature. Undoubtedly uniformity of temperature is essential during the winter.

Fig. 52—2000-Hen Laying Cage Plant. Right wing starting and growing rooms.

[Courtesy E. S. Munro.]

It is of course desirable if "starting from scratch" to consult the cage manufacturers as to specific room requirement for their cages and to construct new buildings specifically designed for the handling of birds in cages so as to utilize labor to the best advantage and secure the desired results in production.

Good Birds:

The second important consideration is good birds, those bred for high egg production, and relatively disease free. No one can hope to get satisfactory production from hens in cages unless the birds have been bred for the production not only of a reasonable number of eggs per bird but also egg size.

Some have mentioned that certain strains of birds are better adapted to the cages than others. Up to the present moment, however, we do not believe this has been convincingly demonstrated. No doubt "strain" differences will show up even more in the cages than on the floor because we have a chance to observe each bird daily if we desire. In our observation, good breeding is more important than breed or variety.

Good Feeds:

The third essential for satisfactory results with laying cages is good feeds and feeding methods. Fortunately there seem to be quite a range of possibilities as to feeds or rations that will give good results in securing high production from hens in cages.

Inasmuch as birds are greatly restricted in the cages, attention to these details is especially important: (1) Feed preferably twice daily regardless of the ration used. (2) Make the birds clean up their feed regularly. If they do not, withhold additional feed. (3) Avoid accumulation of "tailings" or junk or waste feed and material in the bottom of the hoppers. (4) Keep feeders and waterers clean and sweet. (5) If some birds fail to clean up their feed, remove it and replace it with fresh feed, or dump their remaining feed into an empty hopper of a "food hog" and let her clean it up. (6) Keep the birds' appetites keen at all times. This can be done by seeing to it that all hoppers are cleaned out at least once daily by the birds. (7) If birds tend to become too heavy and over-fat, the ration must be modified accordingly. This is the place where strain difference enters into the picture as some fatten more readily than others. In general it is easier to maintain body weight under heavy production in the cages than on the floor. (8) If hens in cages have no direct sunshine it may be necessary to increase the vitamin D in the ration by as much as 50 per cent. (9) If the egg shells are thin and too many cracks and broken eggs result, it may be necessary to increase the shell-forming material in the ration. (10) Grit should be provided at regular intervals. Oyster shell can be fed in the same manner.

Fig. 53—2000-Hen Laying Cage Plant. Starting and growing rooms. [Courtesy E. S. Munro.]

Sanitation:

A fourth factor for consideration in the handling of hens in laying cages is sanitation. The system of management should included a definite schedule for removing droppings and cleaning (1) feeders, (2) waterers, (3) floors of the cages to avoid soiled and stained eggs, and (4) the walls and floors of rooms. Provision should also be made in such a schedule for the regular and systematic cleaning of the entire laying cage units themselves.

Replacements:

A fifth consideration is replacements. In order to utilize laying cages to the best advantage, a definite system must be worked out whereby at regular intervals a new crop of pullets will be coming into production at the proper time to replace those that die, are culled, or prove "duds". There is no place where culls and "duds" can be sorted out quicker than in cages. A cull just doesn't lay; neither does a "dud", but a "dud" looks and handles like a layer, but still never lays. Such a bird on the floor would stay in your pen for months without being detected and removed.

Fig. 54—2000-Hen Laying Cage Plant in operation.
[Courtesy E. S. Munro.]

Farm account records indicate that those poultrymen who keep their houses filled the year 'round nearest to full capacity, in general make the most money. This would apply to even a greater extent to keeping laying cages full to capacity all the time. Consequently a replacement program must be carefully planned.

Disease Prevention:

A sixth important factor is freedom from disease. This has already been mentioned in reference to quality of producers. It is obvious that if we expect to net real returns with laying hens in cages, those birds that are started should be as free from disease as possible. No one can expect to take a flock of pullets infested with parasites, either external (lice or mites) or internal (round worm, tape worm, etc.), place them in cages, and expect to get anything but mediocre results.

The Personal Factor:

There are, of course, many other things that might be listed in trying to cover the high spots, so to speak, of managing laying hens in cages, but we cannot refrain from adding at least a note about the seventh, namely the personal factor. As we have tried to indicate, a large share of the success of any method of handling laying hens is the personal experience and aptitude of the one doing the work; also that in this respect, attention to details is of vital importance. For example, failure to follow some of the suggestions in regard to method of feeding may easily result in 10 to 20 per cent lower production. If you have someone feeding who will not take care of the details mentioned, you can immediately increase per cent production by getting someone who will.

Mortality:

In general hens die from about the same causes in the cages as on the floor; (1) Ruptured egg yolks, (2) enteritis, (3) tumors (chiefly of egg organs), (4) respiratory disease, (5) uremia, (6) leucosis, (7) injury, (8) internal hemorrhage, (9) coccidiosis, (10) nephritis.

Breeding Caged Layers

Confinement of layers in cages has resulted in a detailed study of a method of carrying on breeding work under this method of management. Results of experimental work have indicated that artificial insemination is possible. Also that satisfactory results can be obtained under field conditions when proper technique is used.

Details of Method

Under practical conditions collection of the semen must be considered a two-man job. This also applies equally to the injection of the semen into the oviduct of the hen.

The males should be kept in individual cages and removed as desired. They should be "milked" daily as they respond more readily and yield more semen.

One operator holds the male with the abdomen towards the second operator. The male is held with the legs apart and hang-

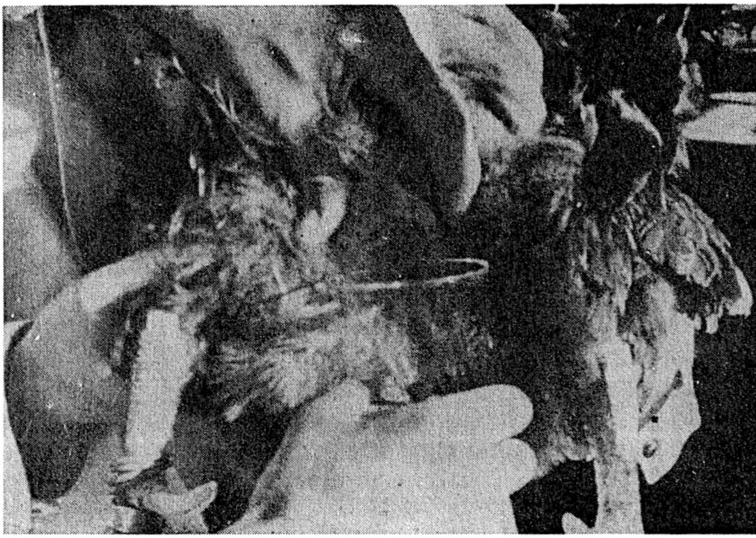

Fig. 55—Method of securing semen from male for artificial insemination.

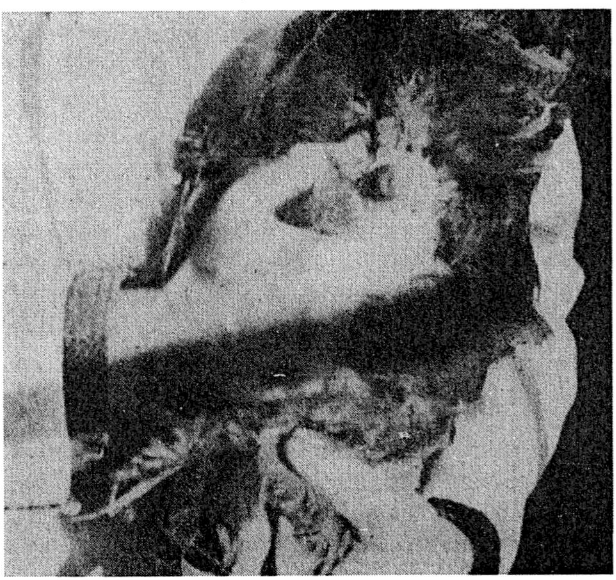

Fig. 56—Position of hen, and hands of operator, for securing eversion of the oviduct for artificial insemination.

ing freely. The second operator is equipped with a glass funnel inserted in a cork, and a paraffin plug in the base of the funnel. Holding this funnel as illustrated warms up the glass to body temperature and helps to preserve the sperms alive and active. The second operator then secures the proper activity of the male by, (1) applying pressure to the abdomen by gentle massage or tickling, or (2) by applying the hand along the back at the base of the tail head and applying suitable pressure and manipulation, or (3) by applying pressure and massaging behind the tail and over the vent. This amounts to more or less a squeezing process with the direction towards the vent. The action should be swift and regular.

The semen which is ejaculated is then collected by the second operator in the glass funnel. Care should be taken that uric acid or other foreign material does not get into the semen. A male in good physical condition will provide enough semen, under average conditions, to inseminate 4 to 10 females when used in .2 c.c. doses.

Artificial Insemination

For most satisfactory results, the semen should be used as soon as collected, and the hens should be inseminated after they have laid.

There are two positions in which the hen may be held for artificial insemination. One position consists of the operator holding the hen under his left arm with the hen's head down, and abdomen towards the second operator. The first operator holds the hen's legs with his left hand and with his right hand slowly and gently applies pressure to the abdomen, causing the hen to evert her oviduct.

The second method consists of letting the hen stand on a bench or platform with her head towards the first operator. By stroking the hen's back (cushion) she is caused to assume a squatting position. Pressure is then applied to the abdomen with the right hand, causing an eversion of the oviduct.

In either case, the second operator having the syringe ready inserts it into the oviduct a distance of 1½″ to 2″ and injects the proper amount of semen. Simultaneously the operator holding the hen releases the pressure on the abdomen so that the oviduct returns to its normal position.

Precautions to Be Observed

1. Only vigorous males should be used.
2. Males should be kept in individual cages.
3. Males should be "milked" daily.
4. If males have been mated, they should be held in isolation two or three days before "milking" for semen.
5. In everting the oviduct of the female, care should be taken not to apply too much pressure.
6. Semen should be fresh and free of foreign material.
7. Dilution of semen may not be as satisfactory as use of undiluted semen.
8. If dilution is necessary to inseminate a certain number of hens, it should not be diluted over 25 per cent. That is, 75 per cent semen and 25 per cent physiological salt solution (raised to the body temperature of the hen).
9. Hens should be inseminated at least once weekly and preferably twice weekly for best results.
10. .2 of a cubic centimeter of semen is recommended for each insemination.

Questions on Chapter IV

1. What are the provisions one should consider in designing a battery brooder room?
2. Ventilation is important in the battery room for what three reasons? Does the requirement depend at all on the size of birds?
3. How would you manage chicks brooded in batteries?
4. Batteries are frequently used as storage brooders. Why?
5. Where should storage brooders be housed?
6. What factors should be borne in mind in considering the purchase of battery equipment?
7. List advantages and disadvantages of the battery system in the management of laying birds.
8. What are the seven essentials in the management of layers in cages?
9. How can caged layers be used for breeding?

CHAPTER V

BROILER PRODUCTION

Subject Matter

Until comparatively recent years broiler production was entirely seasonal, a side line to the annual crop of pullets. This was due to the fact that little knowledge was available on the prevention of leg weakness or rickets during the unseasonal periods of the year.

With the discovery of vitamin D sources, as a preventive, brooding of chicks became practicable the rear 'round. Through improved management practices successful hatching of chicks also became possible throughout the year. These have been incentives to commercial broiler raising, leading many poultrymen in the United States to cater to the year 'round market.

Broiler production is an important phase of the poultry business and for that reason it is discussed as a special phase of the industry.

The Economics of Broiler Raising

The raising of broilers has greatly extended the use of brooding facilities and equipment. Naturally it has aided in carrying the overhead entailed in the purchase of such equipment, and consequently has reduced proportionately the overhead charges debited to the regular seasonal chicks. Many have pointed to the economy of this using equipment that would otherwise remain idle until the spring brooding period. In addition broiler production has aided in the employment of hired help throughout the year.

In some cases plants have been established to make broiler raising a specialty, and many limit their season to the most favorable months for prices, or may cater to a special market, producing broilers every month or even every week in the year. Broilers are reasonably well adapted to the factory method of

Fig. 57—Showing Wire Platform in Use Under Feeders in Rearing Broilers.
[Courtesy Hubbard Farms.]

production. The location of such plants in sections of dense population is desirable to provide an outlet without too great a transportation charge and shrinkage loss and mortality in shipment. Due to the influx of poultrymen into this phase of the business with the resultant lowering of price, a more extensive knowledge of management practices and costs of production is necessary to successfully compete on a smaller margin of profit.

TABLE 11

FEED CONSUMPTION PER 100 CHICKS—NEW HAMPSHIRES

	Pounds Mash	Pounds Grain	Total	Total To Date
First 4 weeks	91.1	21.8	112.9	112.9
Second 4 weeks	245.3	100.3	345.6	458.5
Third 4 weeks	378.8	148.4	527.2	985.7

Conditions under which chickens are brooded and rations employed will influence food intake; so under varying conditions somewhat different figures will be secured on food consumption.

Growth, similar to food consumption, is not constant, but is variable, influenced by many factors besides heredity. The breed and strain of birds, conditions under which brooded, feed and consumption level, all will react on the rapidity of growth.

A study of broiler production in Maryland indicates that there are many factors that influence the net return. The following, however, are the principal ones concerned: (1) gross income and costs; (2) efficiency in the use of feed; (3) efficiency in the use of labor; (4) efficiency in the use of fuel; (5) marketing costs; (6) cost of baby chicks; (7) capital investment; (8) size of the enterprise; (9) weight of birds at time of sale; (10) mortality; (11) feeding practices; and (12) marketing practices.

Considering cost factors, feed represented 54% of the total or $25.62 per 100 birds; baby chicks 22% of the total cost or $10.72 per 100 chicks; labor represented 8% of the total cost, or $3.70 per 100 birds; cost of marketing 5.7% of the total or $2.71 per 100 birds; overhead cost 5.5%, averaging $2.64 per 100 birds; general expenses 5.2% or $2.50 per 100 birds. Combining all these costs represents $47.89 per 100 birds raised, for the production and marketing of broilers as reported in this study of broiler raising in Maryland.

Weight of broilers at time of marketing had a very material influence on profits. Reference to Table 13, indicates that unless special market conditions prevail, that it is best to market birds between the weight of 2.75 to 3.00 pounds or over. This study also indicates that those farms raising between 5,000 and 15,000 broilers made the most satisfactory profit per bird.

Fig. 58—Partial view of a 165,000 broiler plant in Delaware.
[Courtesy Harold M. Morris, Morris Poultry Farm.]

TABLE 12

COMPARATIVE COST OF PRODUCING BROILERS SEASONALLY*

Item	Fall Broilers[1]	Winter Broilers[2]	Spring and Summer Broilers[3]
	Cost Per 100 Birds	Cost Per 100 Birds	Cost Per 100 Birds
General expenses:			
Fuel	$ 2.75	$ 2.24	$ .66
Light	.09	.12	.22
Water	.07	.09	.09
Disinfectants	.16	.15	.05
Total	$ 3.07	$ 2.60	$ 1.02
Feed:			
Mash	$23.70	$24.68	$21.05
Grain	2.46	2.77	2.70
Green feed on farm	.16	.09	....
Litter	.54	.46	.26
Oyster shell	.01	.01	....
Sand, grit, charcoal	.03	.06	.39
Milk	.41	.62	1.50
Codliver oil	.08	.06	.04
Miscellaneous	.02	.02	....
Total	$27.41	$28.77	$25.94
Labor:			
Hired	$ 1.20	$ 1.06	$ 1.33
Family	2.69	2.89	1.92
Total	$ 3.89	$ 3.95	$ 3.25
Total cost of production	$34.37	$35.32	$30.21

[1]Includes Sept., Oct., Nov., Dec.—58 lots, 247,715 birds raised.
[2]Includes Jan., Feb., Mar., April—64 lots, 262,394 birds raised.
[3]Includes May, June, July, Aug.—5 lots, 33,419 birds raised.
*Maryland Bulletin 390

TABLE 13

RELATION OF THE WEIGHT OF BROILERS AT SELLING TIME TO PROFIT*

Item	Average Weight per Bird, Pounds					
	Under 2.24	2.25 to 2.49	2.5 to 2.74	2.75 to 2.99	3 and Over	Average All Farms
Total number of birds	109,030	122,871	322,087	129,257	214,20[illegible]	897,238
Average number of birds per farm	10,903	7,228	7,002	6,463	13,387	8,231
Average per farm:						
Income	$5,765.10	$2,815.06	$3,296.76	$3,627.30	$8,995.6[illegible]	$4,065.27
Cost	6,129.30	3,443.12	3,279.91	2,917.30	8,194.8[illegible]	3,941.72
Profit	—364.20	—628.06	16.85	710.00	800.8[illegible]	123.55
Average per bird, cents:						
Income	52.9	38.9	47.0	56.1	67.2	49.3
Cost	56.2	47.6	46.8	45.1	61.2	47.8
Profit	—3.3	—8.7	0.2	11.0	6.0	1.5

*Maryland Agr. Sta. Bulletin 390.

TABLE 14

TYPE OF FUEL, MORTALITY RATE, AND FUEL COST*

Type of fuel	Number Chicks Started	Mortality Rate	Cost of fuel per 1,000 chicks started
	Thousand	Per cent	Dollars
Wood	74	17.8	9.48
Coal	279	13.5	15.46
Kerosene	51	11.3	24.85
Other (1)	4 (2)	11.4	28.41

(1) One brood with gas at a cost of $10.71 per 1,000 chicks started, one with distillate at a cost of $44.93 per 1,000 chicks.

*Arkansas Exp. Sta. Bulletin 412.

Fuel is an important item of expense in brooding, however it is of less significance than management of the brooder itself. Failure to properly operate the brooding device may result in chilling or overheating the chicks, resulting many times in heavy mortality and uneven growth. Table 14 illustrates the variability of fuel costs in brooding as well as the possible effect on mortality.

Breeds Suitable for Broiler Production:

The price level of the marketable product has some bearing on the selection of a breed or variety for broiler production. Crosses, New Hampshires, R. I. Reds, White and Barred Plymouth Rocks, are satisfactory for broiler production. Leghorns bring a lower price and seldom offer many opportunities for broiler specialization, unless a special market is developed by the individual producer. Many other breeds and varieties may be particularly adapted to meet the requirements of special market demand.

Additional factors to consider in selecting a breed for the broiler business are the uniformity and rate of growth and the rapidity of feathering. Bareback broilers do not present an attractive appearance, are more difficult to dress, and are likely to be cut in price. Selection of rapid-feathering cockerels and pullets as breeders will aid in minimizing trouble from this source.

In the last few years, nationwide attention has been directed to the production of more desirable meat type chickens

Fig. 59—First Carload of Broilers Shipped in New England Leaving Durham, N. H.

[Courtesy N. H. Agr. Exp. Sta.]

for broiler production. This interest is likely to continue and in time may approximate the attainment of turkey breeders in developing broad-breasted turkeys.

Since cockerels as broilers attain a heavier weight than pullets, broiler raisers will find it desirable to purchase straight cockerel chicks where possible. With the gradual development of a demand for sexed pullet chicks, more and more cockerels are available to broiler raisers, frequently at a very attractive price.

Size of chicks seems to be closely related to size of the individual at broiler age. Thus it is evident that a decided advantage is likely in selecting good-sized chicks for the broiler crop. A relationship is also likely between size of chick and size of hatching egg; so much may be gained by selecting a strain that produces large eggs, and by rigorous choice of eggs for hatching purposes.

Management

For continued success in the enterprise the production of "A No. 1" broilers is of prime importance. Little is to be gained by producing an inferior product. The difference in net return will prove to be the disparity between profit and loss. The enterprise does not offer "gold mine" possibilities. To make profits a quality product only must be produced. With quality as the goal, brooding facilities are essential which under your conditions will produce the type of broiler desired.

TABLE 15

FLOOR SPACE REQUIREMENTS FOR BROILER PRODUCTION

Group	Space Allotment per Bird	Weight 100 Birds at 12 Weeks	Weight 100 Males at 12 Weeks	Weight 100 Females at 12 Weeks	Feed per 100 Birds for 12 Weeks	Pounds Feed for 100 lbs. Broilers	Percent Grading U.S. #1
1	1 sq. ft. plus yard	306 lbs.	340 lbs.	270 lbs.	1151 lbs.	378 lbs.	92.1
2	½ sq. ft. plus yard	283 lbs.	308 lbs.	257 lbs.	1151 lbs.	413 lbs.	87.2
3	¾ sq. ft. plus yard	292 lbs.	317 lbs.	266 lbs.	1148 lbs.	399 lbs.	91.2
4	1 sq. ft.	287 lbs.	317 lbs.	259 lbs.	1134 lbs.	400 lbs.	89.3

(Data selected from Florida Expt. Station Bulletin 451)

Some producers raise broilers to a marketable age in batteries, others by floor brooding, and still others by a judicious combination of both systems. Some have received a price premium for broilers which have had access to a good range.

Battery brooding and floor brooding have been discussed at length elsewhere in this text; so the details of management under these systems are omitted here.

The combination of battery and floor systems may be affected by brooding the chicks for a limited period—usually three to four weeks—in the batteries, and then transferring them to the floor system where they are carried to broiler age.

Broiler producers, more and more, are coming to rely on specially compounded rations that result in maximum growth. The Connecticut high energy ration is a good example.

With the raising of large numbers, cannibalism becomes more serious and ways of prevention are pertinent considerations. Crowding, insufficient feeding space and accidents are probably management factors favoring the development of the habit. Obviously it is not easy to stop this pernicious vice once it has gained headway.

In the production of broilers one must decide the time of year the birds will bring the most satisfactory returns. This calls for considerable judgment, as one must consider the market demand, competition from competing areas, the history of seasonal price range during recent years, and the cold-storage holdings of poultry meat, particularly that of broilers.

Marketing Broilers

In planning a broiler enterprise consideration should be given not only to production but also to marketing, so that when the broilers are ready to be sold a satisfactory outlet will have been established. Broilers may be marketed, either live or dressed, through three main channels: wholesaler, retailer, or consumer.

Whether the birds are to be sold alive or dressed will depend on market requirements, distance from market and facilities of the producer to meet these conditions. More and more the consumer is demanding full drawn birds. Where the distance from market will mean excessive shrinkage in transit or where special fattening of the birds is practiced, it becomes necessary to develop a market for dressed or dressed and drawn broilers or for sale alive at the door, in order to avoid this shinkage loss. The importance of shrinkage is shown by the work of E. H. Rinear, New Hampshire Experiment Station. He found this caused a loss of 7,120 pounds or 5.2 per cent in the shipment of 135,495 pounds of broilers from New Hampshire to New York City. Higher shrinkage for such a distance is not uncommon.

In the final analysis the net return actually realized is of prime concern to the broiler producer. The net return or profit is influenced by the cost of production, including overhead and marketing. Marketing costs may be reduced by minimizing shrinkage in transit, by grading birds preparatory to shipment, by careful calculation on means of transportation, and by the decision of selling live or dressed.

The type of transportation—freight, express or truck—will be determined by the rates asked, comparative shrinkage prospects, transportation time, and the time of day that broilers may be put on the active market under each system. Some broiler men,

TABLE 16

COST AND PROFIT IN THE PRODUCTION AND MARKETING OF BROILERS

ITEM—	Av. of All Farms	Calculated Av. Cost per 100 Birds	Av. of 25 Most Profitable Farms	Av. of 25 Least Profitable Farms
Number of Farms	109		25	25
Number of broilers raised	8,231		12,850	13,291
Per cent of mortality	16.8		16.0	18.1
Total broiler investment	$1,417.05	$17.22	$1,964.92	$2,725.92
Per cent of investment for broiler production	20.9		23.3	28.6
Income:				
Value of broilers sold	3,795.33	46.11	7,144.32	5,439.34
Value of broilers consumed on farm	29.30	.36	33.12	34.88
Inventory change	240.64	2.92	176.04	468.60
Total	$4,065.27	$49.39	$7,353.48	$5,942.72
Costs:				
General Expense:				
Fuel	$ 181.59	$ 2.20	$ 265.04	$ 305.96
Light, water, disinfectants	24.50	.30	42.32	40.84
Total	$ 206.09	$ 2.50	$ 307.36	$ 346.80
Feed:				
Mash	$1,764.72	$21.44	$2,561.16	$3,153.72
Grain	237.71	2.88	374.52	357.60
Litter	36.66	.44	57.24	69.04
Milk	44.22	.54	82.92	71.36
Other feeds	25.44	.32	25.36	56.60
Total	$2,108.75	$25.62	$3,101.20	$3,708.32
Labor:				
Hired	$ 97.63	$ 1.19	$ 111.08	$ 282.16
Family	206.87	2.51	251.20	286.44
Total	$ 304.50	$ 3.70	$ 362.28	$ 568.60
Overhead:				
Taxes	$ 17.95	$ .22	$ 23.84	$ 36.28
Interest on investment	71.31	.87	99.32	135.28
Depreciation	65.82	.80	91.96	122.96
Miscellaneous	62.27	.75	98.52	139.32
Total	$ 217.35	$ 2.64	$ 313.64	$ 433.84
Marketing:				
Transportation	$ 43.82	$ .53	$ 46.40	$ 109.04
Shrinkage	79.15	.96	106.68	177.68
Commission	89.64	1.09	107.28	227.68
Miscellaneous	10.14	.12	12.44	27.20
Total	$ 222.75	$ 2.70	$ 272.80	$ 541.60
Cost of baby chicks	882.28	10.72	1,394.72	1,431.52
Total broiler cost	3,941.72	47.89	5,752.00	7,030.68
Total net profit or loss	123.55	1.50	1,601.48	1,087.96
Per cent return on investment	8.7		80.1	—30.9

Material taken from Maryland Agr. Expt. Sta. Bul. No. 390.

Fig. 60—Truck Shipment of Broilers to Market.
[Courtesy N. E. Poultryman.]

after sustaining heavy shrinkage in live shipments, have found it more profitable to dress the birds on the farm. Proper market contacts with a dealer in dressed broilers are then necessary.

Regardless of whether the birds are sold live or dressed, grading is necessary to establish a reputation for quality and to command the best prices. In live broiler shipments, for instance, it is advisable to segregate bareback individuals in separate crates.

To summarize this chapter, it is evident that, in order to make a satisfactory profit in a broiler enterprise, (1) disease-free stock is necessary to keep mortality low, (2) sound management practices must be followed to hold overhead expenses down and produce a satisfactory product, and (3) the right market must be secured and good marketing methods employed.

Questions on Chapter V

1. What was the discovery that made it possible to brood chicks successfully throughout the year?
2. Approximately how much feed will 100 chickens of broiler breeds require up to and inclusive of twelve weeks of age?
3. List breeds and crosses that appear to be suitable for broiler production?
4. What factors will determine whether broilers should be sold alive or dressed?
5. List the important factors that assist in securing a satisfactory profit from a broiler enterprise.
6. What factors besides feed and labor enter into the cost of broiler production and marketing?

CHAPTER VI

CAPONS AND CAPONIZING

Subject Matter

Caponizing, the complete and successful castrating of male chickens, is a practice of ancient times. According to history, the Chinese, Greeks, and Romans all practiced it.

The custom was most extensive in France. Its development in this country has been slow but is gaining fairly rapid favor in northeastern United States. This can be accounted for to a large extent by the nearby markets in the large eastern cities.

The Capon

Capons, since they are docile and quieter in disposition than cockerels, can be confined in pens in larger numbers. They seldom crow. Their appearance is altered considerably from that of cockerels—their comb and wattles lose the brilliant red color, cease to grow and appear to be shrivelled, and the hackle and saddle feathers develop markedly. As the capons grow, they seem to develop frame rapidly and appear lanky. The rates of growth of capons and cockerels are parallel until the cockerels approach maturity, when the former make somewhat better and more economical gains.

The advantages of caponizing are mainly improved quality of flesh and more favorable market price. Capons tend to retain the quality of "broiler flesh", remaining tender and palatable; (as cockerels mature their flesh becomes "staggy"). They do not, however, retain this better texture of flesh indefinitely. They should be marketed when they apparently have attained normal size and condition. The better quality of flesh is reflected in a satisfactory difference in market prices between capons and cockerels.

Breeds Suitable:

Cockerels of any breed may be caponized, but the market demand for heavy capons emphasizes the advantages of general-purpose or heavy breeds. Little success is to be obtained by selecting light breeds, such as Leghorns, due to their lack of suffi-

Fig. 61—Position of Bird for Operation.
[Courtesy N. H. Agri. Exp. Sta.]

cient size. Consequently, they are seldom caponized unless for special trade. Capon production is usually a side line of the poultry business and as a result general-purpose breeds such as New Hampshires, Rhode Island Reds and Plymouth Rocks are largely used. Cross-bred cockerels of the general-purpose and heavy breeds make good capons. The vigor of such individuals is an advantage. In any event, it is well to bear in mind in selecting a breed that the market prefers yellow skin and shanks.

Time and Age:

The time to caponize is important. Usually May, June and July are the months most suitable, inasmuch as nine to ten months are required to properly finish a capon. If birds are to be ready for the Christmas holidays and the market directly following, it is necessary to caponize during these months. Likewise, it will be necessary to operate earlier if the intention is to produce capons for the Thanksgiving trade. Cockerels are suitable for caponizing when they are one and one-half to two pounds in size.

Instruments and Materials:

Several sets of instruments are available for caponizing. Necessities are a good knife, spreaders, a hook, removers (regular or electrical), and tie cords. Absorbent cotton and a disinfecting solution should be used in cleaning the instruments. Some poultrymen also swab the top of the operating table with disinfectant.

Preparing Birds for the Operation:

For a period of 36 to 48 hours before the operation the birds should be starved. It is essential to starve them in batteries or crates, otherwise they are liable to fill up on litter from the floor of the pen. Starving clears the intestines of materials, contracting or deflating them and allowing more room for the operation. Some poultrymen also withhold water during this period, but usually it is better to allow the birds to drink what they want as an aid in clearing the intestines. Select only thrifty, vigorous cockerels for caponizing.

The Operation:

For satisfactory operations it is necessary to perform the work on a bright day. Locate the table in the most suitable position for the sun to light the body cavity during the operation.

The bird is stretched out with the legs and wings securely held. When help is available one person can hold the bird in place, remove from the table after the operation, and replace with another. When one man works alone, tie cords are necessary to hold the birds in position, slowing up the operation.

With the bird in position, remove the feathers directly in front of the hip. Locate the last two ribs with the fingers, draw the skin back over the hip and make an incision between the last two ribs, large enough for the insertion of the tools. The spreaders are then inserted and the ribs spread apart to allow sufficient room for operating. Poultrymen in many cases prefer to make only a small incision and, after spreading, increase its size.

The glass-like membrane now seen can be punctured and torn readily with the hook, revealing the internal organs. If the

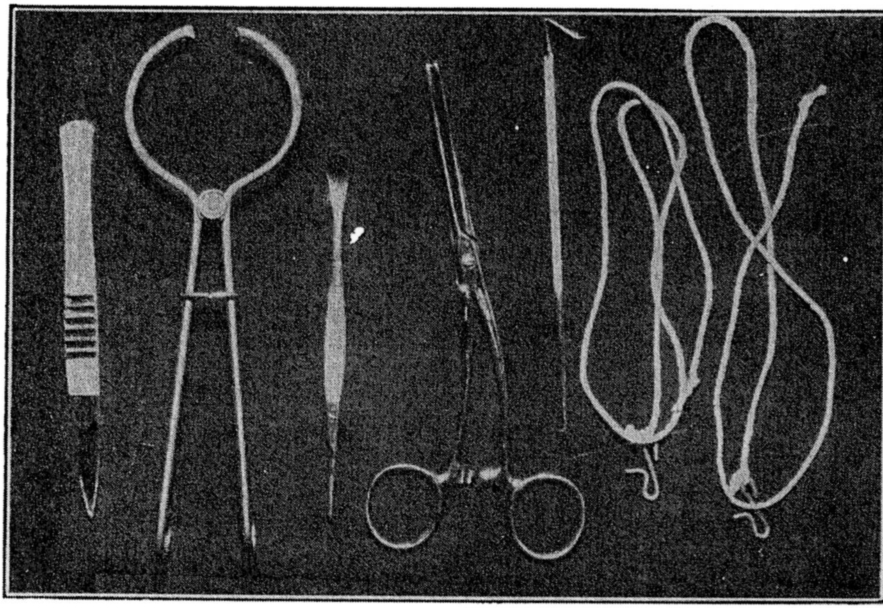

Fig. 62—Caponizing Instruments.
[Courtesy N. H. Agri. Exp. Sta.]

birds have been sufficiently starved the testicle can be located with little difficulty. It is about the size of a navy bean, a light yellow body under the backbone. At times it may be discolored and somewhat more difficult to locate.

The removers should now be inserted, the testicle securely grasped and removed by a pulling, twisting motion. Be careful not to grasp the artery directly above the testicle, or hemorrhage will result. The spreaders can now be removed. If the skin has been stretched as directed, it will resume its original position and cover up the incision in the flesh. The bird is then turned over and the procedure repeated on his other side after the instruments have again been disinfected.

Some operators remove both testicles from one side, but with amateurs this is likely to result in higher mortality.

Occasionally a fold of the intestine may be found protruding through the first incision upon completion of the operation and should be replaced.

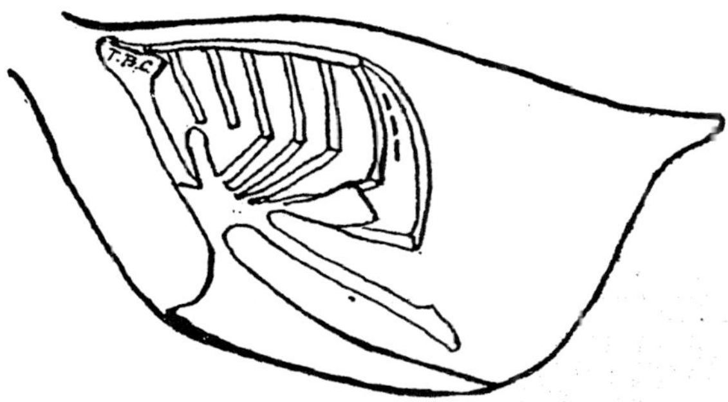

Fig. 63—Dotted Line Illustrates Position of Incision.
[Courtesy N. H. Agri. Exp. Sta.]

Care and Feeding After Operation:

The birds are usually given water and mash immediately after the operation. No hard grains are fed for a few days. Some poultrymen feed scantily on mash at first because the birds have been starved for a considerable period. Others give them all the mash they want.

During this time, the capons may be housed in small groups to prevent crowding, and without roosts to minimize jumping and straining.

During the first week or ten days after the operation the capons will "wind puff." The "wind puff" is the result of an accumulation of air under the skin. The air can be released by cutting the skin with a knife or shears. In some cases the treatment needs to be repeated.

Slips and Mortality:

When a piece of testicle is left in place, the bird will retain its cockerel characteristics and disposition, but is seldom able to reproduce, and is generally termed a "slip." With skilled operators, the percentage of slips will be in the neighborhood of five per cent. With amateurs, it is likely to run considerably higher, 30 to 40 per cent.

Mortality from operations is variable. With unskilled persons, it may approach five per cent or more.

Hormones:

Recently poultrymen have resorted to the use of female sex hormones in order to produce a fleshing condition approximating that of capons in males. Males which are approaching maturity or are matured and are intended for market disposal are those generally treated. The treatment consists of inserting commercially prepared pellets between the skin and muscle of the neck near the skull. This is done four to five weeks before the birds are to be marketed. The effect of this material is to depress the male sexual characteristics and produce a softer meated bird than would otherwise be the case. This depressing effect is of relatively short duration so that gradually the male characteristics reassert themselves.

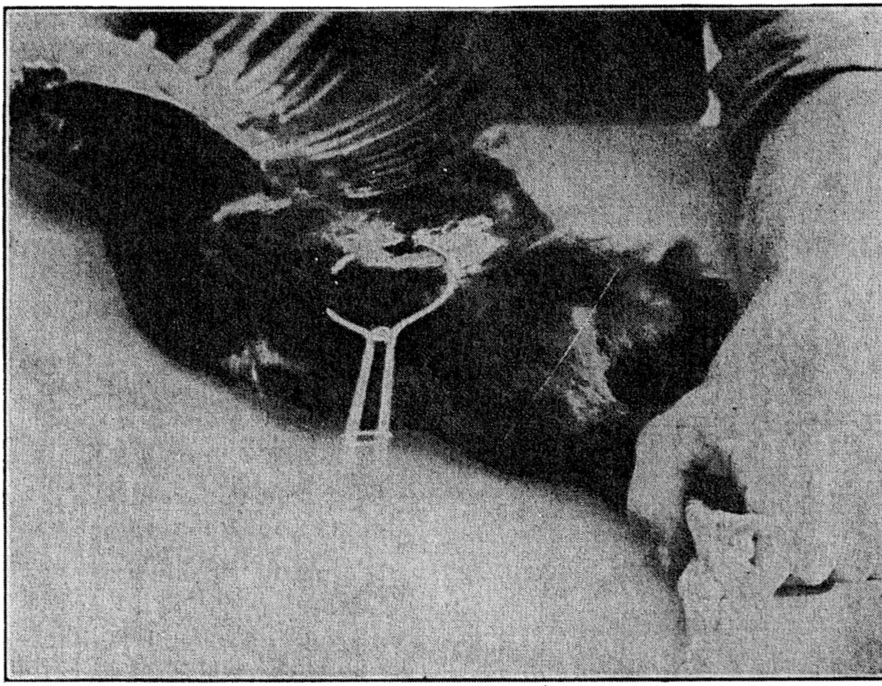

Fig. 64—The Spreaders in Position.
[Courtesy N. H. Agri. Exp. Sta.]

Killing, Dressing and Marketing

It is advisable to starve capons for a day previous to killing so that they may empty their crops and pass as much of the fecal material as possible. The birds keep better when their intestinal tracts are empty.

To kill the birds it is recommended that only the stick method be resorted to. By this method the jugular veins are severed, permitting free bleeding. The knife is then passed upward through the cleft or groove in the roof of the mouth directly between the eyes and slightly back of them in order to pierce the brain and thus release the muscular control on the feathers.

Usually capons are only partially dressed for market. The feathers are left on the neck and head, the last two joints of the wing, the tail and part of the back, and the thighs in the region of the hock joint. This distinguishes capons from stags. Where capons are being retailed to the consumer, however, they should be completely dressed to relieve the buyer of this task.

Capons may either be dry picked, scald picked or picked by use of the slack or semi-scald method. Dry picking or semi-scalding is preferable. For further information on these methods see Chapter XV.

In marketing capons, the general outlet is the large consuming centers. Contacts can be made with wholesalers or retailers who are advantageously located and have a special outlet for high-quality market poultry. Because most capons are disposed of to the wholesaler, they are usually sold alive by the producer. If he sells to retail stores or direct to the consumer, his capons may be most advantageously marketed if completely dressed and drawn. A special retail trade can be developed for this type of poultry meat.

In dressing and drawing capons the loss in weight will be 25 to 30 per cent. These figures can be used in computing the necessary rate for dressed-and-drawn capons over the live-weight price. If live capons are selling for 40 cents a pound, for example, an eight-pound capon will return $3.20. With a 25-per-cent shrink the loss in drawing and dressing would be approximately two pounds. Thus the remaining six pounds would have to bring $3.20, or approximately 54 cents each, plus labor involved.

Capon Commercialization and Its Problems

Not often is an entire farm devoted to capon production, but a few instances are known by the authors. Only with a special favorable market can one successfully specialize in the production of market meat. On the farms where this is known to be done, suitable cockerels of a desirable size for caponizing are purchased. No attempt is made to produce those needed for the business. Due to the skill required in the operation and the limited demand, capon production as a specialization is adapted to only a limited group.

Disease control is probably the primary problem in capon production. Field observations indicate that capons are more susceptible to diseases than pullets and cockerels. Consequently the rate of mortality may be higher and additional precautions are necessary to prevent diseases. In view of this it is well to grow capons entirely separate from other poultry.

Another problem is the relatively high percentage of "slips". These usually sell for about the same price as cockerels, materially reducing possible profits.

The problem of fleshing the birds often arises. They are likely to be framy and lanky and, although of sufficient weight, may not be thoroughly fleshed on the breast. If they are allowed

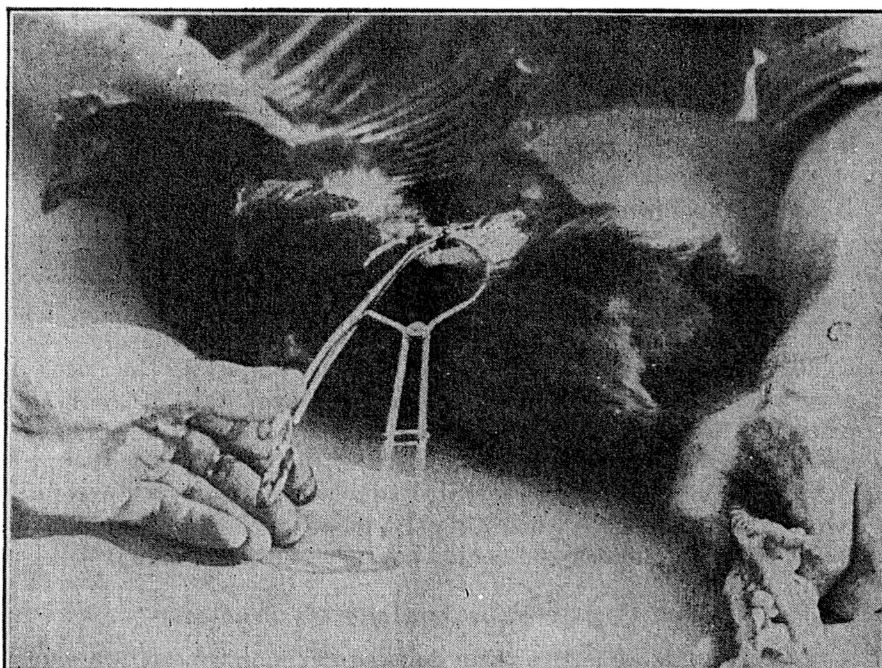

Fig. 65—Testicle Being Withdrawn.
[Courtesy N. H. Agri. Exp. Sta.]

to range it will be of distinct advantage to house them sufficiently early to properly finish fleshing before marketing.

For success good vigorous stock which will make rapid growth is paramount. Little is to be gained by caponizing cockerels showing evidence of disease, such as coccidiosis, or worm infestation.

Questions on Chapter VI

1. What is a capon?
2. What breeds are suitable for caponizing?
3. In preparing birds for the operation, why is feed withheld from the birds for 36 to 48 hours?
4. How is the operation performed?
5. Define a "slip".
6. Is it desirable to give the capons any special care or attention following the operation? If so, what?
7. At what period of the year are birds normally caponized? Why?
8. What are some of the problems of capon commercialization?

CHAPTER VII

PRINCIPLES OF FEEDING

Subject Matter

Today more attention is being given to feeding and its principles than ever before. If climatic situations were always the same, if feeds never varied in composition, if fowls were all alike and if feeders had but a single purpose, feeding might possibly be reduced to a single ration and method. These factors do vary, however, and no system has yet been devised which will relieve the feeder of a little thinking for himself. Thus it is obvious that knowledge of the composition and value of feeds must be gained to secure maximum results, and its application must be accompanied by good breeding and housing and intelligent management in all of its phases.

Object of Feeding

The object of feeding is to produce food useful to man. To attain this goal of feeding it is necessary not only to understand the animal organism and its functions but to be familiar with the sources of requisite food nutrients.

If birds are to perform satisfactorily they must be supplied with water, protein, carbohydrates, fat, minerals and vitamins.

Nutrients and Their Sources

Water—This is vital to digestion and metabolism. It constitutes the largest proportions of both the body and the egg. A continuous supply of it is essential. Without water laying birds will stop production sooner than if they are without feed.

Protein—This nutrient makes up about 20 per cent of the body of the chicken and nearly 14 per cent of the egg. Most feed stuffs used for poultry rations contain some protein, but it is still necessary to give careful attention to providing that the kind and the amount of protein are adequate for the needs. Generally speaking, the protein supplements are the most expensive parts of the ration. Since the kind of protein is as important as the amount, one needs to have a clear understanding of this situation in formulating rations.

By kind of protein, we refer to protein quality. The protein is a very complex substance, made up of many small units known as amino acids. Thus far, 23 naturally occurring amino acids have been discovered. Obviously, with such a substantial number, it is possible to have many kinds of combinations of the amino acids united to form specific proteins. The chicken has certain specific needs for amino acids to formulate the protein of body tissue and of the egg. Thus, proteins, which supply these specific amino acids are considered of good quality for poultry feed utilization; while those which do not, are considered of poor quality.

For convenience, one may think of the amino acids as comparable to the letters of the alphabet. Just as each letter of the alphabet is not equally useful in building a given word, so it is that each amino acid is not equally useful in building a specific protein.

In the past, emphasis has been placed on supplying the proper amount of protein. Today, the emphasis is upon supplying the necessary amino acids in the proper amount.

Carbohydrates—Grains and their by-products are the chief sources of carbohydrates in poultry rations. The starches and sugars present in feed stuffs account for the major portion of the carbohydrates present in a given product. The term "nitrogen free extract" often used by the chemist in feed stuff analysis also refers to the starches and sugars present in the feed product.

Fiber which is poorly digested by poultry serves the same function as carbohydrates. Carbohydrates function as fuel to keep the body operating. They supply the necessary heat and energy which is the largest requirements of an animal organism. Excess carbohydrates consumed over and above immediate needs are stored for future use as fat. For example, in the fattening of poultry for market purposes, rations used are high in carbohydrates.

Fats—Fats occur rather generally in all feed stuff except in the mineral substances; and they serve the same general purpose as carbohydrates in the supplying of heat and energy. Because of their chemical structure, they actually supply two and one-fourth times more energy than an equal amount of carbohydrates.

Because of the value of fats and oils for other purposes than feed, the present day trend is for rations to be lower in fat than formerly. However, it has not yet become a problem in poultry feeding.

Minerals and Grit—The bulk of the mineral matter fed to poultry is used for building the skeletal framework of the bird and the shell of the egg. However, minerals also play an important part in all the tissues and the fluids of the body, even being important to the beat of the heart.

Chickens are known to require calcium, phosphorous, salt (sodium chloride), manganese, potassium, magnesium, iron, copper, sulphur, and iodine. However, it is usually only necessary to give attention to supplying calcium, phosphorous, salt, and manganese, since the other elements are generally supplied in sufficient amounts by the common feed stuffs utilized in poultry rations.

It is important that poultry be supplied with an adequate amount of vitamin D, since this vitamin is necessary to the utilization of calcium and phosphorous. No additional amount of these elements will make up for the lack of vitamin D.

It is also important to maintain a satisfactory ratio between calcium and phosphorous, because an excess of either will reduce the availability of the other.

Meat scraps and fish meal are the main ingredients contributing important amounts of calcium and phosphorous to the ration. Unless the amount of these ingredients used is large, it is usually necessary to add some specific sources of calcium and phosphorous. Calcium is most often supplied as limestone or oyster shell. Both calcium and phosphorous are supplied by bone meal and defluorinated rock phosphate. Care should be exercised in the use of rock phosphates since they are frequently high in fluorine which is harmful.

Salt is not found in the ordinary feed ingredient except perhaps in the case of certain fish meals. Therefore, the common practice is to add ½ of 1 per cent of salt to all poultry mashes. There is some evidence that lack of salt may lead to cannibalism as well as giving poor results in growth and egg production.

Manganese is supplied by all the common feed stuffs but with wheat by-products, limestone, and oyster shells containing the largest amount. In spite of the wide distribution of manganese in feed stuffs, it is still necessary to add manganese supplement to most poultry mashes. Any of the common forms of manganese are satisfactory, with manganese sulphate and carbonate being the most frequently used.

A deficiency of this mineral element results in poor shell strength, lower hatchability, deformed embryos, and poor bone development in growing chicks.

Since chickens have no teeth, it is desirable to feed some hard grit to aid the gizzard in grinding feed. Any hard rock that is not easily shattered may be used. Birds on range generally find their own supply of grit. While it is possible to raise chickens without grit, better feed efficiency results when grit is supplied.

Vitamins—These are necessary food factors known by the effect they exert on the body when the ration is deficient or entirely lacking in one of them. The list of known vitamins is still growing. Poultry require at least 14 of the known vitamins; but since many of them are widely distributed in various feed ingredients, the poultryman definitely needs to be concerned with only a few of them. These are: vitamins A, D, riboflavin, pantothenic acid, and B_{12}.

In the case of deficiency of Vitamin A, any or all of many symptoms may develop, as for example:

1. Lowering of vitality and consequently a lowering of resistance to infection.
2. With lowering vitality, micro-organisms may invade the urinary tract, the alimentary canal, and the eye, ear and nasal passages.

TABLE 17

AVERAGE COMPOSITION OF INGREDIENTS COMMONLY USED IN POULTRY FEEDING

	Protein Percent	Calcium Percent	Phosphorus Percent	Manganese Mgs. Per Pound	Vitamin A Activity I. U. Per Pound	Riboflavin Mgs. Per Pound	Pantothenic Acid Mgs. Per Pound	Niacin Mgs. Per Pound	Thiamin Mgs. Per Pound
Alfalfa meal, dehydrated	16	?	?	12	7–242,000	6.73	17.42	?	2.0
Alfalfa meal, sun cured	16	1.44	.21	?	6–118,000	5.36	12.05	17.70	1.35
Alfalfa leaf meal, dehydrated	20	?	?	14	15–300,000	9.15	18.0	?	4.0
Alfalfa leaf meal, sun cured	20	1.42	.25	?	17– 90,000	7.18	12.6	24.1	2.0
Barley	12.8	.07	.32	7	0	.55	2.84	30.44	2.71
Barley, Pacific coast	9.7	?	?	?	?	?	?	?	?
Bone meal, special steamed	6.5	27.0	13.2	0.9	0	0	0	0	0
Brewers yeast, dried	46.5	1.26	1.21	0.9	0	20.27	59.60	216.	131.20
Buttermilk, dried	34.3	1.32	.93	?	0	15.0	19.80	7.70	1.24
Corn, yellow	9.3	.01	.26	2.3	1– 4,000	0.6	3.36	6.72	2.06
Corn gluten meal	43.0	.10	.47	1.8	2.5–15,000	.39	6.26	13.40	?
Cottonseed meal	41.0	.19	1.11	8.2	?	4.08	6.35	20.60	6.13
Fish meal, Menhaden	57.8	6.50	3.6	20.2	?	2.13	?	40	?
Fish meal, sardine	68.1	4.73	2.63	18.2	?	3.17	?	?	?
Limestone, high calcium	0	39.2	0	91.0	0	0	0	0 ?	0
Meat scrap	55.0	8.70	4.30	8.2	?	2.78	3.54	?	?
Meat and bone scrap	50.4	10.5	5.21	4.5	?	2.60	?	30.53	?
Oats	12.5	.10	.40	15.4	0	.58	4.50	6.50	3.43
Oats, Pacific coast	9.6	?	?	?	?	?	?	?	?
Peanut meal	45.1	.17	.55	?	0	2.35	24.10	77.50	3.27
Rice bran	12.8	.10	1.84	127.4	0	1.38	10.33	129.10	10.32
Soybean oil meal, expeller	41	.29	.67	13.6	0	1.87	6.27	17.60	2.62
Soybean oil meal, solvent	44.0	.34	.71	13.6	0	?	?	?	?
Skimmilk, dried	35.8	1.34	.99	?	0	9.33	15.20	7.31	1.59
Wheat	12.0	.05	.38	17.0	0	.51	5.62	26.74	2.10
Wheat, Pacific coast	9.9	?	?	?	?	?	?	?	?
Wheat bran	16.4	.10	1.14	54.1	0	1.34	11.33	139.97	3.24
Wheat, std. middlings	17.0	.09	.90	54.1	0	.74	7.10	52.80	7.0
Whey, dried	12.8	.73	.66	6.4	0	12.92	24.0	8.31	?

?—Indicates data incomplete.

(Data of the National Research Council)

3. The slowing up of the growth rate with eventual cessation of growth. The period preceding the cessation of growth will probably depend on the amount of vitamin A previously stored up in the body.
4. Ophthalmia. A nutritional disease of the eyes may develop along with perforating ulcers.
5. A loss or repression of the appetite.
6. Egg production and hatchability may be markedly reduced.
7. In severe cases of long duration, pustules form in the roof of the mouth and throat.

Good sources of vitamin A activity are found in yellow corn, alfalfa meals, feeding oils and most all green leafy plants and grasses.

In a deficiency of vitamin D, symptoms such as the following may develop:

1. Rickets, or bone disorders resulting from failure to utilize the calcium and phosphorous of the feed.
2. Muscular weakness.
3. Possible diminished excretions of calcium by the kidneys.
4. Retardation or failure of growth, hatchability, fertility, egg production and shell formation.
5. Decreased resistance to infection.

Vitamin D is always deficient in poultry rations unless a special source is added to the feed or the chickens have access to direct sunlight. Direct sunlight acting upon the surface of the skin produces vitamin D which the chicken can utilize. Good sources for addition to the feed are fish oils and activated animal sterols.

A deficiency in riboflavin is considered to result in the following abnormalities:

1. Low hatchability.
2. Lack of vigor.
3. A development of curled toes in growing birds in which the toes always curl inward.
4. Impairment of growth.
5. Nervous and digestive disturbances.
6. Loss of body weight.

Good sources of riboflavin are found in dried milk byproducts, yeast and fermentation byproducts. Moderately good sources are alfalfa meal, fish meal, meat scraps and distillers byproducts.

Pantothenic Acid deficiency results in:

1. Retarded growth.
2. Granular incrustation on eyelids and in corner of mouth.
3. Dermatitis of the feet frequently evidenced as cracked or wart-like lesions.
4. Lowered hatchability.

Good sources of pantothenic acid are yeast, dried milk by-products and alfalfa meal.

Vitamin B_{12} deficiency results in:

1. Retarded growth.
2. Low hatchability.

Sources of this vitamin are fish meal, meat scraps and dried skim milk.

Composition of Feeding Stuffs

Feeding-stuff analyses deal with the following constituents: water, ash, protein, ether extract, crude fibre and nitrogen-free extract.

How Each Is Determined:

Water: The loss in weight which a substance undergoes at a temperature at the boiling point of water. It is not absolutely accurate because other volatile substances escape.

Ash: The residue left after complete incineration of the feed.

Crude protein: The total nitrogen content multiplied by 6.25 determines the crude protein. This is based on the assumption that all protein contains 16 per cent nitrogen and that all the nitrogen in the feed is in protein form.

Crude fat: Determined by dissolving the fat with a common solvent such as ethyl ether. The extracted material is then referred to as ether extract.

Crude fibre: Material that is nonsoluble in either dilute acids or alkalies.

Nitrogen-free extract: The percentage of this equals 100 minus the summation of all the previous determinations.

A knowledge of feeding-stuff analyses will enable a person to intelligently digest and interpret the material presented annually in the various states on commercial feed stuffs. It is also of value to know not only the analysis of feed but what becomes of food material that is ingested. Food material is of no value to livestock until it has passed through the digestive processes and the resultant products are carried by the blood stream.

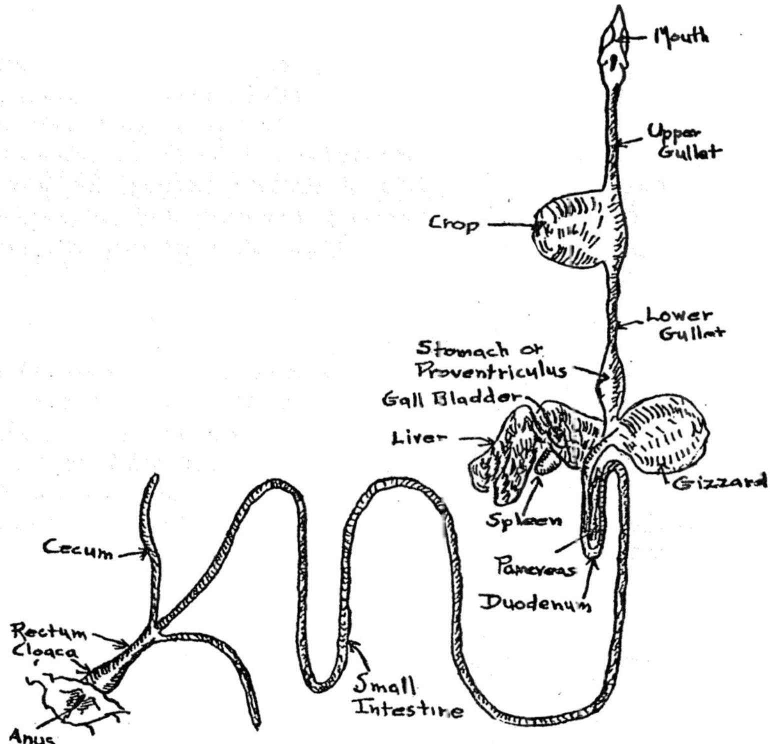

Fig. 66—Diagram of Alimentary Tract.

TABLE 18

OUTLINE OF DIGESTION

KINDS OF DIGESTION	ACTIVE AGENT	ACTS UPON	PRODUCED
Salivary	Ptyalin	Starch	Maltose
Gastric	Pepsin Rennin	Protein Milk	Proteosis and peptones Ca caseinate and whey albumin
Pancreatic	Amylase Steapsin Trypsin Nuclease	Starch Fats Protein Nucleic acid	Maltose Glycerol and fatty acids Proteosis, peptones and amino acids Phosphoric acids (Carbohydrates and purine and pyramidine bases)
Intestinal	Maltase Sucrase Lactase Erepsin	Maltose Sucrose Lactose Proteosis and peptones	Glucose Glucose and levulose Glucose and galactose Amino acids

Digestion

Digestion is a breaking down process whereby the nutrients of the food are reduced to forms which can be absorbed. These are absorbed through the walls of the alimentary tract into the blood stream and transported throughout the body for utilization. For instance, in the formation of protein tissues, the amino acids are reassembled in necessary amounts and proportions into a complex protein form. In Table 16 a general skematic digestive outline is presented.

Metabolism

By metabolism is meant the sum total of all changes which the absorbed nutrients undergo within the body. It may be expressed as (1) anabolism or the building up of the bodily reserve or (2) katabolism or the tearing down of bodily material. Both processes are going on practically continuously and call for constant replacements through feeding of the necessary nutrients if optimum results are to be secured.

Carbohydrate Metabolism:

When carbohydrates are ingested in the feed and pass through digestion processes, they are broken down into simpler forms of sugar. These are absorbed by the blood capillaries of the intestines and carried by the portal vein to the liver. Here the carbohydrates pass into the general circulation system.

The liver serves as a storehouse for carbohydrates after they undergo a chemical reaction and form glycogen. Some quantity of the carbohydrates is stored in the muscles. The amount of sugar in the blood is practically constant in the normal individual.

The storage of glycogen is limited. When a surplus of carbohydrates is digested it may form fat tissue, particularly in the abdominal region.

Protein Metabolism:

Protein is made up of small units commonly called amino acids. Through digestion protein is broken down into these small units. Amino acids in excess of requirements are deaminized into excretory products. By deaminization is meant the cleavage of the amine or nitrogen-containing group from the molecule.

In the deaminization of proteins by poultry, ammonia is formed and execreted as uric acid. The non-nitrogenous material remaining closely resembles both the carbohydrates and fats. This may be utilized for energy.

The protein in the feed is used for body building as well as for reproduction in egg form.

Fat Metabolism:

Fat material is broken down into fatty acids and glycerol. It is to be noted that in the digestion of all the nutrients, the main consideration is to reduce them to their usable forms. In this manner they are assimilated and built up into the complex forms found in the body of the individual. Where the feed contains a proportionately large amount of fat material, considerable of it may be stored in the fat tissue of the body.

The energy derived from food material depends on the oxidization of its carbon and hydrogen. Fat contains an abundance of carbon and hydrogen and as a result carries a large amount of energy. In fact it is usually considered to possess about two and one-quarter times as much energy as carbohydrates. Because they are consumed in such large proportions carbohydrates are distinct energy carriers.

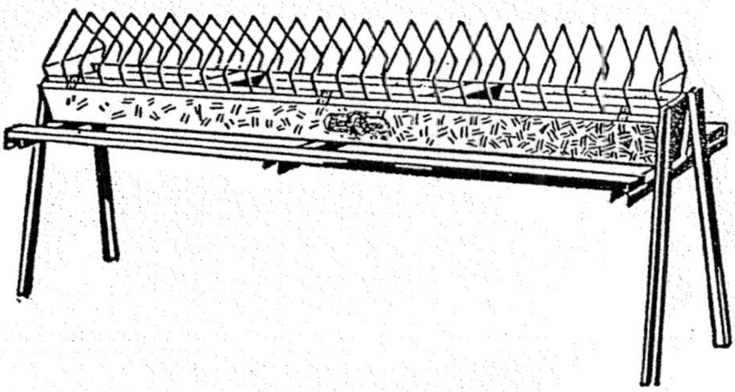

Fig. 67—Metal Flock Feeder with Stand.
[Courtesy Buckeye Incubator Co.]

Mineral Metabolism:

The mineral compounds found in feeds are split into the various elements such as calcium, phosphorus, sodium, chlorine, and many others. The elements needed are then absorbed through the intestinal wall into the blood stream and carried to all parts of the body for use by the cells. Except for the bones which do represent a reserve of calcium and phosphorus there is little storage of minerals in the body. However, withdrawal of calcium and phosphorus from the bones for other use cannot be tolerated for any lengthy period of time.

Balanced Ration

A balanced ration would be one containing all the necessary nutrients in the proper proportion to meet the requirements of the particular purpose intended such as growth, egg production, fattening, or maintenance. Actually, however, a definitely balanced ration is impossible until all the requirements are fully understood. A ration may be considered fairly well balanced for the requirements of the birds when it is compounded in accordance with present-day knowledge and produces results that satisfy.

Fig. 68—Automatic fountains, of various types, provide a constant supply of water and eliminates drudgery of carrying water.

[Courtesy of Doughboy Industries, Inc.]

Animal Versus Vegetable Protein

The ideal source of protein in a poultry ration would be poultry meat because it contains the correct amino acids in the proper proportion. Obviously poultry meat as a protein source for poultry is impossible. As near an approach to this as economically possible is made by supplying protein from animal sources.

The proteins of the plant kingdom generally will not effectively replace animal protein because their make-up is considerably different. The use of suitable minerals and vitamins, however, will enhance the utilization of vegetable proteins. These proteins naturally are an integral part of rations containing grain and grain products.

Among the vegetable protein concentrate sources, one that appears to have considerable merit is soy bean meal. From various viewpoints it would appear that the amino acids contained in the protein of soy beans are of a particularly desirable type and

Fig. 69—Showing Comparative Growth of Pullets at 12 Weeks of Age, on 20%, 17%, 14% and 11% Protein Levels, Respectively.

may more closely approach the amino acid make-up of animal protein concentrates. Thus soy beans as the source of vegetable protein concentrate are being looked upon with more favor in the formulation of poultry rations. As research work progresses in determining the role of minerals and vitamins it is being found possible to use larger quantities of vegetable proteins and less animal protein.

Formulation of a Ration

There are many things that must be taken into consideration in the formulation of a ration: i.e., adaptability, palatability, mechanical and physical condition, effect on product, variety, and cost. Accordingly knowledge is needed covering these factors and the effect the ration will have on the birds and the product, if one is to design feed formulas.

Adaptability:

It would appear obvious that a ration to produce satisfactory results must be adapted to the class of poultry to which it is fed. A ration designed for laying hens will not be satisfactory for day old chicks because the nutrient requirements differ. Rations must be designed to meet the known requirements of the poultry to which it will be fed.

It is not always necessary to change a formula to make a ration adaptable. Sometimes it may mean a change in the way a ration is fed. This is often true in the feeding of the scratch and mash type of ration. Frequently, a change in the proportion of scratch and mash makes it possible to secure the desired results in growth, production or physical condition of the birds.

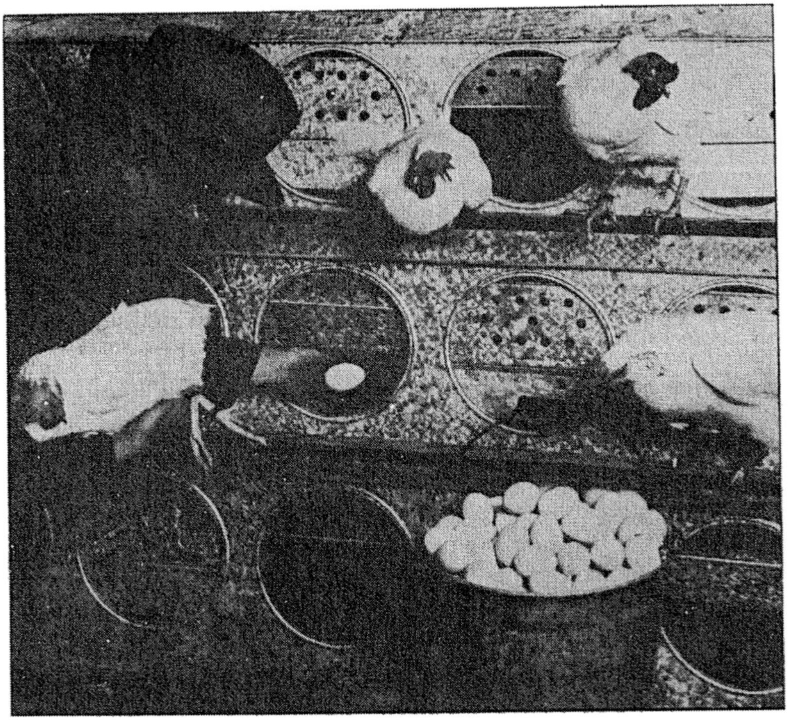

Fig. 70—Birds Must be Adequately Fed for Heavy Production.
(Courtesy James Mfg. Co.)

Palatability:

For maximum egg production, growth or fattening, high feed consumption is necessary first of all. This is impossible unless palatability was sought in the formulation of the ration. Unpalatable feeds must be used in very limited amounts if a mixture is to be appetizing enough to lead to satisfactory consumption.

The chemical analysis of an ingredient is not an index of its palatability. Two ingredients fairly close in their analysis will not always be found to possess the same appetizing effect. Rye, for instance, is fairly comparable in its analysis to other common grains but is not eaten readily by poultry. Catering to the appetites of the birds will be necessary for best results.

Fig. 71—Palatable Rations Encourage Consumption.
[Courtesy N. H. Agr. Exp. Sta.]

Mechanical and Physical Condition—Both mechanical and physical condition are important to securing a high total feed intake. Poultry prefer cracked and whole grains to finely ground feeds. Most systems of feeding recognize this in attempting to get first, a high intake of mash; after which grains or pellets are fed to make up the total feed needs.

Finely ground mashes are not generally as relished as those coarser ground. For best results, the more granular a mash can be made, the better it can be consumed. Research work has shown important differences in growth and egg production attributed only to the fineness to which the mash was ground.

The physical condition of the feed refers mainly to its bulk and to its stickiness when wet. Because the capacity of the digestive tract is limited, fibrous and bulky feeds not only reduce the total feed intake in a given period of time, but also reduce the actual amount of nutrients the bird may obtain. Therefore, more emphasis is being placed on reducing the amount of fiber and bulk in poultry rations.

Stickiness in a feed is important because of the drinking habits of poultry. They tend to drink small amounts frequently. Ingredients of mash which are inclined to become sticky and pasty when wet will tend to stick to the beak and mouth of the bird, making it more difficult for the bird to eat; and more feed is transferred by the bird to the drinking fountains causing undue waste.

Affect on Production—Certain feed stuffs are known to have an affect on the product produced. Research work, however, has not progressed along this line as far as along other lines. The color can be influenced materially by the amount of pigment in the feed. Such materials as yellow corn and alfalfa meals increase the depth of the color of the yolk, while wheat and oats have but little effect. Cotton seed meal fed at high levels tend to produce olive green yolks and a pink coloration in the albumen. At lower levels, the effect may not show up until after the eggs have been in storage for a long time.

Corn gluten meal tends to increase pigmentation in the shanks and the skin, while high levels of animal proteins have a depressing influence on shank pigmentation. Certain feed stuffs have been reported to affect the taste and the aroma of the meat. It is, therefore, evident that one should carefully consider all aspects of feed ingredients before utilizing them for feeding purposes.

Variety—A certain amount of variety in a ration is desirable and necessary. Obviously no feed ingredient alone will constitute a complete ration for poultry, and until such a time as more complete information is available concerning the nutritive requirements of poultry as well as the exact composition of feed ingredients, it will be desirable not to limit the formulation of a ration to a few ingredients.

Poultrymen generally feel that a grain mixture gives better results if it is composed of more than one cereal. Mash mixtures usually require 6 or 7 ingredients in order to meet the known nutritive requirements, and obviously one of these ingredients should be a vitamin D supplement, and another a source of animal protein. It is, of course, not necessary to go to the extreme of utilizing 20 or more ingredients just for the sake of variety. A little common sense would dictate that this would be unwarranted. Common sense would also dictate that careful consideration should be given to the cost.

Cost—The cost of 100 pounds of a feed or of a ration is not the most important consideration. It is much more important to judge this factor by measuring the results of feeding. A ration carefully formulated and utilizing quality ingredients will, of course, cost more per bag than others. However, the cost of producing a pound of meat or a dozen of eggs is the real measurement of the value of a feed. Thus the price asked for a feed and the results which are secured must be carefully weighed. Generally the poultryman gets just about what he pays for.

Rations.

At the present time rations are furnished in various forms, namely: mash and scratch, all-mash, and pellets. There are nearly as many rations or combinations of rations as there are poultrymen, and surely as many different ways of feeding. But there is no best ration for all conditions. Good results come with a wide variety of formulas, providing the feeders are skilled. For special formulas see Chapters 8, 9, and 16.

Results should be the basis for any change in formulas. It is unwise to make a change in a mixture when it is proving satisfactory, and particularly is this true at critical times when a revision would be likely to put the birds out of condition. When birds are in a high state of production, for instance, any change in the ration should be made gradually so as not to throw them out of production and into a molt.

The foundation for feeding practices has been laid by experimental work. Future fundamental research will undoubtedly very materially affect the formulation of rations and feeding management.

Feeding Experiments by Poultrymen

Many poultrymen are prone to experiment in feeding their birds. Many are too easily susceptible to high pressure sales arguments and will give almost anything a tryout. This is costly and most of us are not financially situated to afford it. Work of this type should be left to experiment stations which have been established for research purposes. If in doubt as to the practicability of any ingredient or feeding practice it is well to consult the poultry department of your own state university or other reliable agencies.

If one must experiment it is well to use only a portion of the flock for the purpose so that if a loss is incurred the entire business will not be jeopardized. Many do not weigh the consequences seriously enough. They fail to take into consideration all of the possible factors involved.

Factors Influencing Feeding Results

Some of the major factors influencing feeding results are housing, breeding, management, the health of the birds, and the skill and experience of the operator. For example, a laying flock improperly housed or from poor production parentage or in poor health or improperly managed through lack of experience and aptitude of the attendant, will not respond satisfactorily even though its ration is suitable.

On the other hand healthy birds bred for high production, satisfactorily housed and managed by a skillful operator, may be expected to give good results when fed the identical ration. This emphasizes the importance of giving due consideration to all factors influencing feeding results.

Questions on Chapter VII

1. What is the primary object of feeding?
2. List the essential nutrients?
3. What is meant by the term "good quality protein"?
4. What vitamins must the poultryman be concerned about in formulating rations? List sources of these.
5. What determinations is the chemist concerned with in feedstuffs analysis?
6. Outline digestion?
7. How can one enhance the utilization of vegetable proteins in poultry rations?
8. What factors should we be concerned with in the formulation of rations?

CHAPTER VIII

FEEDING CHICKENS

Subject Matter

Time to Feed
What to Feed
How to Feed
Where to Feed
Varying Nutrient Requirements in Floor Versus Battery Brooding
Feed Requirements of Poultry
Hoppers and Hopper Space
Size of Grain
Clean Feed and Water
Grading Pullets According to Feeding Needs
Feed Consumption and Growth
Rations

The nutritional requirements of chickens are met with rations that have been carefully compounded. Yet a ration adequately formulated for the purpose intended is not alone sufficient. Satisfactory management and good judgment are also essential in any feeding program.

Time to Feed

Several years ago it was common practice to recommend withholding feed from chicks until they were 48, 60, or 72 hours old. The reason usually advanced was that this gave opportunity for the assimilation of the yolk which had been drawn into the body cavity of the chick during the end of the incubation period. No feed was deemed necessary for the first two or three days.

Chicks do not hatch uniformly, however, so many are relatively old by the time they are removed from the incubator. Experimental work has shown no advantage in withholding feed from them. The sooner it is placed before them the better. Those hatched first will then have an opportunity to satisfy their appetite; the others will still be a little weak and will be resting and, therefore, will not eat much immediately.

Obviously when chicks are placed under the brooder without feed they are likely to fill up on indigestible material, such as litter. The result is digestive disturbances and mortality. After a digestive upset a chicken is likely to be stunted and have less chance of making satisfactory growth.

As soon as chicks have thoroughly dried off in the incubator they should be moved to the brooder and supplied with feed.

What to Feed

The feed requirements of chicks are even more exacting than those of laying birds. If chicks are to make rapid gains in weight they must be provided with a ration carrying all the necessary nutrients. Their ability to utilize protein is more pronounced than that of adult birds. This ability decreases as they approach maturity. Thus it is evident that for the first few weeks their ration should provide ample protein as well as other nutrients. They may be fed nothing but the chick mash for the first four to six weeks. Then in addition grain of desired size may be placed in separate hoppers, or scattered in the litter, thus automatically reducing protein intake.

At times difficulty may be experienced with chicks "pasting up" around the vent even though all conditions appear satisfactory. When experienced it is most troublesome during the first week. The feeding of finely cracked grain the first two days seems to be helpful in preventing this condition. Some make a practice of starting the chicks on cracked grain while others will start them on chick mash and utilize only two or three feedings of cracked grains at the first appearance of pasting.

Regular feeding of grain in addition to the mash may be inaugurated at four to six weeks of age but if maximum growth is desired it may be delayed till the chickens are eight to ten weeks of age. This is particularly true if the birds are to have free choice grain as well as mash.

With the all-mash system of feeding grain feeding is not practiced but because of lower nutrient requirement at eight to ten weeks of age, chickens are transferred to all-mash growing ration.

How to Feed

With the mash constantly in front of the birds the routine of feeding is relatively simple during the first few weeks. However, since it is an important period in the life of the chicks close observation is necessary to get them off to a good start.

In addition to small hoppers filled with feed it is desirable for the first few days to place small handfuls of feed on egg cup flats well distributed around the brooding area. When placed at or even under the edge of the hover the chicks find it readily and when feed is readily available there is less danger of the chicks eating litter. Corrugated cardboard or other rough surfaced material may be used in place of egg cup flats. Smooth cardboard or other smooth surfaced materials are not satisfactory as they are too slippery for the chicks to walk on.

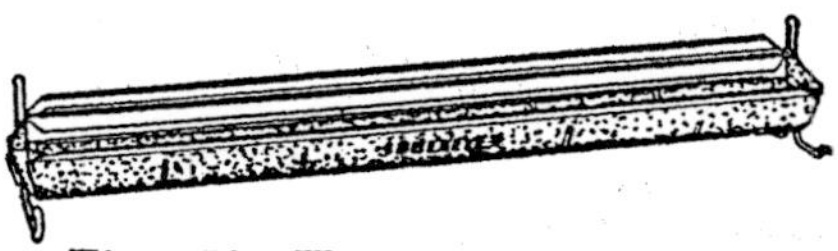

Fig. 72—Hopper for Starting Chicks.

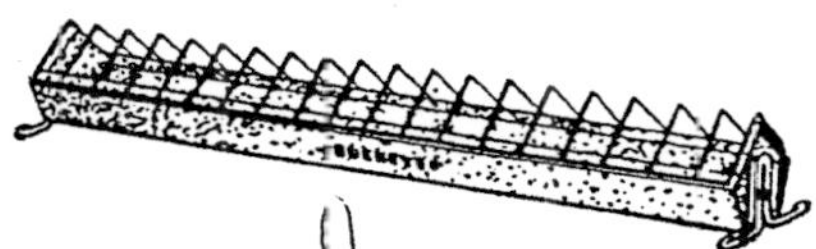

Fig. 73—Intermediate Feeders.

[Courtesy Buckeye Incubator Co.]

Water is as important as feed at all stages of life. Care should be exercised to see that all chicks can get to the water supply. It is better to have a surplus of water fountains than too few. Sometimes litter gets into the fountains to such an extent that it is impossible for the chicks to drink. This should be watched and avoided as much as possible. Many poultrymen prefer to give water the first few days that has had the chill removed.

Chickens will soon outgrow the small feeders and water containers originally provided so larger equipment must be provided as the chickens grow. Feed hoppers designed to prevent excessive feed waste and contamination with fecal material should be used. Sometimes excessive feed wastage is due to the attendant filling the hoppers too full.

Where to Feed

During the early brooding period it is evident that the chickens must be fed indoors. When the birds are permitted outdoors, either on a runway or on a wire platform, some of the hoppers should be moved out to attract them outside for considerable periods. When the birds are placed on range the placing of all hoppers entirely outside of the buildings will be found advantageous. These hoppers may be moved at regular intervals to new locations to encourage better use of the range. In hand feeding scratch grain on range it is considered good practice to avoid bare areas which are likely to harbor contamination.

TABLE 19

RECOMMENDED NUTRIENT ALLOWANCES FOR CHICKENS

	Total Protein	VITAMINS									MINERALS				
		Vitamin A Activity[1]	Vitamin D	Thiamin	Riboflavin	Pantothenic Acid	Nicotinic Acid	Pyridoxine	Biotin	Choline	Calcium	Phosphorus[2]	Salt[3]	Manganese	Iodine[4]
	Percent	I. U. per lb.	A.O.A.C. Units per lb.	Mg. Per lb.	Mg. per lb.	Mg. per lb.	Mg. per lb.	Mg. per lb.	Mg. per lb.	Mg. per lb.	Percent	Percent	Percent	Mg. per lb.	Mg. per lb.
Starting Chicks, 0 - 8 weeks	20	2000	180	0.9	1.6	5.0	8.0	1.6	0.045	700	1.00	0.60	0.5	25	0.5
Growing Chicks, 8 - 18 weeks	16	2000	180	?	0.9	?	?	?	?	?	1.00	0.60	0.5	?	0.5
Laying Hens	15	3300	450	?	0.9	2.5	?	1.6	?	?	2.25[5]	0.75	0.5	?	0.5
Breeding Hens	15	3300	450	?	1.3	5.0	?	1.6	0.070	?	2.25[5]	0.75	0.5	15	0.5

[1]May be fish oil vitamin A or Provitamin A from vegetable sources.
[2]Inorganic Phosphorus should constitute 0.2 percent of the total feed.
[3]This figure represents added salt or sodium chloride.
[4]This allowance is larger than that specified in National Research Council Reprint and Circular Series No. 111, May, 1942, "Iodine — Its Necessity and Stabilization," because of changes in the feed supply which increase the need for iodine.
[5]This amount of calcium need not be incorporated in the mixed feed inasmuch as calcium supplements fed free choice are considered as part of the ration.

Data of the National Research Council

Varying Nutrient Requirements in Floor vs. Battery Brooding

Under different environmental conditions chickens are likely to have varying nutritional requirements and will generally respond differently. It is apparent that they will not stand as heavy a mineral intake under close confinement, such as in batteries, as when brooded under less-restricted conditions. The so-called perosis or "slipped-tendon" disease has been more prevalent in batteries than in floor brooding where identical rations have been fed.

It is also quite possible that vitamin requirements may vary. Research work within the next decade will probably point out definite differences in the nutritional needs under varying environments.

TABLE 20

QUANTITIES OF FEED REQUIRED TO OBTAIN CERTAIN SELECTED AVERAGE LIVE WEIGHTS WITH DIFFERENT KINDS OF POULTRY*

Average live weight (pounds)	Kind of poultry and quantity of feed required per bird				
	White Leghorn chickens (males and females)[1]	Cross-bred[2] chickens (males)	Rhode Island Red chickens (males)	Turkeys[3] (males)	White Pekin ducks (males and females)[1]
	Pounds	Pounds	Pounds	Pounds	Pounds
0.5	1.20	1.29	1.12	0.95	0.83
1.0	3.18	2.91	2.53	2.20	2.01
1.5	5.27	4.65	4.05	3.49	3.28
2.0	7.75	6.52	5.69	4.83	4.66
2.5	10.80	8.56	7.49	6.21	6.17
3.0	14.75	10.78	9.46	7.65	7.83
3.5	20.39	13.24	11.66	9.15	9.68
4.0	....	15.97	14.13	10.70	11.77
4.5	....	19.07	16.95	12.32	14.17
5.0	....	22.62	20.25	14.02	16.99

[1] For a group containing approximately the same number of birds of each sex.
[2] The male offspring resulting from mating Barred Plymouth Rock females to Rhode Island Red males.
[3] Several different breeds from parent stock which had been selected for small size.
*U.S.D.A. 1939 Yearbook.

Hoppers and Hopper Space

Feed hoppers should be constructed to check wastage and prevent chickens from contaminating the feed with their droppings. Covers are needed to prevent feeds from wetting and spoiling when the hoppers are outdoors.

During the first few weeks one linear inch of feeding space can be allotted to each chick. If the cockerels are removed in six weeks this will give the pullets increased feeding space per bird. If the cockerels must remain for a longer period additional hoppers will be necessary. At least half of the chicks should be able to eat simultaneously. Plenty of hopper space will help prevent cannibalism, and will give more gratifying results.

Size of Grain

The grain mixture may be modified as the chickens grow by changing from fine to medium-cracked corn, and later to coarse-cracked. During the later part of the growing season, it is advisable to start feeding whole corn on the range, if the birds are to get it in the grain mixture after attaining laying age. Begin sufficiently far in advance of housing to thoroughly accustom the birds to it and thus avoid delay later.

Clean Feed and Water

Only clean, wholesome, high-quality ingredients are recommended for the feed mixture for growing chickens. The use of questionable, musty or moldy feedstuffs is dangerous. Only a small amount of poor feed is required to throw chicks out of condition and start serious digestive disorders and mortality. Purchase feed ingredients on the basis of classification and analysis, giving quality rather than price first consideration.

Another requirement is plenty of fresh, clean water in fountains that can be easily cleaned, yet not be contaminated by fecal material. For young chicks tepid water is best. Cold water is too chilling and will disturb digestion. During the summer when the chicks are older and on range provide cool water by burying the pipe line.

Brook water is usually considered dangerous where it receives the wash of land that is contaminated or thought to be contaminated. Fence off stagnant pools and swampy areas from the chickens.

Grading Pullets According to Feeding Needs

Too little attention is normally given to grading birds. Grading can be started when the chicks are taken from the incubator and should be continuous throughout the entire brooding and ranging cycle. Inspect chicks for size, deformities, fluffiness, and healing of the navel.

When they are sufficiently developed to identify as pullets, systematic grading can be practiced by segregating them into two or more groups as determined by their physical development.

Grade them according to size, rapidity of feathering, health, vigor, and freedom from physical defects.

The larger and better pullets can then be maintained on the usual feeding system. Those in the below standard group will have a chance to feed without competition from the bigger birds, and can be given supplementary feedings to promote growth, in addition to the regular ration. Any which do not respond satisfactorily and prove undesirable to retain to maturity can be disposed of for meat purposes at broiler age.

Feed Consumption and Growth

There is a strong relationship between the amount of feed consumed by chickens and the growth rate. If the consumption is subnormal, growth will obviously be subnormal for the strain involved. This makes it imperative that the chickens have access to plenty of palatable feeds at all times for maximum growth.

Only a part of the feed consumed by chickens can be used for growth since part of the feed is not digested. Another part is used to maintain the body and all the various functions necessary to life. This is referred to as the maintenance requirement. And as the chickens grow and become larger the quality of feed necessary for maintenance becomes larger. Thus a decreasing proportion of the feed consumed can be used for growth. This is the same as saying that the efficiency of feed utilization for growth decreases as chickens increase in weight. Early rapid growth therefor is the most economical growth. This relationship is illustrated in the following table:

TABLE 21

Period in Weeks	Pounds of Feed to Produce One Pound of Gain
0-2	2.86
10-12	4.92
22-24	15.24
0-8	2.97
8-16	5.85
16-24	10.92

TABLE 22

CHICKEN RATIONS*

Feeding Ratio—Mash to Grain Mash Ingredient	1/0 Starter	2/1 Grower
Ground yellow corn	750	700
Ground barley or pulverized oats	400	400
Ground wheat or standard middlings	400	400
Alfalfa meal (17%)	100	100
Soybean oil meal	300	300
Fishmeal (55%) or meat scrap (50% (1)	200	100
Steamed bonemeal or equivalent (2)	—	50
Calcium carbonate (3)	25	40
Iodized salt	10	20
Fort. oil (400D - 2000A) or equivalent	2	3
Manganese sulfate	.25	.25
Riboflavin supplement, mgs. (4)	600	—
Total (5)	1987	2013

(1) Either fishmeal or meat scraps may be used alone, or in any combination of the two.

(2) Defluorinated rock phosphate or di-calcium phosphate may replace the bonemeal. Raw rock phosphate containing not more than 3.5% fluorine may replace one-half of the bonemeal in the starter and grower and one-fourth in the layer and breeder formulae.

(3) Oyster shell flour or feeding limestone (98% calcium carbonate).

(4) Sources of riboflavin are: (1) dried milk powders 9 -15 mgs. per pound, (2) dried whey 12.9 mgs. per pound, (3) yeast 11.4 - 18.2 mgs. per pound, (4) dried distillers solubles 6.8 mgs. per pound, (5) fermentation products 36.3 - 3.600 mgs. per pound, and (6) synthetic riboflavin supplements 16.000 mgs. per pound. To determine amount to use divide number of milligrams desired by manufacturers guaranteed potency. Example - Desire to furnish 600 msg. using a product having a guaranteed potency of 15 milligrams per pound. Solution 600 divided by 15 equals 40.

(5) If an even 2000 pounds is desired adjust by adding or removing ground grains or middlings.

*New England Conference Recommendations.

TABLE 23

CHICK-GROWING MASHES TO BE FED WITH GRAIN*

Ingredient	Percent	Percent
Ground yellow corn	25.0	25.8
Ground oats or wheat	10.0	—
Ground corn, wheat, barley, or grain sorghum	11.8	—
Wheat middlings	—	15.0
Wheat bran	—	10.0
Alfalfa leaf meal	5.0	5.0
Soybean meal	21.0	17.0
Cottonseed meal, peanut meal, corn gluten meal, or soybean meal	10.0	10.0
Meat meal or fish meal	5.0	5.0
Riboflavin supplement (20 micrograms riboflavin per gram)	6.0	6.0
Steamed bone meal, defluorinated superphosphate, or other low-fluorine calcium phosphate	3.0	3.0
Ground limestone or oystershell	2.0	2.0
Manganized salt	1.0	1.0
Vitamin A and D feeding oil (400 A.O.A.C. units vitamin D, 2000 International units vitamin A per gram)	.2	.2

*Two of a number of formulas recommended in U.S.D.A. Circular 788.

TABLE 24

ALL-MASH CHICK-STARTING RATIONS*

Ingredient	Percent	Percent
Ground yellow corn	25.0	25.0
Pulverized oats or ground wheat	10.0	—
Ground corn, wheat, barley, or grain sorghum	20.4	13.4
Wheat middlings	—	10.0
Wheat bran	—	10.0
Alfalfa leaf meal	5.0	5.0
Soybean meal	16.0	14.0
Cottonseed meal, peanut meal, corn gluten meal, or soybean meal	10.0	10.0
Fish meal	2.0	2.0
Meat meal	4.0	4.0
Riboflavin supplement (20 micrograms riboflavin per gram)	5.0	4.0
Steamed bonemeal, defluorinated superphosphate, or other low-fluorine calcium phosphate	1.0	1.0
Ground limestone or oystershell	1.0	1.0
Manganized salt	.5	.5
Vitamin A and D feeding oil (400 A.O.A.C. units Vitamin D, 2000 International units Vitamin A per gram)	.1	.1

*Two of a number of formulas recommended in U.S.D.A. Circular 788.

TABLE 25

ALL MASH CHICKEN RATIONS*

Mash Ingredient	Starter	Grower
Ground yellow corn	750	800
Ground barley or pulverized oats	200	400
Ground wheat or standard middlings	400	500
Alfalfa meal (17%)	100	50
Soybean oil meal	300	100
Fishmeal (55%) or meat scrap (50%) (1)	200	100
Steamed bonemeal or equivalent (2)	—	20
Calcium carbonate (3)	25	30
Iodized salt	10	10
Fort. oil (400D - 2000A) or equivalent	2	2
Manganese sulfate	.25	.25
Riboflavin supplement, mgs. (4)	600	—
Total (5)	1987	2012

(1) Either fishmeal or meat scraps may be used alone, or in any combination of the two.

(2) Defluorinated rock phosphate or di-calcium phosphate may replace the bonemeal. Raw rock phosphate containing not more than 3.5% fluorine may replace one-half of the bonemeal in the starter and grower and one-fourth in the layer and breeder formulae.

(3) Oyster shell flour or feeding limestone (98% calcium carbonate).

(4) Sources of riboflavin are: (1) dried milk powders 9-15mgs. per pound, (2) dried whey 12.9 mgs. per pound, (3) yeast 11.4 - 18.2 mgs. per pound, (4) dried distillers solubles 6.8 mgs. per pound, (5) fermentation products 36.3 - 3,600 mgs. per pound, and (6) synthetic riboflavin supplements 16,000 mgs. per pound. To determine amount to use divide number of milligrams desired by manufacturers guaranteed potency. Example - desire to furnish 600 mgs. using a product having a guaranteed potency of 15 milligrams per pound. Solution 600 divided by 15 equals 40 pounds per ton of feed.

(5) If an even 2000 pounds is desired adjust by adding or removing ground grains or middlings.

*New England Conference Recommendations.

TABLE 26

ALL-MASH CHICK-GROWING RATIONS*

Ingredient	Percent	Percent
Ground yellow corn	25.0	25.0
Ground oats or wheat	10.0	—
Ground corn, wheat, barley or grain sorghum	34.9	23.9
Wheat middlings	—	15.0
Wheat bran	—	10.0
Alfalfa leaf meal	5.0	5.0
Soybean meal	12.0	8.0
Cottonseed meal, peanut meal, corn-gluten meal, or soybean meal	5.0	5.0
Meat meal or fish meal	3.0	3.0
Riboflavin supplement (20 micrograms riboflavin per gram)	2.0	2.0
Steamed bonemeal, defluorinated superphosphate or other low-fluorine calcium phosphate	1.5	1.5
Ground limestone or oyster shell	1.0	1.0
Manganized salt	.5	.5
Vitamin A and D feeding oil (400 A.O.A.C. units Vitamin 2000 International units Vitamin A per gram	.1	.1

*Two of a number of formulas recommended in U.S.D.A. Circular 788.

Questions on Chapter VIII

1. Why is it desirable to feed chickens as soon as they are removed from the incubator or, if purchased, as soon as they are received?
2. Are feed requirements of chicks as exacting as those of laying birds?
3. Do chicks require the same or higher protein intake per units of feed consumed as laying birds?
4. What is the advantage gained in moving hoppers on range to new locations frequently?
5. How can you determine whether a sufficient amount of hopper space has been provided for a group of chicks?
6. In general, when is the most economical growth secured?

CHAPTER IX

FEEDING ADULT BIRDS

Subject Matter

To successfully feed birds one must have the ability to observe their condition and the effect of feeding. No "cut-and-dried" method has yet been devised that will relieve the feeder of responsibility of making necessary changes to meet varying conditions throughout the year.

It is the goal of this chapter to point out many of the successful practices in the feeding of adult birds.

The Grain and Mash Method of Feeding:

This method is one of the most widely used at the present time. It is based on feeding a mash containing about 20 per cent protein with about an equal amount of grain. The mash is kept before the birds at all times and grain fed once and sometimes twice daily. When fed twice daily, one third the allowance is fed in the morning and two thirds in the late afternoon. The grain and mash method gives excellent results in the hands of experienced poultrymen.

In some sections poultrymen utilize a highly concentrated mash containing from 32 to 40 per cent protein. In such cases mash and grain are kept in separate hopper before the birds at all times thereby permitting free choice. High consumption of grain results with the birds consuming as much as 75 to 80 per cent of their ration as grain. To properly balance the total intake, the mash must not only contain a high percentage of protein but must contain adequate minerals and vitamins to make up for the deficiencies of grain.

Hopper Feeding of Grain:

Some poultrymen advocate the feeding of grain in hoppers where grain and mash systems are employed and where the mash contains the usual amount of protein. They point out that it is not necessary for the birds to scratch in the litter to obtain

Fig. 74—Place Hoppers Where They Will Be Well Lighted. Note Easily Constructed Laying Mash Feeder For Hens.

exercise and that the system is more sanitary. From the results available it would appear to make little difference where the grain is fed. The important point is the amount of grain fed and how the nutrient deficiencies of the grain is offset in the use of the mash.

Litter Feeding of Grain:

The feeding of grain in the litter has long been common in the handling of laying birds. One must remember the impossibility of accurately designating a definite amount to feed, as the requirements of the birds vary from time to time, depending on their physical condition and rate of production and on climatic conditions. To feed in the litter intelligently the feeder must be skillful, observant as to the appetites of the birds, *and sure that the grain is entirely cleaned up at least once daily.*

This method of feeding induces exercise by encouraging the birds to scratch for their feed in the litter, but may be a questionable advantage, for probably the birds will exercise sufficiently even if they are hopper-fed both grain and mash. It may help to prevent or control cannibalism, however, by keeping the birds busy. It undoubtedly aids in airing out the litter and keeping it drier.

The All-Mash Method of Feeding:

The all-mash method of feeding laying birds has seemingly gained popularity. In this system both the mash- and scratch-feed ingredients are combined into a ground mash mixture. The advantages claimed by its advocates are that it is simpler, and more sanitary and that it reduces labor, gives a definite ration, and yields as good results as other feeding systems. It is recommended to the inexperienced feeder.

Where feed consumption is not at a sufficiently high level for the purposes of maintaining a satisfactory production rate, a wet mash feeding shortly before the birds go on the roosts or a feeding of laying pellets may be desirable. Should a reduction in body weight become a factor during the laying season, these additional feedings may consist of a wet fleshing mash instead of all-mash, or fleshing pellets in place of laying pellets.

The Pellet System of Feeding:

A general description of pellets may be that they represent a feed ration that has been compressed by machinery into small cylindrical units. They may represent a formula designed to supplement other systems of feeding such as the all-mash method and it is perhaps best exemplified by the fattening pellets. Here they are only supplementary to the regular ration used. In other cases, the entire ration of birds may consist of laying pellets. However, this system has not proven entirely, satisfactory because of canabalistic outbreaks. They may also be used as supplementary feeds to other methods. Feed in pellet form is relished by the birds and has a claimed advantage of reduction in waste.

Supplementary Feedings

Moist Mash:

Dry mash should always be in front of the birds, but where production does not respond satisfactorily a moist mash may also be provided for greater stimulation. About three pounds of dry mash moistened—or about what they will eat in 15 or 20 minutes—will be sufficient for 100 birds daily.

Morning seems to be a natural dry-mash-consumption period; so it is preferable to feed the additional moist mash just after the noon hour. In this way the dry-mash consumption is not seriously interfered with and considerable time remains before the afternoon grain feeding.

Fig. 75—Summer Laying Shelter.
[Courtesy Prof. J. C. Graham.]

The primary object of a moist mash is to stimulate production. It should be held in reserve until needed. After birds have been maintained through the summer period for production and are nearing the completion of their laying year, a stimulant often becomes necessary to hold production at a suitable level.

Fleshing Mash:

During the laying season it is desirable to maintain the body weight of the birds. If for any reason they are losing weight, the use of a fleshing mash will be profitable. This mash serves as a supplementary feeding and may be composed of (1) one part of the regular laying mash and one part of corn meal, or (2) a special fleshing mash formula of reduced animal-protein content.

Fleshing Mash

1 part dry mash
1 part yellow corn meal
1 part ground oats

In either case the ingredients should be mixed with milk of some form as it adds to the palatability and feeding value of the mixture.

It is considered good policy to examine a sufficient number of representative birds monthly or at some other regular period to determine the need for such a feeding before trouble begins. By carefully watching the fleshing on the breast various slumps in production can be avoided and molts experienced by many poultrymen with their flocks can be prevented.

The feeding of a fleshing mash should be gradually discontinued when the condition of the birds justify it, providing best results in production are desired.

Artificial Illumination

During fall and winter months, days in some sections are not sufficiently long to provide the necessary light stimulation to the pituitary gland in order to obtain maximum egg production. To remedy this condition and thereby stimulate egg production, artificial lights are commonly used by many poultrymen. It is generally advocated that not more than a 12- or 14-hour day be resorted to. There are several methods by which the length of day may be increased:

(1) morning lighting period
(2) evening lighting period
(3) morning and evening lighting combination
(4) all-night light.

The morning lighting period is perhaps the most popular because it is easier to get the birds off the roosts in the morning than to keep them off in the evening. An additional advantage of morning lighting is that the use of a dimming device is eliminated.

Amount of Light:

In determining the amount of light necessary for a pen it is well to allow at least 40 watts for each 200 square feet of floor space. For all-night light this should be reduced to 15 watts. The light should be hung so as to illuminate not only the floor, feed hoppers and waterers, but also to the roosts. If used regularity of operation is essential.

Length of Day:

If artificial illumination is contemplated it should be inaugurated as a part of the management program in the fall, when the length of the hens' natural working day has shortened to

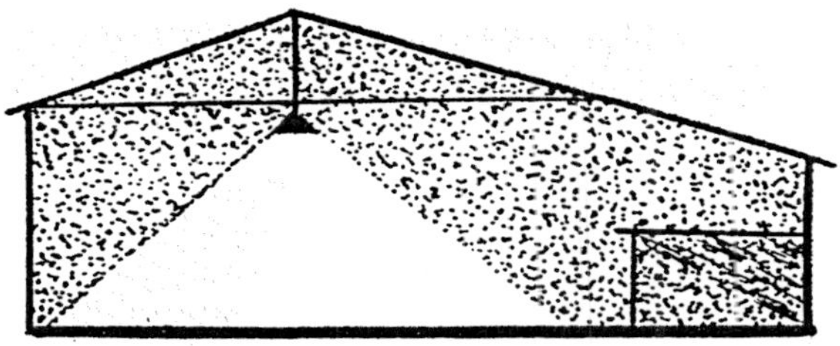

Fig. 76—Poor Lighting. Roosts Are Not Illuminated.
[Courtesy N. H. Agr. Exp. Sta.]

this period, and continued until spring when daylight again covers a similar length of time. With all-night light the lights may be turned on at 8 or 9 in the evening and left until morning. When to start lights in the fall will depend on the latitude in which the farm is situated. It will vary from September 1 to October 1 in the northern states. Economic studies indicate that artificial lights do influence egg distribution of production but not total annual production.

In the management of hens for breeding purposes lighting ought to begin about four weeks before one intends to start saving eggs for hatching purposes. Hens given an opportunity to molt normally will respond very quickly. This will provide a volume of eggs at the time desired.

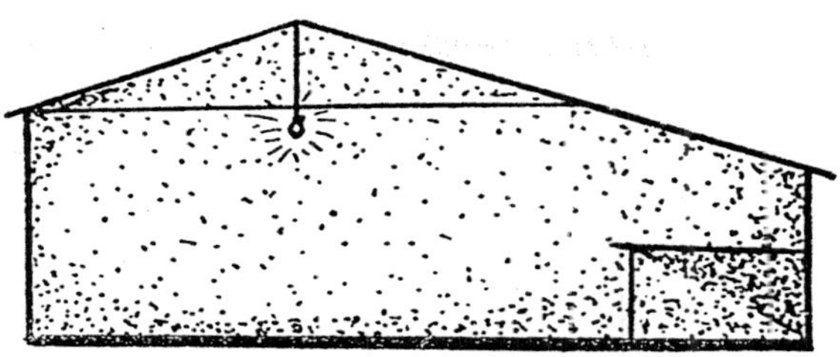

Fig. 77—Faulty Lighting. Lights Too Diffused.
[Courtesy N. H. Agr. Exp. Sta.]

Maintain Body Weight:

In forcing layers for production it is possible to stimulate them too heavily for eggs without maintaining sufficient body weight. In general an increased grain allowance will aid materially in keeping up their weight. If they are losing in weight, resort to a daily feeding of a fleshing mash may be necessary.

Many poultrymen are successful in securing continuous high production, and at the same time maintaining the weights of the birds. They are also obtaining a very high hatchability percentage throughout the fall, winter and spring season.

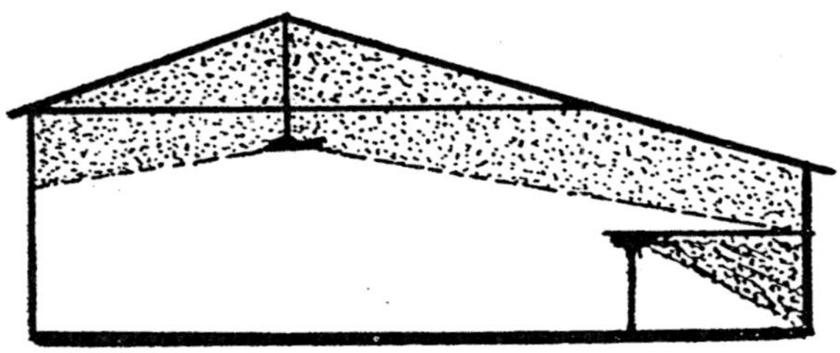

Fig. 78—Proper Lighting. Depends on Position and Shape of Deflector.

[Courtesy N. H. Agr. Exp. Sta.]

Factors Affecting the Success of Artificial Lighting:

1. Inauguration of lighting in season.
2. Regularity of lighting and all management practices.
3. Housing of birds in good physical condition.
4. Segregation of birds according to sexual development.
5. Maintenance of body weight of birds.
6. Proper installation of lights of sufficient intensity.
7. Available water and feed when lights are on.
8. Lights not discontinued too early in the spring.
9. Regular check on time switch to see that it is keeping proper time and that it is in working condition.
10. Good production breeding back of the stock.

Feeding Molting Birds

Hens which have completed their production year and are going into their annual molt should not be fed materially different than laying birds. The old practice of removing mash and feeding them solely on grain is not good practice. The making of feathers requires considerable protein. Since mash is the portion of the ration possessing the protein concentrates, it is obvious that one should continue feeding mash. At the same time the birds should receive all the scratch grain they will readily consume.

During the molting season of the year many birds and flocks which have laid heavily and have largely depleted their body reserve may require fleshing mash to return them to normal physical condition in the shortest space of time.

Feeding Breeding Birds

As production is essentially reproduction breeding birds should be handled by the best-known methods, similar to laying birds. The ration should contain all of the nutrients known to be necessary and should promote reasonable production. As to differences in feeding pullets and hens, the latter may need to be managed by more intensive methods to stimulate production of volume of eggs at the season desired.

TABLE 27

TABLE OF FEED REQUIREMENTS

The following table represents data from Cornell University with White Leghorns and an estimate of the amounts of feed required for modern production-bred Rhode Island Reds, Plymouth Rocks, or New Hampshires:

% Production	Feed Required Per 100 Layers Per 24 Hours	
	Leghorns	Rocks or Reds
—	18–19 lbs.	21–26 lbs.
10	19–20	22–28
20	20–21	24–30
30	21–22	25–31
40	22–23	26–32
50	23–24	27–33
60	24–25	28–35
70	25–26	30–36
80	27–28	32–39

Normal continuous production and even continuous high production does not appear to be a hindrance to good hatchability and livability. There is evidence that it actually enhances the breeding value of the stock, for when birds are maintained in low production our experience has been that fertility and hatchability does not reach its optimum. It is desirable to bring the breeders along into production uniformly to avoid poor fertility, which results when birds come into production irregularly.

Confining breeders to the house places the feeding in the attendant's control. Breeders need not have access to yards or range in order to produce desirable eggs of high hatching power. Failure to secure results where birds are confined may be charged to faulty feeding, housing or management.

Feed Consumption

The amount of feed consumption, particularly mash, is directly reflected in the egg production of layers. High mash consumption indicates a proportionately high egg yield. Increased feed consumption precedes increased production by two or three weeks, and, conversely, decreased feed intake precedes a decrease in egg flow. By keeping consumption records the trend of production can be predicted. Where feed consumption is dropping, supplementary feedings can be used to stave off declines in production, as they help to maintain the level of consumption. The records aid in computing the feed cost of producing eggs and ought to form a part of every poultry account system.

In general, laying birds consume in the neighborhood of 1¾ to 2¼ pounds of feed per bird per week. This amount includes approximately equal amounts of grain and mash and the figures may be used as the basis for computing probable feed requirements for a flock of birds.

The appetite of hens for different ingredients of the scratch-grain mixture vary from time to time, and it is desirable to cater to their tastes by changing the formula of the mixture in accordance with their choice. Differences in the nutrient values of many of the common cereals used in scratch mixtures are not great enough to interfere with a desirability to provide flexibility in the formula to meet the preferences of the birds. The relative costs should be given consideration.

Frequently the intake of an adequate quantity of clean, wholesome water is not sufficiently emphasized in regard to its direct effect on production. Daily cleaning of water containers will materially increase the amount of water ingested. To continue a desirable level of intake during freezing weather, non-freezing fountains should be provided.

Effect of Feed on Product

Certain feeds have a decided effect on the color of yolk, on the flavor, and in other cases on the nutritive value of the product. The feeding of onions, rape and other strong odoriferous

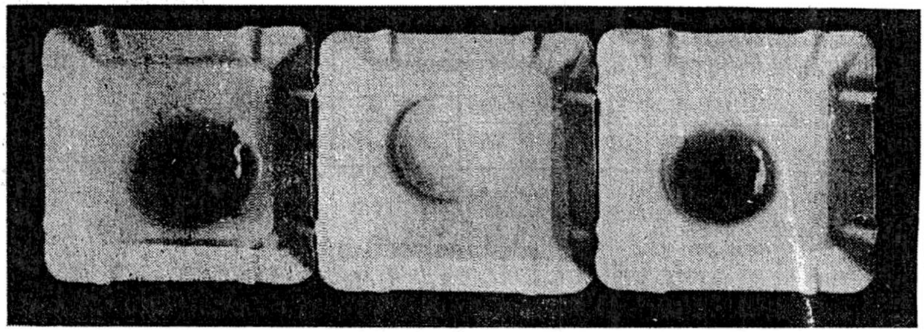

Fig. 79—Showing Effect of Cottonseed Meal on Yolk Color. Center Yolk is Normal.

[Courtesy L. N. Berry.]

products is likely to add an undesirable taste to eggs. It has been demonstrated that the use of ingredients of high free-fatty acid content will produce fishy-flavored eggs and meat.

Foods carrying large amounts of xanthophyll, such as green foods and yellow corn, are conducive to a highly pigmented yolk color. Such products as oats, wheat and white corn tend, on the other hand, to produce pale yellow yolks. Birds fed cottonseed meal produce eggs which when stored develop typical olive green or black yolks. Such eggs are very undesirable for holding or cold-storage purposes. Restriction of the amount of cottonseed meal in the ration is necessary.

The vitamin nutritive value of eggs is dependent upon the vitamin content of the ration. It is possible, therefore, to influence the nutritive value of the product.

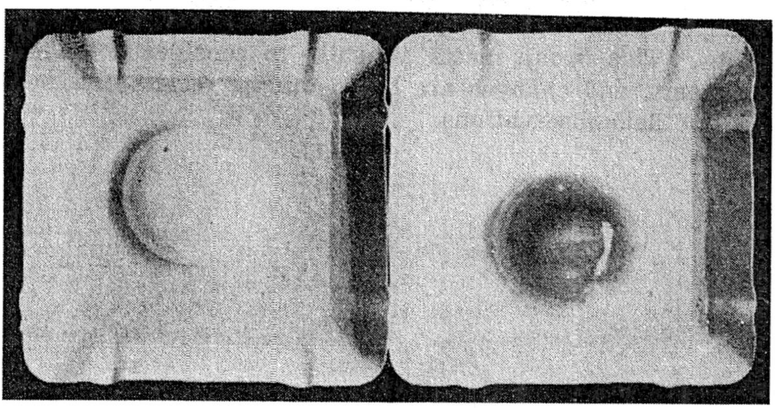

Fig. 80—Effect of Excessive Green Feed Consumption on Yolk Color. Egg on Left is Normal.

[Courtesy L. N. Berry.]

SUGGESTED NUMBER OF POUNDS OF SCRATCH FOR 100 LAYERS*

Month	N.	D.	J.	F.	M.	A.	M.	J.	J.	A.	S.	O.
Light Breeds	12	12	12	12	12	12	10	10	8	8	8	8
Heavy Breeds	15	16	18	16	16	15	14	14	12	12	12	12

*These amounts should be varied to meet specific conditions.

Don'ts in Feeding

1. Don't over-feed on scratch grain.
2. Don't over-feed green feeds.
3. Don't give up a definite schedule of feeding once it is in operation.
4. Don't be inattentive to details.
5. Don't make any abrupt change in feeding; if a change is necessary make it gradually.
6. Don't forget to provide plenty of feed-hopper space—half the birds should be able to feed at one time.

Rations

The rations that follow are selected as fairly representative. The poultryman is not limited to these, however. He may adopt any one of a wide range of formulas in addition to a large number of commercial mixtures.

New England College Conference Ration:

This first ration was formulated co-operatively by the poultry departments of the six New England state universities and colleges. This group meets annually to consider possible improvements, and changes are based on the results of research work and field observations.

TABLE 28

Feeding Ratio—Mash to Grain Mash Ingredient	1/1 Layer	1/1 Breeder
Ground yellow corn	600	600
Ground barley or pulverized oats	300	300
Ground wheat or standard middlings	350	350
Alfalfa meal (17%)	100	100
Soybean oil meal	300	300
Fishmeal (55%) or meat scrap (50%) (1)	200	200
Steamed bonemeal or equivalent (2)	60	60
Calcium carbonate (3)	40	40
Iodized salt	20	20
Fort. oil (400D - 2000A) or equivalent	10	10
Manganese sulfate	.5	.5
Riboflavin supplement, mgs. (4)	—	1540
Total (5)	1980	1980

(1) Either fishmeal or meat scraps may be used alone, or in any combination of the two.
(2) Defluorinated rock phosphate or di-calcium phosphate may replace the bonemeal. Raw rock phosphate containing not more than 3.5% fluorine may replace one-half of the bonemeal in the starter and grower and one-fourth in the layer and breeder formulae.
(3) Oyster shell flour or feeding limestone (98% calcium carbonate).
(4) Sources of riboflavin are: (1) dried milk powders 9 -15 mgs. per pound, (2) dried whey 12.9 mgs. per pound, (3) yeast 11.4 - 18.2 mgs. per pound, (4) dried distillers solubles 6.8 mgs. per pound, (5) fermentation products 36.3 - 3,000 mgs. per pound, and (6) synthetic riboflavin supplements 16,000 mgs per pound. To determine amount to use divide number of milligrams desired by manufacturers guaranteed potency. Example—Desire to furnish 600 mgs. using a product having a guaranteed potency of 15 milligrams per pound. Solution 600 divided by 15 equals 40.
(5) If an even 2000 pounds is desired adjust by adding or removing ground grains or middlings.

TABLE 29

ALL-MASH BREEDING RATION*

Ingredient	Percent
Ground yellow corn	25.0
Ground oats or wheat	10.0
Ground corn, wheat, barley, or grain sorghum	31.2
Alfalfa leaf meal	5.0
Soybean meal	7.0
Peanut meal, corn-gluten meal, or soybean meal	5.0
Fish meal	2.0
Meat meal	4.0
Riboflavin supplement (20 micrograms riboflavin per gram)	5.0
Steamed bonemeal, defluorinated superphosphate, or other low-fluorine calcium phosphate	2.0
Ground limestone or oystershell	3.0
Manganized salt	.5
Vitamin A and D feeding oil (400 A.O.A.C. units vitamin D, 2,000 International Units vitamin A per gram)	.3

*One of the rations recommended by U.S.D.A. Circular 788.

TABLE 30

ALL MASH RATIONS

Mash Ingredient	Layer	Breeder
Ground yellow corn	800	800
Ground barley or pulverized oats	400	400
Ground wheat or standard middlings	500	500
Alfalfa meal (17%)	50	50
Soybean oil meal	100	100
Fishmeal (55%) or meat scrap (50%) (1)	100	100
Steamed bonemeal or equivalent (2)	20	20
Calcium carbonate (3)	30	30
Iodized salt	10	10
Fort. oil (400D—2000A) or equivalent	5	5
Manganese sulfate	.45	.25
Riboflavin supplement, mgs. (4)	—	760
Total (5)	2015	2015

(1) Either fishmeal or meat scraps may be used alone, or in any combination of the two.

(2) Defluorinated rock phosphate or di-calcium phosphate may replace the bonemeal. Raw rock phosphate containing not more than 3.5% fluorine may replace one-half of the bonemeal in the starter and grower and one-fourth in the layer and breeder formulae.

(3) Oyster shell flour or feeding limestone (98% calcium carbonate).

(4) Sources of riboflavin are: (1) dried milk powders 9-15 mgs. per pound, (2) dried whey 12.9 mgs. per pound, (3) yeast 11.4 - 18.2 mgs. per pound, (4) dried distillers solubles 6.8 mgs. per pound, (5) fermentation products 36.3 - 3,000 mgs. per pound, and (6) synthetic riboflavin supplements 16,000 mgs. per pound. To determine amount to use divide number of milligrams desired by manufacturers guaranteed potency. Example - desire to furnish 600 mgs. using a product having a guaranteed potency of 15 milligrams per pound. Solution 600 divided by 15 equals 40 pounds per ton of feed.

(5) If an even 2000 pounds is desired adjust by adding or removing ground grains or middlings.

TABLE 31

LAYING MASH WITH WHICH AN EQUAL AMOUNT OF GRAIN IS TO BE FED*

Ingredient	Percent
Ground yellow corn	25.0
Ground oats or wheat	10.0
Ground corn, wheat, barley, or grain sorghum	9.0
Alfalfa leaf meal	5.0
Soybean meal	24.0
Peanut meal, corn-gluten meal, or soybean meal	10.0
Meat meal or fish meal	4.0
Riboflavin supplement (20 micrograms riboflavin per gram)	4.0
Steamed bonemeal, defluorinated superphosphate, or other low-fluorine calcium phosphate	5.0
Ground limestone or oystershell	2.5
Manganized salt	1.0
Vitamin A and D feeding oil (400 A.O.A.C. units vitamin D, 2,000 International Units vitamin A per gram)	.5

*One of the rations recommended by U.S.D.A. Circular 788.

Questions on Chapter IX

1. What are the different methods or systems of feeding laying birds?
2. If a wet or moist mash is being fed supplementary to the regular ration, at what time of day is it most desirable?
3. When should a supplementary fleshing mash be used in the feeding of laying birds?
4. What different lighting systems are employed by poultrymen?
5. What amount of illumination should be provided for 200 square feet of floor space?
6. What factors affect the success of artificial lighting?
7. How should birds be fed during the molting period?
8. How much feed do laying birds normally consume per week?
9. Does feed have any effect on the eggs produced?

Fig. 81—A Permanent Laying Shelter with Sides Closed for Late Fall Use.

[Courtesy Prof. J. C. Graham.]

CHAPTER X

REARING AND RANGE MANAGEMENT

Subject Matter

Successful management practices must be continued after the chicks are past the brooding stage. Too often a poultryman considers his task completed upon finishing the brooding operations, and allows the birds on range to shift for themselves with little personal attention. It is particularly important during the hot summer months to meet the needs of the growing birds for water, feed and the general details of supervision. Special attention should be given to watching for the appearance of mites. If they appear, prompt treatment is advisable. Painting the roosts and roost rails and supports with one of the heavy coal-tar disinfectants is usually effective.

The care given the birds on range is quite often the determining factor as to the quality of pullets raised and the cash income after they are placed in permanent laying quarters. There is likelihood of a distinct relationship between the quality of pullets housed and the mortality of adult birds during the laying year. This may spell the difference between success and failure. The complete cycle of good management should be followed.

Range Rotation

It is advisable to allow sufficient land for ranging purposes so that a three-year rotation system may be followed. A three-year rotation implies that the range has not been used for poultry for two successive years and has not had poultry manure applied to it.

Fig. 82—Showing Use of Both Colony Houses and Shelters on Spacious Range.
[Courtesy Hubbard Farms.]

With plenty of land available, a four-year rotation may be advantageous.

In laying the plans, allow one acre for each 300 to 350 birds—certainly not over 500, after the cockerels are removed this would mean not over 250 pullets per acre. In other words, to provide proper range for the production of 1,000 pullets, four to six and one-half acres must be available annually, or a total of approximately 15 acres for a three-year rotation, such as follows:

TABLE 32

	1st Range	2nd Range	3rd Range
1st year	chickens	hay	hay with oats
2nd year	hay with oats	chickens	hay
3rd year	hay	hay with oats	chickens

The crops best suited for a chicken range are clover, grass and alfalfa in the order named. Chickens are grown on a portion one year, being carefully excluded from the other two sections. The second year the area on which the chickens were grown is plowed and seeded to a cereal crop, such as oats, together with a hay crop. About two bushels of heavy oats to the acre are sufficient. This should be seeded with alfalfa or clover, if land is prepared for either, or with a mixture of grass seed composed of ten pounds timothy, six red clover, four alsike clover and four red-top, per acre. This is sown preferably in the spring after the land has lain over winter. The oats may be cut and stored for use as litter in the hen houses.

The third year the hay crop is cut for cattle or horse feed, and the fourth spring the chickens are again brooded on this area where a good sod has developed.

Where natural shade is lacking, suitable protection from the sun can be provided by plowing a narrow strip of the range and planting sunflowers or corn or a combination of both. The chickens should be fenced out of the strip until the plants have sufficiently developed.

Location of Range

The range should be conveniently located near the center of operations, yet sufficiently separated to prevent contamination from spreading from the mature birds to the growing stock. In laying out the range its accessibility to water, feed and other supplies must be considered. Select relatively high tillable ground, sloping preferably to the south or southeast. It needs to be well drained.

Fig. 83—A Well-sodded Range with Natural Shade.
[Courtesy Roy Jones.]

Protection from prevailing winds by natural or artificial wind breaks is wise. In most instances thorough fencing will also be found advisable, not only to restrict the pullets to the range, but to protect them from natural enemies, such as foxes, dogs and other large predatory animals.

In locating the range, precautions against theft of birds may be necessary. If the range is along a main highway and very noticeable to passersby, considerable thievery may be encountered. In some cases where the size of the enterprise will justify it, employment of a night watchman may become necessary.

Summer Shelters

The summer shelter has come into quite general use for housing pullets on range from the time they can safely be removed from the brooding house, whether it be of the colony or permanent-type system. These shelter the pullets conveniently without great expenditure of capital. Of light construction, they can be easily moved from range to range succeeding years. The general feeling is that birds develop better where shelters are used than where colony houses are resorted to for range houses.

Many innovations have been made by various poultrymen in the arrangement of the roosts, the wire floor, and the door. Tar paper, galvanized sheets and some types of wall board have all been used for roofing material. The tendency is to use stiffer roofing than tar paper, material that is not easily punctured, for it is placed directly on the rafters.

Fig. 84—Temporary Yard Trains Chicks to Location of House.
(Courtesy of H. D. Munroe)

Colony Houses

Colony brooder houses will also serve for range quarters after the chicks are past the brooding stage. Due to this fact they have been a very popular type of brooder. With a system of range rotation which necessitates moving them, however, more labor is required and they do not remain sufficiently tight over a period of years to satisfy brooding requirements.

Where they are used they should be of a size readily movable by team or tractor. This permits range rotation just as with range shelters. Due to their size and weight it is advisable to move them in winter when the ground is frozen.

When used on the summer range special ventilation will be necessary. It is wise not to crowd them with pullets for this reason, although the birds do not appear to develop individually to the same extent that they do where shelters are used.

As a precautionary measure the foundation of the house may be screened to prevent pullets from congregating under it, thus avoiding intense soil contamination. Litter should be used on the floor and be frequently renewed.

To give the chickens easy access to the house an inclined approach of soil or boards may be constructed. This will prevent them from huddling outside when they are first turned on the range and before they become adept at reaching the trap door.

Roosting Capacity

Regardless of the type of range house, ample roosting space needs to be provided. Six linear inches of roost per bird at 12 weeks of age is adequate. If this is the gauge for estimating housing capacity, overcrowding is not likely. With the heavy breeds additional perch space is demanded as the birds approach maturity.

Fig. 85—Range Shelters in Use on a Well-sodded Range.
Note Covered Hoppers.

[Courtesy New England Poultryman.]

Separation of Sexes and Culling

Separating the cockerels from the pullets as soon as they can readily be identified is an advantage. In many cases a large percentage of the males will be evident at six weeks of age and practically all may be distinguished at eight weeks. The reasons for separation are mainly: (1) If both sexes are permitted together the cockerels are bothersome to the pullets. (2) The birds increase in size so that the capacity of the house soon becomes severely taxed. Reduction in numbers is almost necessary. (3) By removing some of the competition around the hoppers, greater food consumption is possible for the remaining birds.

Why cull at this time? Inasmuch as it is necessary to handle many or all of the birds for sex identification, a culling at this time will entail no extra handlings. Culls at this time are likely to be culls at a later date; so they might just as well be removed and disposed of a few weeks later along with the broilers. More can also be realized on them on a normal broiler market than can be secured later on at fowl prices. In making this culling, select for vigor, rapid feathering and good size. Cull any deformed specimens.

Only vigorous individuals should remain after the culling process. Those that appear thin or scrawny or show evidence of disease should have been removed. Similarly, individuals that have feathered slowly should not remain for use as breeders. If retained they may find their way to the breeding pens as mature individuals. Many cases of slow feathering are traceable to genetical make-up. If such birds are used for breeding purposes they may throw numerous slow-feathering offspring.

Seldom do undersized birds ever develop rapidly enough to make up for lost ground. The rapid-growing individuals are usually the most efficient in the utilization of feed. Attention must be paid to size to secure sufficient weight at maturity.

Tattooing

A development to help prevent thieving and to make positive identification of poultry possible is tattooing. This consists of stamping into the web of the wing a suitable number which is permanently preserved by impregnating it with indelible ink.

Tattooing can be done any time, as soon as the wing web is large enough to take the marking. The most satisfactory time, however, is when the birds are moved onto the range at from eight to twelve weeks of age. Chickens or other poultry partly grown, have very few feathers under the web of the wing, the web of the wing is thin, so that the markings come out plainly. Also the work can be done more rapidly.

The process of tattooing consists of piercing the under side of the web of the wing with the tattooing pliers. The indelible ink is then "patted" into the marks left by the needles. Rubbing the ink in may result in smearing the tattoo. Some operators

Fig. 86—An Automatic Range Waterer with Dry Well.
[Courtesy New England Poultryman.]

prefer to put some ink on the skin first, then to pierce the tattooing needles into the area covered with ink, and after making the mark to "pat" the area lightly with more ink. (It is probably preferable to use the bare finger in applying the ink, rather than to use a brush.)

Suffice to say, that unless you can read the tattoo immediately after you have completed the process, and before you release the bird, it will not be legible later.

Types of Waterers

Two types of waterers are suitable for birds on range, the movable and the stationary. If movable, the water containers should be rotated around the range with the feed hoppers. The stationary ones should stand over dry wells constructed to care for the surplus water. The dry well may be built merely by digging a sufficient-sized hole in the ground and filling it with loose rock. This may be covered by a wooden frame overlaid with inch mesh hardware cloth to assist in keeping conditions sanitary around the waterers.

The drinking vessels should have ample capacity and be so constructed as to prevent the birds from getting in them. Daily cleaning is advisable as it will increase water consumption.

If water is piped to the range, a furrow may be plowed for burying the pipes. This will help to provide cool water and prevent damage to the pipe by farm implements, and trucks which may be used to haul supplies to the range.

Hoppers

A variety of hoppers are in use for feeding pullets on range. These vary from the large stationary type, with a capacity of 200 to 300 pounds of feed, to the small movable kind with a capacity of 10 to 30 pounds.

Small hoppers are preferable because they can easily be rotated on the range. They should be moved regularly. If shifted daily they may be located a few feet from the previous site, but if the shift is made only weekly greater distance between locations is needed. By continuous moving many bare spots on the range may be avoided.

With these small-capacity hoppers requiring more frequent filling, the attendant sees the birds oftener and is able to keep a more accurate check on developments. And with only a small amount of feed in the hopper, spoilage due to rainy weather is greatly reduced and the chance of the birds developing digestive disturbances due to moldy feed is much smaller.

The hoppers should provide sufficient linear feet of feeding space and should be covered for protection from rain and designed to keep the birds out of the feed.

In some instances, poultrymen have eliminated the use of hoppers by feeding pellets scattered daily over the range.

Feed Consumption and Growth

To make good growth birds must eat a normal amount of feed. Less will mean slower growth and retarded maturity. Sweet wholesome feed must be available at all times. Keep the birds eating and you will keep them growing.

TABLE 33

NEW HAMPSHIRES. FEED CONSUMPTION, FEED COST, AND AVERAGE WEIGHTS OF MALES AND FEMALES TO 16 WEEKS*

Week	Mash			Scratch			Total Feed			Mash to Date			Scratch to Date			T. F. to Date			Cost of Feed to Date			Av. Weight Records in Lbs. by Weeks	
	Males & Females	Males Only	Females Only	Males & Females	Males Only	Females Only	Males & Females	Males Only	Females Only	Males & Females	Males Only	Females Only	Males & Females	Males Only	Females Only	Males & Females	Males Only	Females Only	Males & Females	Males Only	Females Only	Males	Females
1	10	---	---	--	--	--	10	---	---	10	---	---	0	--	--	10	---	---	$ .24[1]	$ .24[1]	$ .24	.0873	.0873
2	17	---	---	4	--	--	21	---	---	27	---	---	4	--	--	31	---	---	.73	.73	.73	.1218	.1218
3	25	---	---	12	--	--	37	---	---	52	---	---	16	--	--	68	---	---	1.56	1.56	1.56	.1962	.1962
4	31	---	---	10	--	--	41	---	---	83	---	---	26	--	--	109	---	---	2.51	2.51	2.51	.3025	.3025
5	41	---	---	11	--	--	52	---	---	124	---	---	37	--	--	161	---	---	3.72	3.72	3.72	.4350	.4350
6	61	---	---	13	--	--	74	---	---	185	---	---	50	--	--	235	---	---	5.44	5.44	5.44	.6431	.6431
7	102	101	104	21	24	19	123	125	123	287	286	289	71	74	69	358	360	358	8.31	8.36	8.36	.9075	.9075
8	104	106	102	22	23	20	126	129	122	391	392	391	93	97	89	484	489	480	10.85	11.37	11.17	1.353	1.203
9	103	110	96	30	30	32	133	143	128	494	502	487	123	127	121	617	629	608	14.32	14.61	14.12	1.750	1.510
10	116	103	129	37	40	34	153	156	163	610	605	616	160	167	154	770	772	770	17.84	17.90	17.89	1.960	1.695
11	114	121	100	42	48	36	156	169	136	724	726	716	202	215	190	926	941	906	21.41	21.76	21.02	2.390	2.050
12	148	148	148	50	41	58	198	189	206	872	874	864	252	256	248	1124	1130	1112	25.97	26.14	25.74	2.815	2.320
13	127	140	113	58	70	46	185	210	159	999	1014	977	310	326	294	1309	1340	1271	30.24	30.91	29.39	3.165	2.730
14	160	145	173	67	76	58	227	221	231	1159	1159	1150	377	402	352	1536	1561	1502	35.43	35.94	34.71	3.480	2.930
15	145	152	138	42	41	41	187	193	179	1304	1311	1288	419	443	393	1723	1754	1681	39.76	40.41	38.85	3.950	3.350
16	147	170	128	51	39	63	198	209	191	1451	1481	1416	470	482	456	1921	1963	1872	44.31	45.27	43.19	4.345	3.520

*Poultry Dept. N. H. Agr. Exp. Station.
[1]Mash @ \$2.40; Scratch @ \$2.02.

To maintain maximum feed consumption it is advisable to cater to the appetites of the birds. As weather conditions and their appetites change, make adjustments in the ration. The grain mixture is more adaptable and should be modified whenever the birds evidence a desire for an additional amount of certain grains.

As the birds approach production they may, in many cases, be consuming as much as two parts of grain by weight to one part of mash. This is excellent if they show a desire for this additional grain.

At 12 weeks of age birds consume about equal amounts of grain and mash. From then on to maturity they will eat a slightly increasing proportion of grain. The poultryman who plans to feed whole corn in the grain mixture when his birds are laying, should start replacing part of the cracked corn with whole corn while they are still on range. Difficulty may otherwise be experienced in getting laying birds to accept the whole grain readily.

The tendency on the part of many poultrymen is to cheapen feed mixtures by including ingredients of questionable value or by removing part of the more expensive elements of the mash formula. It is economy to purchase the best grades of material rather than attempt to cheapen the mash by including poor-quality feedstuffs.

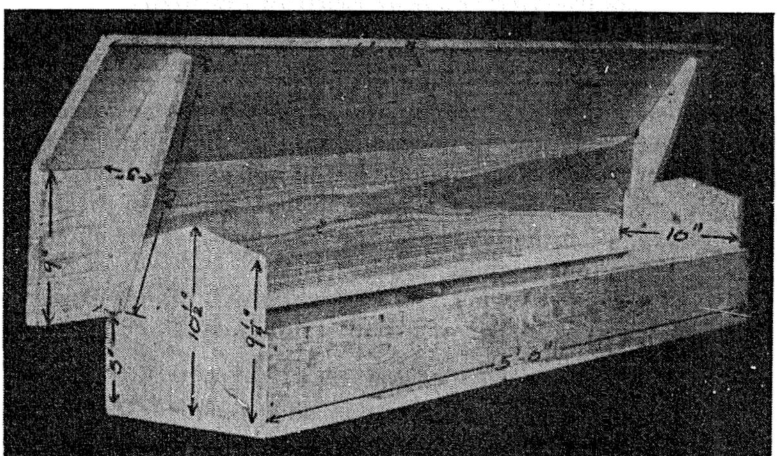

Fig. 87—Range Hopper with Hinged Top.
[Courtesy N. H. Agr. Exp. Sta.]

Shade

Probably the ideal range would be one having no natural shade or other obstruction to prevent complete tillage. Where it is possible, however, natural shade will permit the birds to get out of the sun during hot summer days. Too much shade may react as a disadvantage in view of soil contamination and lack of sufficient direct sunshine for the birds. Artificial shade may be supplied by summer shelters, sun shelters, corn, sunflowers and the like. The sun shelter may be moved to a new location periodically to avoid accumulation of contamination through droppings on one site.

Table 88—Teach Birds on Range to Use Nests.
[Courtesy New England Poultryman.]

Nesting Provisions

Birds should be trained on the range to use nests so that when they are changed to permanent laying quarters they will give little of the usual trouble about laying in the litter of the floor. Ventilate the nests on range and locate them in the shade, convenient for both the birds and the attendant. They should be similar to those in the laying house.

Nests on range prevent loss of eggs which otherwise would be deposited in stolen nests and inaccessible places.

Mites and Lice

Regular inspection of the range shelters or colony houses will help to prevent severe infestations of mites. Examine roost rails and braces. If mites are found apply treatment immediately by painting the roosts and their supports with a good mite

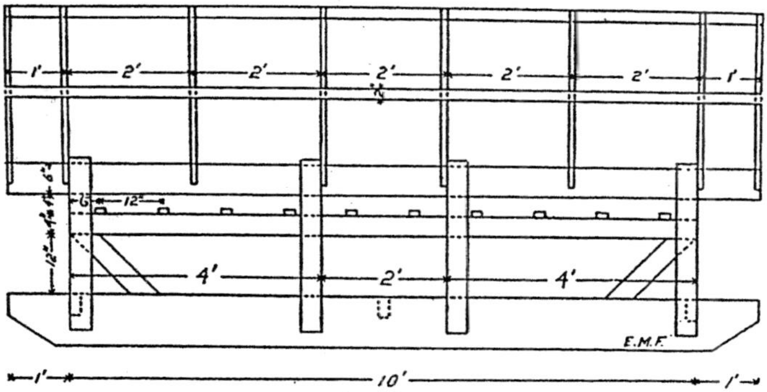

SIDE VIEW

Fig. 89—Penn State Range Shelter.

[Courtesy Prof. H. C. Knandel.]

BILL OF MATERIAL—(Penn State Shelter)

	Detailed Bill	Order Bill
Skids	2—2″ x 10″ x 12′	2—2″ x 10″ x 12′
Sills	2—2″ x 4″ x 8′ 8″	11—2″ x 4″ x 12′
Floor joists	3—2″ x 4″ x 9′ 8″	
Spacer	2—2″ x 4″ x 2′ 7″	
Spacer	2—2″ x 4″ x 1′ 4″	
Studs	8—2″ x 4″ x 2′ 10″	
Studs	4—2″ x 4″ x 4′ 6″	
Braces	4—2″ x 4″ x 3′ 3″	
Braces	4—2″ x 4″ x 2′	
Rafters	16—1″ x 3″ x 6′	8—1″ x 3″ x 12′
Rafter spacers	2—1″ x 2″ x 12′	9—1″ x 2″ x 12′
Perches	20—1″ x 2″ x 4′	
Ridge beam	1—1″ x 4″ x 12′	
Ridge cap	2—1″ x 4″ x 12′	5—1″ x 4″ x 12′
Door	2—1″ x 4″ x 4′	
Door	2—1″ x 4″ x 2′ 3″	
Door	1—1″ x 4″ x 4′ 9″	
Tie beams	2—1″ x 4″ x 2′ 2″	
Perch supports	4—1″ x 4″ x 10′	4—1″ x 4″ x 10′
Plates	2—1″ x 6″ x 12′	2—1″ x 6″ x 12′
Base boards	2—1″ x 6″ x 8′ 8″	2—1″ x 6″ x 10′
Roof	2—6′ x 12′ wall board	2—6′ x 12′ wall board

Hardware

Wire	55 ft. —1″ mesh 2′ wide
Wire	30 ft. —1″ mesh 3′ wide (16-gauge)
Nails	3 lbs.—20d
Nails	2 lbs.— 8d
Nails	2 lbs.— 8d finish
Nails	3 lbs. 1½″ galvanized roofing
Staples	2 lbs.—wire
Hinges	4 pair—3″ strap

Paint

Asphalt paint 2 gal.

exterminator. On the range, mite infestations are more common than lice, but both offer a problem.

Handling of some of the birds on range at regular intervals will serve as a check-up of lice infestation. Lice are most readily observed below the vent where moisture conditions are conducive to their growth.

When found they may be controlled by applying nicotine sulphate to the perches shortly before the birds go to roost at night. For further information see Chapter XIX.

Transferring Stock to Laying Quarters

Previous to housing make all necessary preparations, such as cleaning and getting the pens ready for occupancy. It will be well to bring the pullets in from range at night when they may be placed on the roosts.

When they are transferred during the day, it is difficult in many cases to get them to use the roosts. Sometimes they have to be placed on the perches repeatedly.

The time of housing will depend on the condition of the range, the climate and the maturity of the birds. If the range is poor, or the season late and climatic conditions unfavorable, it will undoubtedly pay to house the birds in regular quarters earlier, as they will develop better and come into profitable production sooner. At the very latest they should be housed before or shortly after they come into production. Not much is gained by permitting them to lay on the range for a prolonged period.

In housing, birds undesirable because they show evidence of disease or weakness should be eliminated. The pullets should be segregated into different pens according to sexual maturity. This will aid in intelligently feeding and managing them.

Birds toe-punched as to hatch can be checked as to maturity. This is a way of identifying for elimination the very slow maturing individuals.

The transition from range conditions to restricted room in laying quarters is so abrupt that special attention given the birds the first few days after housing will assist in avoiding many of the troubles and pernicious vices that are likely to develop.

Frequent trips through the pens will show how they are responding to their new environment and give an opportunity to check trouble at its inception. Attend the birds cautiously and quietly to avoid exciting and frightening them and they will respond more rapidly.

In recent years the use of a built up litter in laying pens has become common practice. A few inches of litter is used to cover the floor when the pullets are first housed. As this is broken up additional layers of fresh litter are added so that by the time really cold weather arrives a deep, dry, highly absorbent litter covers the floor. Ordinarily this will not need to be removed until the following spring or summer and may not be replaced until the house is being prepared for the next crop of pullets. Hydrated lime added to the litter will help to maintain it a dry condition. Generally three fourths of a pound of lime per bird capacity is sufficient. However, it should not be added all at once but added gradually as the litter is being built up.

Fig. 90—Range Hoppers Cleaned, Disinfected, Stacked and Ready for Use Next Year.

[Courtesy N. E. Poultryman.]

Questions on Chapter X

1. What is meant by a range rotation system?
2. What factors should be taken into consideration in locating the range?
3. Do range shelters make a more desirable method of housing stock on range than do colony houses?
4. Should the foundations of colony houses be screened to prevent birds from congregating under them?
5. Ordinarily, how early should the sexes be separated? Why?
6. When separating sexes, should one also cull? Why?
7. How does tattooing help to prevent poultry thieving? What does the process consist of?
8. Are small hoppers more desirable for range use than large ones? If so, why?
9. Is it desirable that nesting provisions be made for birds on range?
10. In housing pullets, should birds be segregated into different pens in accordance with sexual maturity?

CHAPTER XI

SELECTION

Subject Matter

- Time to Select
- Points to Consider in Selection
  - Vigor
  - Breed Type
  - Head
  - Body Conformation
  - Proportionate Measurements
  - Quality
  - Pigmentation
  - Molt
  - Temperament
  - Broodiness
  - Condition of Comb, Vent and Abdomen
  - Precocity or Early Sexual Maturity
  - Physical Maturity
  - The Use of Trapnest Records in Selection
  - Disqualifications and Standard Qualifications
- Selecting Male Birds
- Culling
- Comparison of Characteristics of Good and Poor Producer
- Glossary of Selection Terms

Selection is the choosing of individuals of desired characteristics. These characteristics may be either ornamental or productive or a combination of both. The production breeder may choose entirely on a production basis while the exhibition breeder may select for conformation to standard requirements. Still others attempt to combine both production and standard qualifications. The basis for selection is the tendency for individuals to beget like individuals. This tendency is of great value to the skillful breeder.

Careful selection of the birds to be housed is valuable in the following ways:

1. Number of necessary removals during the laying year is reduced.
2. Culls are eliminated, thus increasing the square feet of floor space per bird.
3. Average egg production is higher.
4. Profit is greater.
5. Less labor is necessary.

6. Segregation of best birds for breeding purposes is made possible.
7. Appearance of flock improved from the standardization viewpoint.
8. Late-maturing individuals can be excluded.
9. Production of subsequent generations of higher-quality individuals is aided.
10. The spread of disease is retarded through the elimination of diseased individuals.

Time to Select

Selection is a continuous process, starting with the baby chicks at hatching time and continuing during the lifetime of the individual. Consequently it may be practiced at any time. Baby chicks showing any abnormality or deformity may be eliminated at time of removal from incubator. Pay attention to their growth and feathering throughout the brooding period and grade carefully at broiler age, when the undesirables may be marketed along with the surplus cockerels. Generally more can be realized from their sale at this time than at a later date.

The next systematic selection may come at housing time, or when the birds are in 10 to 20 per cent production. This is a logical period, for the late-maturing birds may be more easily detected.

Points to Consider in Selection

To select birds on the basis of one consideration without due regard for the others is unwise. Best results come from the proper balancing of all points of selection. Those to keep in mind are:

1. Vigor.
2. Breed type.
3. Head type.
4. Body conformation.
5. Proportionate measurements.
6. Quality.
7. Pigmentation.
8. Molt.
9. Temperament.
10. Broodiness.
11. Condition of comb, vent and abdomen.

12. Precocity or early sexual maturity.
13. Physical maturity.
14. Trapnest records if available.
15. Disqualifications and standard qualifications.

Vigor:

The first essentials to note in the practice of selection are health and vigor. Vigor is the inherent ability of a bird to withstand the ravages of disease and other abnormal conditions, and to stand up under intensive poultry methods. The beneficial effects of hybrid vigor have been clearly illustrated, denoting that the consideration of vigor is of paramount importance in a selection program.

Mismanagement may be responsible for undermining the vitality of inherently vigorous individuals. Others may be heritably weak. Neither type should pass inspection. Evidence of vigor is primarily performance, chiefly with respect to activity, growth and reproductive ability.

Birds of high constitutional vigor are active and alert, and likely to excel the phlegmatic in the production of eggs under identical conditions of management and environment. They are also likely to possess bright outstanding eyes, grow and feather rapidly and show good carriage, type and body conformation.

Breed Type:

Birds to be maintained as breeders should be representative of the breed and variety to which they belong. A breed is a domestic race possessing, and reproducing with general regularity, certain more or less conspicuous pecularities of size, shape and plumage by which individuals of the breed are readily identified. A variety is a subdivision of a breed used to distinguish from other groups of the same breed, fowls differing in color of plumage, shape of comb, and other characteristics.

The production breeder in the past has paid scant attention to breed type, disregarding it largely in his attempt to secure high egg yields. On the other hand, the exhibition breeder has been prone to lay too little emphasis on egg production. During the past few years, however, demand has been increasing for birds combining both production and reasonably good breed type. There is a real opportunity for the breeder to develop his business along these lines. Many states in cooperation with the

Federal Government are fostering breeding programs, giving due consideration to type and production qualities.

Head:

The head can be used as an indication of vigor, intelligence, physical condition, disposition and laying condition. In applying any of the various so-called systems of judging by head points, it is advisable not to study these alone but to supplement them with the other considerations. Arbitrary standards in this regard do not allow for strain differences.

A good individual should have bright, prominent, well-placed eyes. They should not be depressed nor show evidence of physical debility. The so-called "pearly" eye and the retracted pupil are commonly associated with fowl paralysis.

Fig. 91—Body Conformation and Type is an Essential Consideration.
[Courtesy N. H. Agr. Exp. Sta.]

The beak should be relatively short, strong, well-curved, and in proportion to the head of the bird. The mandibles should meet properly and not overlap each other nor be deformed in any other manner. Such deformities are likely to interfere with the food consumption of the bird and consequently with production.

The face of the individual should be clean cut and lean, yet should not present a pinched-in appearance. If, in facing the front of the bird's head, the face of the fowl is narrow and the eyes appear to be crossed, this is undesirable and is usually an indication of low vitality. The comb and wattles should be of reasonable size, well placed and fine in texture. A smooth, velvety feel indicates good texture.

Females showing the following head types are usually poorer producers: Crow head, fat or beefy head, masculine head and bullet head. The crow head is elongated and narrow in proportion to its length, and presents an appearance typical of that of the crow from which it derives its name. Lack of vigor is to be expected in such an individual.

A thick, coarse head, usually accompanied by overhanging eyebrows with sunken eyes denotes a fat or beefy type. Such a bird usually has a skin "thick to the feel," coarse texture of comb and wattles, wrinkled face and an abundance of fatty tissue. It is generally phlegmatic.

Masculinity in females and femininity in males are extremely undesirable. A bird should possess a head characteristic of its sex. Masculine heads in females are usually accompanied by coarseness of face and adjuncts and increased size of comb and wattles. Abnormalities of the sex organs are sometimes responsible for this condition.

Bullet head refers to the type notably pointed and relatively short in proportion to the size of the bird. Such birds are likely to be poor producers and should be eliminated from the breeding flock.

Birds with refined and overly refined heads are medium to good layers. Refined heads are those possessing the previously mentioned desirable characters. Overly refined heads are those with refinement evident to an excessive degree. Such a head is likely to be proportionately small for the bird's body. Very precocious individuals many times fall within this classification.

The head is an index of health and condition. Production, disease and various abnormalities may be reflected by its appearance. In disease the evidence thus presented may be noticed long before definite diagnostic symptoms are observable.

Body Conformation:

To lay a large number of eggs a bird should have a good body with the following characteristics:

1. A flat, broad back that carries well out towards the pubic bones.
2. A wide heart girth.
3. Good depth of body that increases toward the abdomen in a bird in laying condition.

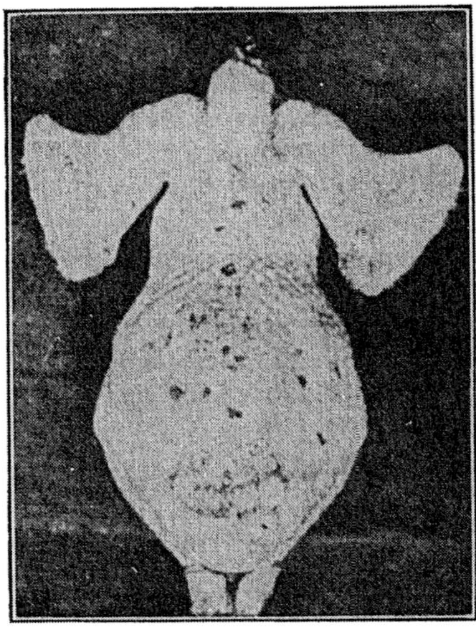

Fig. 92—Illustrates Good Width of Heart Girth.

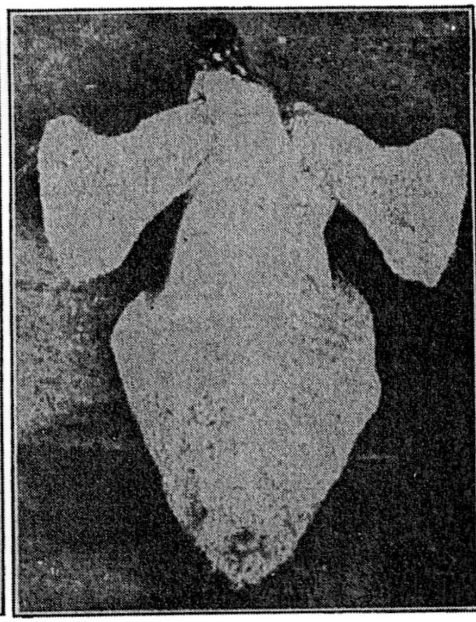

Fig. 93—Illustrates Narrow Heart Girth.

[Courtesy N. H. Agr. Exp. Sta.]

4. Good span between the pubic bones and keel bones.
5. Thin, straight pubic bones set well apart.
6. Good quality of skin.
7. Body wedge-shaped from back to keel bone.
8. Legs set well apart.
9. Good fleshing in all parts of the body.
10. Reasonably long keel bone.
11. A wide spread of lateral processes.

Proportionate Measurements:

The error is sometimes made of judging birds for productive qualities on the basis of actual measurements. Body measurements should be considered from the standpoint of their relationship to the size of the individual. A bird of the Asiatic or American class, for instance, may actually have larger body measurements than one of the Mediterranean class, but in proportion to the size of the individuals the former may have a much smaller

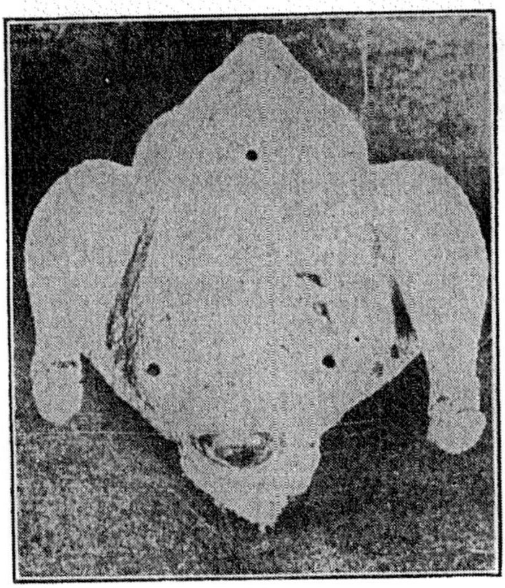

Fig. 94—Dots show position of keel and pin bones. Note distances and dilated vent of a producing bird.

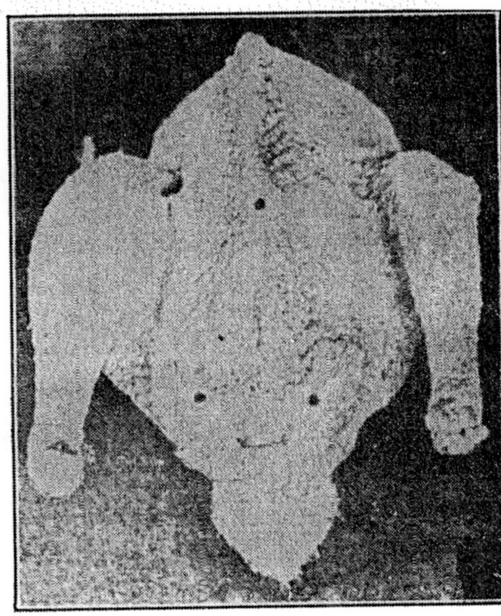

Fig. 95—Note shrunken abdomen and vent of a bird out of production.

[Courtesy N. H. Agr. Exp. Sta.]

proportionate capacity and consequently be a poorer producer. Each breed should be judged on the basis of the characteristic breed size. A bird should be well proportioned and have skeletal development consistent with its size.

It is a general practice of some poultrymen to judge the productive capacity of their birds on a basis of depth of abdomen. This is not a reliable method. As a matter of fact birds having four-fingers' depth may surpass those five fingers deep on a basis of proportionate measurements consistent with the size of the birds.

As an aid in arriving at comparative measurements for different individuals, it is necessary to handle all birds in an identical manner. If a bird is allowed to draw its legs up toward the body when being handled, the apparent width of the back across the hips may be misleading. The bird's legs should be distended for more accurate determination of the width. If one bird's legs are distended and the next one's are not, a false comparison of the two results.

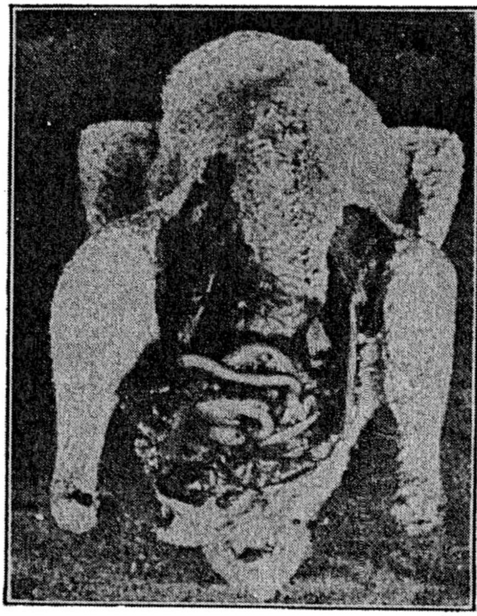

Fig. 96—Shrunken Intestinal Tract of a Non-producer.

Fig. 97—Dilated Intestinal Tract of a Producer. Tissue Cut to Show Egg in Oviduct.

[Courtesy N. H. Agr. Exp. Sta.]

Quality:

In selection, the quality factor should not be overlooked. Quality is denoted by the character of the skeleton of the bird, as evidenced by the examination of the shanks, keel and pubic bones. The bone formation should be sufficiently rugged for the size of the individual but should not tend towards coarseness. The shanks should appear wedge-shaped. Rounded shanks are indicative of coarseness and beefiness. The pubic bones should be pliable, pointed and thin when the bird is in production. The keel ought to be sufficiently long to support the abdomen and should be of refined nature.

In a good laying bird the skin is loose, pliable, and of a velvety quality. These characteristics are also shown in the flesh, face and adjuncts of the head. In a poor laying bird the skin is likely to be tightened, less pliable and coarser in texture.

The scales of the shanks are likely to be similar in quality and texture to the other adjuncts of the skin. In case of diseased condition, such as scaly-leg mite, favus and other troubles, it is difficult to determine quality in the sections affected.

Plumage is sometimes used as an index in selection. It is particularly applicable when the hen is nearing the end of her laying year. At this time the feathers of a high producer are usually brittle, rough, broken and soiled. Normally in such birds, due to the pliability of the skin, the feathering is generally loose.

Thus far the discussion has been limited to handling and quality and to the evidence used in the determination of the latter. The quality of the bird should be viewed, however, with reference to standard requirements. Particular attention should be paid to developing a flock uniform as to breed type, characteristic plumage, size, and freedom from standard disqualifications. The problem of combining production and standard characteristics is one worthy of the best efforts of our breeders.

Pigmentation:

A lack of pigmentation in yellow-skinned breeds is a guide to past and not to future production. It should not be confused with the results of parasitic infestation.

Yellow pigment is bleached from a bird during egg production in the following order: (1) Vent loses color rapidly and shows a marked paleness after two or three eggs. (2) Eye-ring bleaches almost as fast as the vent. (3) Ear-lobes, when fairly well bleached, generally indicate two or three weeks of laying. (4) Beak starts bleaching at the corner of the mouth and bleaches outward toward the tip. When well bleached it indicates six to eight weeks of fairly continuous production. Generally the lower mandible bleaches somewhat faster than the upper one. (5) Well-bleached condition of the shanks indicates fairly continuous production for four to six months although this is apparently dependent upon the intensity of laying.

It is significant that the removal or bleaching of pigment in this sequence is in the direction from the softer to the harder tissues and that replacement is made similarly. The speed of disappearance of yellow pigment may be influenced by the amount of xanthophyll contained in the ingredients of the ration.

1. *Return of pigment*—When production ceases the pigment-forming elements of the ration are diverted from yolk formation to tissue building in various parts of the body. The visible reappearance of pigment can be observed in the same order of its disappearance, as follows: (1) vent, (2) eye-ring, (3) earlobes, (4) beak, and (5) shanks. The replacement is usually more rapid than the withdrawal. In like manner, the body, face, comb and wattles are affected by pigment removal and replacement.

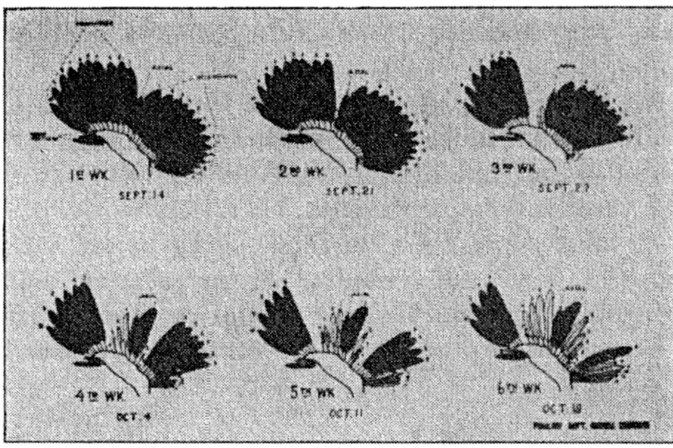

Fig. 98—Illustrates the Order of the Molting of the Primary Flight Feathers.

[Courtesy Prof. James E. Rice.]

2. *Value of Pigmentation Tests in Selection*—Because pigment fading and egg production take place simultaneously, it is obvious that pigmentation tests cannot be accurately applied to pullets coming into production. They are most applicable at the end of the laying year. In those breeds not having yellow-pigmented tissues, the other factors considered in selection must be relied upon.

3. *Effect of Rest Periods on Pigmentation*—Intermittent layers and birds taking vacation periods can be detected by the unbalancing of the pigmentation factor. For instance, a band of yellow color around the beak indicates that the bird has been in production long enough to entirely bleach the beak, has then taken sufficient vacation to give time for the reappearance of pigment in a portion of the beak, and is again back in production, removing this deposition. In instances where the tip of the beak is yellow and the rest bleached, this indicates either a rela-

tively short period of production or that the bird went through a sufficiently lengthy rest period to completely replace pigment in the mandibles and is again in production. Where both ends of the mandibles are yellow, with a bleached band in the center, the explanation is that the bird has been in production a short time, followed by a rest period.

Molt:

Molting or renewing of feathers in the domestic fowl is usually an annual occurrence. Occasionally by skillful breeding and management some birds can be carried through without a complete molt. Partial or complete molts periodically may be due to

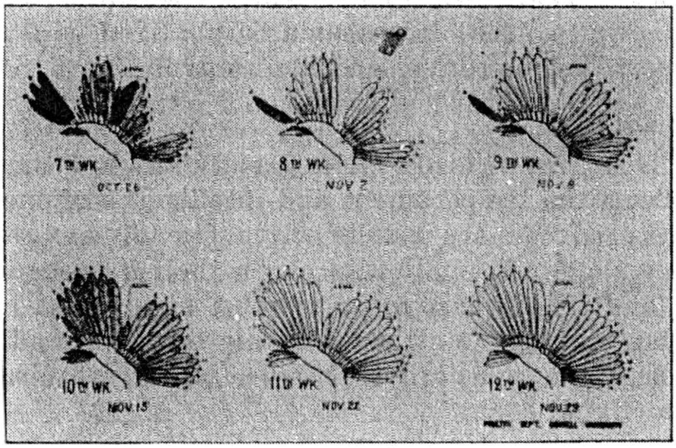

Fig. 99—Molting (Continued)
[Courtesy Prof. James E. Rice.]

management and environmental conditions. Normally the order of the molt is as follows: (1) head and neck, (2) breast, (3) body, (4) tail and secondary flight feathers, (5) primary flight feathers.

Of these the wing molt of the primary feathers is most indicative of productive ability, because of the normal sequence of replacement. The small axial feather divides the primary and secondary flight feathers. The primary one next to the axial feather is the first to be shed. In a slow molting bird approximately two weeks elapses before the next primary feather in line is dropped.

Approximately six weeks are required for a primary feather to grow to full size. Thus, allowing six weeks for the first feather and two more weeks for each additional primary, about 24 weeks would be necessary for a slow molter to replace the primaries of the wings. If two or more feathers are shed simultaneously, they should be counted as one on the time allowance. Rapid molters, in many cases, drop several feathers at once and thus greatly shorten the molting period.

If a bird loses some of her primary feathers during a vacation molt in the summer, she will start her primary wing molt in the fall where she left off.

Forcing or hastening a molt generally is not advisable. It is better to permit the hen to continue laying rather than to force the molt. Many poultrymen have a tendency to restrict birds to a grain ration during the annual molt. Protein is needed for feather renewal; so the birds should be provided with sufficient mash to meet their needs, in addition to grain.

Temperament:

The disposition of birds varies greatly and to a considerable degree reflects the temperament and intelligence of the individuals. Good producers are usually active, friendly and easily handled. They show a friendly interest in their surroundings, are energetic, and are the last to go to roost at night and the first down in the morning. Contrasted to this the poor producers are flighty, suspicious and hard to handle. They are also elusive, hard to catch and squawky.

Light and heavy birds vary considerably in their temperament. Meat classes of birds are likely to be sluggish, dull and inactive in disposition, while egg-production classes may be energetic and friendly.

Broodiness:

Broodiness in fowls is the external evidence of the maternal instinct and the inherited tendency to reproduce the species. The percentage of broodiness in a flock may be materially decreased by selection. For this purpose records should be kept of the number of broody periods a hen has had.

It is a fairly common practice to leg-band birds with a spiral band at each broody period. These indicate the number of such occurrences. Another general practice is to remove from breeding consideration hens that have been broody on two occasions. In general, breeds of the Mediterranean class, such as the Leghorn, do not exhibit this tendency to such a marked degree as many of the heavy breeds.

Broodiness during the winter period is sometimes due to mismanagement. The hen should not necessarily be penalized for it.

It is evident from experiment station reports that broodiness has little effect on winter production.

Condition of Comb, Vent and Abdomen:

Condition, in reference to these three anatomical factors, means the dilation or retraction in size with the approach or discontinuance of production. The comb of a non-producer is shriveled, dry, hard, cold and scaly. With the return of laying condition, size increases and the comb becomes soft and warm and exhibits a bright red color which tends to fade as production progresses. The change in the condition of the wattles accompanies that of the comb. The comb and wattles reflect the functional condition of the ovary.

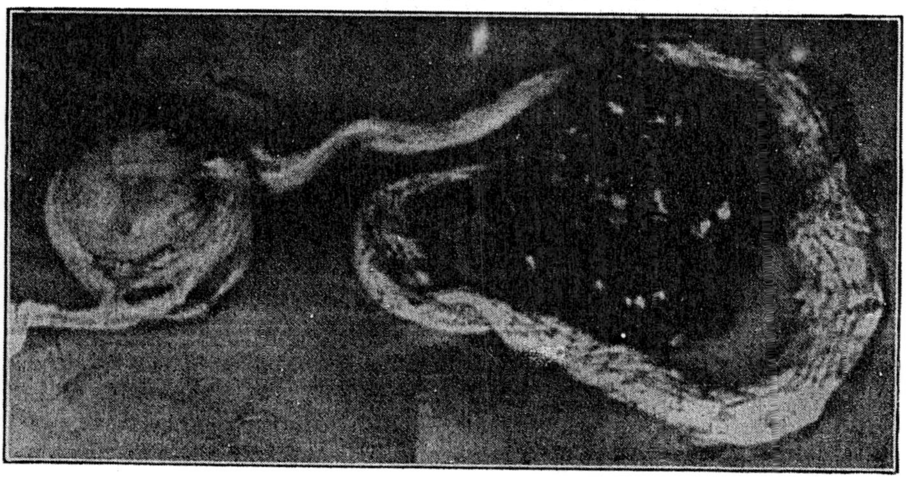

Fig. 100—A Dilated Oviduct of a Producing Bird.
[Courtesy N. H. Agr. Exp. Sta.]

In a non-producer the vent is small, dry, puckered and yellow, while in a laying bird it is enlarged, moist and faded in color. In individuals producing intensely the vent takes on a bluish appearance after a prolonged laying period.

The abdomen, the area between the pubic and the keel bones, tends to increase in size with the approach to laying condition. The greater the depth in proportion to the size of the body, the better. The expansion of the abdomen is largely due to the downward spread of the posterior end of the keel bone. This allows more room for the functioning organs of reproduction and increases the body capacity for the digestion and assimilation of large quantities of food material. The intestines, in turn, dilate to handle more food to meet the increasing requirements of the hen.

It is well to bear in mind that production in reference to poultry is actually reproduction and that a hen in laying condition is comparable to a cow in calving condition, in that all tissues are pliable and yield readily to the labors of reproduction. In view of this, birds should be carefully handled to avoid internal injury. Conveniently located nests and perches will aid materially in preventing rupture conditions.

Precocity or Early Sexual Maturity:

Precocity refers to the ability of birds to mature sexually at an early age within the bounds of the breed that they represent. Physical as well as genetical make-up determines the age of sexual maturity.

Inherent maturity may be affected by management during rearing, indicating that precocity cannot be used as an absolute index of production. Yet under normal conditions of rearing, the percentage of the flock maturing early usually represents most of the birds in the higher classification of egg producers. In this general manner precocity can be used in the selection program.

Very precocious individuals mature sexually at an abnormally early age for the breed and strain they represent and should be discriminated against. These birds are likely to show intensity of production but generally will not stand the strain of continued laying and are liable to be producers of small eggs of low marketable value. Naturally these birds represent only a small percentage of the pullet flock.

Late-maturing individuals within a strain are usually equally undesirable because of the lower annual egg production and the cost of maturing the bird over this longer period of time. Due to the length of the maturing period such birds are not in production during the favorable egg-marketing period of late summer and early fall.

Physical Maturity:

By environmental and management conditions birds may be forced into early production before they are physically mature. If they are to cover a prolonged laying period satisfactorily the pullets must carry sufficient body flesh. If either sexual or physical maturity is out of proportion unsatisfactory results will be forthcoming. Both sexual and physical maturity must be balanced for maximum efficiency.

Too often maturity is spoken of as the time the pullet commences egg production, and has little or no reference to her physical condition. In many cases the occurrence of unseasonable molts is traceable to neglect of this detail. If a bird is physically matured at an early age it is not detrimental to permit laying.

Attempts have been made to establish favorable maturity dates for the different breeds; but, unfortunately, strains within a breed differ widely in this regard, emphasizing the inability to establish arbitrary limits.

Physical maturity, as here referred to, means a suitable size and weight for the bird when laying commences. It is to be understood that birds do not attain their maximum weight until a later date. Maintenance of body weight is one of the most important factors in egg production.

Trapnest Records in Selection:

From all considerations trapnesting is the only means of obtaining a definite check on the productivity of an individual. It also provides the opportunity to study egg size, color, texture and shape, and production intensity and persistency. Naturally it gives a material index for use in the selection of birds. Many poultrymen have a tendency to pay considerable attention to "trapnesting records" and not enough to physical make-up. A wise procedure is to first select the birds according to handling qualities and then resort to the trapnest for conformation of the selection. Neither method of selection is as efficient as the combination of both.

Disqualifications and Standard Qualifications:

Certain defects considered serious enough to effectively mar the appearance of the bird, or those of heritable nature, are sufficient to disqualify a bird in competition. In addition, more minor defects, while they need not debar a bird from competition, should be considered in the selection, exhibition or judging of the individuals.

Since it is desirable to produce birds that approach the standard of the breed to which they belong, it is wise to remove defective individuals from breeding consideration so as to minimize defection in the flock and in subsequent generations. It is as important to observe external defects in production-bred fowl as it is to observe deficiencies in productive ability in birds of otherwise standard quality. Some of the major, general disqualifications for breeds and varieties according to the American Standard of Perfection are as follows:*

"Specimens unworthy of a score or lacking in breed characteristics.

Faking in any manner.

Crooked or otherwise deformed backs.

A split wing.

A slipped wing.

Wry tails.

A squirrel tail, in any breed except Japanese Bantams.

In Leghorns and Anconas, males falling more than one and one-half pounds, and females more than one pound, under standard weight.

Combs foreign to the breed.

Side sprigs or sprigs on all single-comb varieties.

Absence of spike in all rose-comb varieties.

Positive enamel white in the face of Mediterranean cockerels and pullets, except White-Faced Black Spanish.

Positive enamel white in the ear-lobes of males or females of Chantecler, Dorkings and Redcaps."

"In all breeds required to have unfeathered shanks, any feather or feathers, stub or stubs, or feather-like growth on shanks, feet or toes; or unmistakable indications of feather, feathers, stub, stubs, or down having been plucked from same.

Shank, shanks, foot, feet, or toes of color foreign to the breed."

In addition to the general disqualifications, others are specific for certain breeds. The following table indicates these for

*The above is reproduced by permission from the American Standard of Perfection (according to which all standard-bred poultry is judged) published and copyrighted by the American Poultry Association, Fort Wayne, Indiana.

New Hampshires, S.W.C. Leghorns, Barred Plymouth Rocks, Rhode Island Reds, White Wyandottes, and White Orpingtons:

"New Hampshires—One or more entirely white feathers showing in outer plumage."

"Single Comb White Leghorns—Shanks other than yellow.

Barred Plymouth Rocks—Red or yellow in any part of plumage. Two or more solid black primaries, or two or more solid secondaries, or two or more black main-tail feathers. Shanks other than yellow.

Rhode Island Reds—One or more entirely white feathers showing in outer plumage. Shanks and feet other than reddish.

White Wyandottes—Red, buff, or positive black in any part of plumage. Shanks other than yellow.

White Orpingtons—Shanks other than white or pinkish white."

In addition to the above list of serious defects which are sufficient to disqualify or disbar a bird from competition in a showroom, certain minor defects should not be overlooked. Certain numerical values are allocated to different sections of the bird and listed under shape and color. When judging by score card, defects should be cut on a comparative basis, considering the deviation from the normal. In practical competition, birds are universally judged by the comparison method. It is necessary, therefore, for the judge to be intimately acquainted with the standard requirements and be able to judiciously compare values in all sections of the specimens.

Selecting Male Birds

In choosing male birds watch particularly for vigor, breed type, pedigree, body conformation, quality, and freedom from standard disqualifications. Above all, the male should be vigorous, active and rugged, and should possess distinctive masculinity. The so-called "weak sister" should be eliminated at an early age and never find his way to the breeding pens. Docility in males is undesirable.

Fewer factors of body conformation are affected, of course, by sexual activity in males and consequently we base judgment on a fewer number of considerations than is the case with females. The male should possess a wide heart girth, broad back well carried, a deep body and a long keel. The bird should be set well on its legs, and be neither bow-legged nor knock-kneed.

A wise policy is to make a rigid selection of cockerels for breeding purposes, beginning at broiler age, and to follow this with continued culling of the poorer individuals during the growing season. Selection at this early date gives added information for basing the choice on rapidity of growth and feathering.

A mistake often made in the marketing of broilers is the disposal of all males as they develop. The result is that the

Fig. 101—Pigmentation Removal is Effected by Production.
[Courtesy N. E. Poultryman.]

owner ends with only the late-maturing and inferior individuals remaining for breeding purposes. A poultryman should advisedly maintain more males during the rearing season than he will actually need so that he can be certain of enough desirable individuals for breeding purposes.

Culling

Culling refers to the removal of undesirable birds from the flock. Birds may be rated as undesirable on the basis of egg production, plumage and other external characteristics, and disease or disease epidemics. Many times in view of limited housing facilities, it becomes necessary to decrease the size of the flock to be maintained.

There are several types of culls from a production viewpoint. From a market standpoint, however, a cull is a bird of inferior quality, particularly one that is thin, emaciated, or diseased. Fowls like this should never be marketed but should be disposed of on the farm. Only those in good condition physically and well fleshed should be sold as meat.

TABLE 34

A COMPARISON OF CHARACTERISTICS OF GOOD AND POOR PRODUCERS

	Good Producer	Cull
Vigor	Strong	Weak, unthrifty
Breed type	Wedge-shape body	Shallow and tendency toward rocker keel
Head	Neat, refined	Beefy, crow head
Body Conformation:		
1. Back	Broad, carried well out	Narrow, tapering
2. Breast	Broad, full	Narrow, shallow
3. Abdomen	Deep, soft	Short, coarse
4. Pelvic arch	Wide, thin, straight	Narrow, thick, curved
5. Vent	Large, moist	Smaller, dry, puckered
6. Legs	Wide apart, well set	Close together, knock-kneed
Quality		
1. Skin	Fine, velvety	Coarse
Pigmentation:		
1. Vent	Bleached, white to pink	Yellow
2. Eye-ring	" " " "	"
3. Ear lobe	" " " "	"
4. Beak	" " " "	"
5. Shank	" " " "	"
6. Comb and wattles	Fading to waxy red	Coarse, dull red
Molt	Late and rapid	Early and slow
Temperament	Alert, active, friendly	Elusive, squawky

Glossary of Selection Terms

Adjuncts—The fleshy growths on the head.

Axial feather—The short feather in the middle of the wing, dividing primaries and secondaries.

Beak—The horny, projecting mouth parts.

Beefy head—Denotes a condition of coarseness, usually accompanied by a ring of fat around the eye.

Bullet head—A small, short, pointed head.

Comb—The red, fleshy growth on top of the fowl's head.

Crow head—A long, narrow head presenting in appearance a downward curve of the beak typical of the head of a crow.

Dilation—The stretching of tissues to increased size.

Disqualification—A deformity or defect of major importance.

Ear lobe—The fleshy, unfeathered formation of skin below the ear.

Eye ring—The edge of skin surrounding the eye ball.

Epidermis—The outer layer of the skin.

Heart girth—The spread of ribs in the region of the heart, measured by placing the fingers of the hand on each side of the body directly back of the point of attachment of the wings.

Keel bone—The breast bone of the bird extending along the under side of the body from the region of the crop to the abdomen.

Mandible—Either division of the beak.

Masculine head—The head of a female having comb and wattles approaching in coarseness the size of the male-type head.

Pubic or pelvic bones—The two bones found on each side of the vent.

Posterior—Rear.

Rocker keel—A keel bone having a decided upward curve toward the rear.

Shanks—The part of the fowl's leg between the hock and the foot.

Sternal processes—Two bones connected onto and extending outward from the keel bone along the under side of the abdomen.

Vent—The posterior opening of the alimentary tract.

Wattles—The red fleshy growths extending from the base of tl beak.

Xanthophyll—A yellow pigment.

Questions on Chapter XI

1. In what ways is the selection of birds to be housed valuable?
2. List the points to consider in selection of birds?
3. Describe the head of a good producer.
4. Describe the head of a poor producer.
5. Why are proportionate measurements more important than actual measurements?
6. What characteristics does a good body conformation consist of?
7. What is the order of pigmentation bleaching? What is the order of its replacement?
8. What is the condition of the comb, vent and abdomen in a bird that is in laying condition?

9. Should small, very precocious and also very late maturing pullets be eliminated particularly from breeding consideration?
10. Why is it desirable that in the selection of birds for breeding purposes, all birds showing standard disqualifications be eliminated?
11. If trapnest records are available, should selection on physical conformation also be practiced?
12. How would you select male birds for breeding purposes on the basis of physical conformation?

CHAPTER XII

BREEDING AND SEX DISTINCTION

Subject Matter

Poultrymen doing breeding work are primarily concerned with inherent characteristics of poultry, realizing at the same time that environmental conditions play an important role. Unless satisfactory environmental conditions prevail—i.e., balanced diets, satisfactory quarters, good management—maximum results from intelligent breeding work may not be displayed. On the other hand, birds of poor breeding background will not give satisfactory performance regardless of the environmental advantages that are supplied.

The theory that "like begets like" is not proven by breeding practice. "Like" only tends to produce "like." This has definitely been established in connection with high egg production. Not all or possibly any of the daughters of a high-producing dam may equal or excel their mother's production. It is evident that the breeder, to make the most progress, must understand and appreciate the principles of genetics and some of the limitations of breeding.

Principles of Inheritance

To get a picture of the principles of genetics, let us examine some of the basic facts involved in inheritance. All cells of an animal are made up of a portion called a nucleus, which when properly stained will show thread-like material called chromosomes. These chromosomes may be thought of as pieces of a pearl necklace, the pearls representing the genes or hereditary factors. The chromosomes contained in body cells always occur in pairs and the number is constant for the type of animal involved, although varying according to sex. The male chicken has two chromosomes, called sex chromosomes, while the female chicken has only one such chromosome. These are so named because whether an offspring develops into a male or female depends on whether it receives one or two of these particular chromosomes.

During cell division prior to the formation of the sperm and ova, a separation takes place whereby only one member of each pair of chromosomes is incorporated into a given sperm or ova. Therefore, the sex cells or reproductive cells will have only one-half the chromosome number of the body cells. Each sperm of the chicken will have a given number of autosomes, a term used for all chromosomes other than the sex chromosome, and one sex chromosome. One half of the ova or female reproductive cell of the chicken will have a given number of autosomes and one sex chromosome, while the other half of the ova will have the given number of autosomes but no sex chromosomes. The former ova upon uniting with a sperm will develop into the males and the latter ova will develop into the females. Inherited factors or genes born on these sex chromosomes are said to be sex-linked.

As a result of mating, the sperm and ova unite. This union restores the original chromosome number of the body cells and results in an offspring with the same chromosome number as the father if the offspring is a male, and as the mother if the offspring is a female.

Certain factors in poultry are termed dominant because they have the power of suppressing the expression of the alternate or recessive character. This is perhaps best illustrated in the following cross of black and white bantams:

Let A = black   a = recessive white
(AA) black x (aa) white
First generation or F_1 (Aa) all black
Mating of F_1 males and females (Aa) black x (Aa) black
Second generation or F_2 1 (AA) black pure breeding for black)
2 (Aa) black (not pure breeding for black or white)
1 (aa) white (pure breeding for white)

The result indicates a simple three-to-one Mendelian ratio in the second generation. The individual (AA) contains two dominant characteristics or genes for black and is thus homozygous for black. A similar condition is evident in (aa) for the recessive factors or genes, and is therefore homozygous for white. On the other hand, the two individuals of the (Aa) nature contain two dissimilar genes, one for black and one for white, and are therefore termed heterozygous.

We refer to the appearance of an individual as the phenotype, which in this case is black and white. The genetic make-up or in this case (AA), (Aa), or (aa) is termed the genotype.

From this brief discussion it can readily be seen that as the number of factors increase, the number of possible combinations will also increase and consequently the more complicated the breeding problem will be. For the purpose of clarity, an attempt will be made wherever possible in this text to utilize crosses involving but one factor at a time. For the more complicated combinations of characters, it is advisable to consult a standard book on genetics or poultry breeding, as the aim here is to briefly outline some of the principles of breeding.

Aim in Breeding

The aim in poultry breeding is to combine as many desirable characteristics as possible, and eliminate the detrimental and undesirable factors. All factors that we work with are present in the parents, no new ones being created, but we do attempt to recombine the factors in the most desirable manner. We are interested in combining factors for high egg production, large egg size, high hatchability, good fleshing, a rapid rate of growth, good livability, and other desirable broiler and egg characteristics. If we select one of these problems and work on it, we can make considerable progress on this one character, but if we try and select for all of them simultaneously, then we can make some progress in all characters selected for, but less for any given one. Which method a breeder uses depends largely upon his own needs.

Types of Matings

Single-male matings:

Individual matings or single male matings are frequently used by breeders who want to identify the sire of off-spring, or who want to test the fertility, hatchability, etc., of a certain male. One male generally is mated to 15 to 18 females.

Multiple-male matings:

This is the most common type of mating. In such matings, two or more males are mated with a group of females at the rate of 7 males per 100 females. Many breeders prefer to house 9 or 10 males per 100 in order to allow for culling or mortality and still permit good fertility without adding extra males.

Artificial insemination:

It is especially useful in cases where hens are kept in batteries or when old proven sires are used. One male can fertilize more females when artificial insemination is practiced, although considerable work is involved. A male produces about 1 cc. of semen per day, and 0.1 cc. twice a week has been found to give good fertility.

Collecting Eggs for Hatching

Actually a fertile egg should be produced on the second day after mating occurred, but due to preferential mating and the inability for a male to mate with all females the first day, it is advisable to start saving eggs about one week to ten days after the birds are mated. After a mating is broken up, it is still possible to save the eggs for hatching for one week.

Methods of Breeding

Physical selection:

This consists of the mating of birds solely on a physical basis. No records are kept, but birds are thoroughly selected for egg production, meat qualities, etc., by handling each bird individually.

Pedigree breeding:

By this term, we refer to the mating of birds the offspring of which are identified by wing banding or toe punching at hatching time. In addition to a knowledge of the ancestry of a bird, records of growth, hatchability, egg production, etc., are usually noted. This is an extremely expensive and time-consuming task. It should be noted that in physical selection, we consider the individual bird only; while in this method of breeding we know the ancestry. We can thus base our selection on the performance of the individual as well as the performance of the ancestors.

Progeny test:

Real progress in poultry breeding can be made when this method is employed. We base our opinion of the breeding value of a bird on the performance of the offspring; or, if this is not possible, upon the performance of a whole family of close relatives, If the offspring or the sisters or the aunts of a given bird are consistently good producers, it may be assumed that this family including the given bird carry the desirable genetic factors. This method also involves the identifying of chicks and observing the performance of the offspring or families rather than the performance of the individual parents.

This is the basis of real improvement, for breeding from families of chickens that are consistently good will provide us with a better chance for good results than where a given bird is one good bird in a variable family.

Family A		Family B	
A1	200 eggs	B1	70 eggs
A2	195 eggs	B2	280 eggs
A3	240 eggs	B3	112 eggs
A4	260 eggs	B4	130 eggs

For example, bird A2 of Family A will probably be a better breeder than B2, even though the former laid 85 less eggs. Males that have sisters that consistently perform well also are more desirable as breeders than those that have poor sisters.

Systems of Breeding

Crossbreeding:

The mating of individuals of different breeds or varieties is practiced, usually with the hope of increasing the vigor of the stock. Such vigor is termed hybrid vigor, but is sustained only for the first generation. Numerous experiments have shown advantages in rate of growth, feathering, livability, hatchability, etc. in crossbreds over the same strains of birds used as purebreds. For broiler production, crosses are more common than purebreds; but the parents used in a cross are of the greatest importance. Certain crosses are used extensively for sex differentiation at hatching time.

Fig. 102—First-Generation Hybrids from the Cross of Jersey Black Giant Male by White Leghorn Female.
[From Kansas Agr. Exp. Sta. Bulletin 252, by special permission D. C. Warren.]

Inbreeding:

This is the mating together of closely related individuals. The mating of brother and sister, or of parent and offspring is considered close inbreeding, while the mating of cousins would be considered as mild inbreeding. An inbred line refers to a group of birds with three generations of brother-sister mating or its equivalent. It is used to increase the homozygousity of an individual by making more and more factors or genes homozygous every generation and thereby increasing the uniformity. The undesirable as well as the desirable characters are purified however, so that great care must be taken to select for and maintain high hatchability as well as other economic qualities. Where practiced for a few generations, the stamina and vigor of the stock may be undermined and fertility, hatchability, livability and egg production lowered. When two highly inbred lines are crossed, the results may be very good provided proper selection has been made. The extreme uniformity resulting from crosses of inbred lines is one of the great advantages.

Linebreeding:

This is a form of inbreeding in that one individual regarded as unusually desirable is related to a good many of the individuals in the flock. Many of the breeders used will be related to the "idol" but will be related to each other as little as possible. Its purpose is to perpetuate and intensify in the offspring the good qualities of individuals, and avoid the difficulties often encountered from close inbreeding.

Outcrossing:

The mating of individuals of a different strain of the same variety is termed outcrossing in breeding practice. It is done with the intention of increasing desirable characteristics, such as vigor, fertility, hatchability, size of bird, etc. Often the results are discouraging and difficult to forecast, as the blood lines of different strains do not always combine to the best advantage. It is best to try but a small unit first and observe the results. Then if warranted, the entire flock may be outcrossed.

Inheritance of Egg Production

Egg production is a complicated phenomenon influenced by environment and inheritance. There are at least five inherited characteristics which contribute towards the total annual egg production of a bird, and each one is governed by one or more pairs of genes.

Fig. 103—First-Generation Hybrids from the Cross of White Leghorn Male by Jersey Black Giant Female.
[From Kansas Agr. Exp. Sta. Bulletin 252, by special permission D. C. Warren.]

Rate of sexual maturity:

This refers to the age in days at which a bird lays her first egg. This is influenced by both autosomal and sex-linked factors. For most heavy breeds, it is desirable to have the birds mature between 160 and 200 days. Birds maturing earlier are apt to be small and lay small eggs. Leghorns and other light breeds mature somewhat earlier.

Persistency:

This is influenced by at least one factor and refers to the length of laying to the first annual molt. It has been shown that early-maturing birds usually have a higher persistency than late-maturing birds. of course, those birds that lay for a long period prior to their first annual molt are expected to lay more eggs in their pullet year than those laying for a short period.

Intensity:

Rate or intensity of production is a measure of the number of eggs laid in a certain given time. A bird laying clutches of four or five eggs with a one day skip between clutches would be laying at a good rate, while a bird laying clutches of one or two eggs would be of low intensity. The intensity with which a bird is laying is inherited and affected by several genetic factors.

Broodiness:

Broodiness is inherited, but is rather difficult to eliminate, especially when breeds, varieties, or strains are crossed. A genetically broody bird will carry two factors for broodiness, but neither one of the factors themselves would produce broodiness. Consequently, it is possible to mate two non-broody birds, and produce broody offspring. Removing broody birds from the breeding flock, and not using breeders who have previously produced broody offspring, or who had close relatives that were broody is the most effective way to eliminate broodiness.

Winter pause:

Winter pause may be defined as a rest period between winter and spring production periods. This pause may be due to either inheritance or environmental conditions, or a combination of both.

TABLE 35

PRODUCTION INCREASE IN CONNECTICUT FARM FLOCKS*

Twenty-two Year Contest Summary

Contest Year	Total No. of Farms	Total No. of Birds	Av. Size of Flocks	Av. No. of Eggsper Bird
1919-1920	166	26,361	158	131.04
1920-1921	194	29,036	149	147.67
1921-1922	282	55,131	192	144.82
1922-1923	252	64,053	254	143.11
1923-1924	350	88,297	252	142.58
1924-1925	273	72,673	266	143.24
1925-1926	236	56,433	239	147.93
1926-1927	223	56,514	253	153.88
1927-1928	179	52,168	291	161.45
1928-1929	153	46,162	301	168.22
1929-1930	231	103,715	449	161.96
1930-1931	200	84,694	423	163.20
1931-1932	187	71,707	383	165.79
1932-1933	208	74,680	359	161.69
1933-1934	114	46,701	410	162.08
1934-1935	77	31,004	403	169.62
1935-1936	100	52,772	528	170.56
1936-1937	106	64,273	606	175.77
1937-1938	102	78,489	628	180.00
1938-1939	111	79,582	717	184.77
1939-1940	93	83,820	901	186.43
1940-1941	76	63,046	830	193.01
1941-1942	58	58,295	1,005	196.23
1942-1943	59	64,447	1,092	194.78
1943-1944	34	48,927	1,233	197.29

Inheritance of Egg Characteristics

Egg Size:

The egg size is often related to the body size of a bird, the smaller birds laying smaller eggs. Research workers have also found that frequently those birds that lay at a high rate produce smaller eggs than those that lay at a low intensity. To increase the egg size of a flock, the females should be carefully selected for good egg size, and the males should be selected on the basis of the egg size of the female ancestors and the egg size of the sisters or daughters.

Egg shell color:

Shell color is inherited and is probably due to several genetic factors. When dark brown shelled birds are mated to white shelled birds, the egg shell color of the offspring is intermediate, being neither white nor deep brown. As for all charcaters that

*Conn. Home Egg Laying Contest, Sept., 1944.

are inherited and due to several gene pairs, selection on a family basis is most fruitful.

Inheritance of Rate of Growth

This characteristic is inherited and apparently to a very high degree. Rapid growth rate is one of the major objectives of broiler producers, and consequently is receiving a lot of attention at present. By selecting and identifying those birds that are the heaviest at five weeks and again at 8 to 12 weeks, rapid progress can be made at improving the growth rate. Males grow faster than females, but should be given special attention since any one male is the parent of more offspring than any one female.

Inheritance of Rate of Feathering

The dressing plant operator especially is interested in a well-feathered bird at the time of dressing. Sex-linked as well as autosomal genes influence the rate of feathering in chickens. To select the most rapid feathering birds from a flock several methods are employed.

1. The number of secondaries at hatching time may be counted, and those chicks with eight or more secondaries at hatching time are considered as fast feathering. The reader should be cautioned that all chicks should be of the same age and newly hatched. Many poultrymen are using this method, although it is a somewhat tedious task.

TABLE 36

PULLET VERSUS HEN PRODUCTION

Connecticut Home Egg-Laying Contest—1944[1]

Months	Pullets	Hens
	Eggs per Bird	Eggs per Bird
1934-1935 Summary	180.65	131.39
1935-1936 Summary	181.47	134.86
1936-1937 Summary	188.91	140.25
1937-1938 Summary	193.61	144.98
1938-1939 Summary	194.22	143.09
1939-1940 Summary	197.02	141.35
1940-1941 Summary	200.09	143.62
1941-1942 Summary	201.21	157.38
1942-1943 Summary	197.76	155.60
1943-1944 Summary	210.34	164.47

[1]Courtesy, R. E. Jones.

2. Taking observations on the length of tail at ten days of age is a simple and effective method for improving feathering. Those chicks with a tail of one-half inch or more at 10 days are rapid feathering, those with no evidence of a tail would be slow feathering.

3. At 8 to 12 weeks, or slightly prior to or about the age at which the birds are normally sold as broilers is a good time to make a final check on the feathering of the birds. Those birds that are well feathered at that time, and also had a ½ inch tail at 10 days and 8 or more secondaries at hatching should make good breeders for rapid feathering.

Inheritance of Livability

Livability is one of the main objectives of a poultryman, whether he is producing meat or eggs. The possibility of breeding for resistance or susceptibility to a specific disease has been demonstrated. However, to breed for resistance to all diseases would be extremely difficult, if not impossible. The elimination from the breeding flock of all males and females from families with high mortality is certainly desirable.

Inheritance of Other Characteristics

Recently it has been shown that the efficiency of feed utilization is inherited. It has also been demonstrated that males used their feed more efficiently than females, and that the rapid growing birds used feed more efficiently than slow growing birds.

Many poultrymen know that crowding chicks, providing roosts at an early age, or a deficient diet may cause crooked keels in chickens, but many do not realize that inheritance also contributes towards this undesirable characteristic. At Kansas, Dr. Warren has been able to produce strains of chickens by selection that when raised under identical conditions 88 per cent of the males and 82 per cent of the females of the crooked keel strain developed this condition, against 3 per cent of the males and 4 per cent of the females in the straight keel strain.

Fertility

At the present time, there is no information that fertility is inherited, but there are a number of non-inherited factors that do affect fertility. Lower fertility is likely when combs are frozen, when they are light in weight, when they are molting, or when they are infected with lice. Restricted feed consumption, too many or too few males, abnormal sperm, or no sperm production may be other causes for low fertility.

Hatchability

Hatchability refers to the number of chicks obtained from a given number of fertile eggs. Inherited factors affecting egg quality probably also affect hatchability. A few of these would include the size, shape, shell texture, and the thickness of the albumen. Approximately twenty genetic conditions which cause death of the embryos and called "lethals" have been reported.

Since the hen manufactures the egg which will develop into the chick, it can readily be understood why the female seems to affect hatchability more than the male. Undoubtedly a large number of genes determine the hatchability of the hen's egg and

Fig. 104—A Black Female from the Cross of Rhode Island Red Male by Barred Rock Female.

[From Kansas Agr. Exp. Sta. Bulletin 252, by special permission of D. C. Warren.]

consequently the application of the progeny test seems to be the most effective means for improving hatchability.

Sex Distinction

The distinguishing of sex at hatching time has been growing in importance during the past few years. There are advantages in separating the sexes at hatching time both to the egg producer and the meat producer. Sexing makes it possible for the buyer to secure males or females only, depending upon whether he is producing meat or eggs. Broiler producers secure more uniformity, more rapid growth, and a greater efficiency of feed utilization by getting males. On the other hand, the egg producer can reduce his requirements for space and equipment materially, and reduce his labor and overhead costs considerably by rearing pullets only. It is also possible to raise better pullets when cockerel interference is eliminated.

Sexlinkage:

Those genetic factors born on the sex chromosome are used in certain crosses to differentiate sex at hatching time. Whenever the dominant characteristic is introduced through the female in such a sex-linked cross, the males will resemble the mother, since she passes her sex chromosome to her sons, but not to her daughters.

Plumage and down color:

This is the most common characteristic of the sex-linked group. The barring factor found in Barred Plymouth Rocks may be used. A cross of a New Hampshire or Rhode Island Red male on a Barred Plymouth Rock female will result in progeny the male chicks of which are black with a head spot and yellow shanks, in contrast to the female chicks which are black without a head spot and with black shanks and beak. As adults, the males will be barred and the females will be black with varying amounts of red in the hackles.

To illustrate:

Let B = barring; b = non barring; and — indicate the absence of a sex chromosome

New Hampshire male bb (non barred) X Barred Plymouth Rock female B— (barred)

First generation offspring Bb = males and all will have a head spot, as chicks and will be barred as adults
b— = females and all will be black as chicks and will be non-barred and black as adults

Any gold male mated to Barred Plymouth Rock females will give a similar result, while a gold male mated to Columbian or

Fig. 105—Slow Feathering Chick Four Weeks of Age.
(Courtesy New Hampshire Agr. Exp. Sta.)

Silver females will result in chicks the males of which are white and the females red. As adults, the males from the latter cross are of the Columbian pattern or at least exhibiting considerable white, while the females are fundamentally red.

To illustrate:

Let S = silver; s = gold; and — indicate the absence of a sex chromosome

Rhode Island Red male ss X Columbian Plymouth Rock female S—

First generation offspring Ss = males, all white down with black primaries and Columbian pattern as adults
s— = females, all red down and red adult plumage

Suggested Crosses

Golds	Silvers
Rhode Island Red male	Columbian Rock female
New Hampshire male	Columbian Wyandotte female
Dark Cornish male	Delawares
Buff Plymouth Rock male	Light Sussex female
Red Sussex male	Light Brahma female
Buff Orpington male	Silver-Laced Wyandotte female
Buff Wyandotte male	Dark Brahma female
Brown Leghorn male	White Wyandottes (in most cases)
Partridge Plymouth Rock	

Fig. 106—Rapid Feathering Chick Four Weeks of Age.
[Courtesy New Hampshire Agr. Exp. Sta.]

Rate of feathering:

Certain breeds and strains vary materially in the rapidity of feathering. For example, Leghorns are usually rapid feathering while Jersey Giants, Brahmas, etc., are often slow feathering. Slow feathering is dominant to rapid feathering. Consequently, by crossing a fast feathering male with a slow feathering female, the male progeny will be slow feathering and the female progeny fast feathering. The difference between the sexes is noted by the number and length of primaries and secondaries at hatching time.

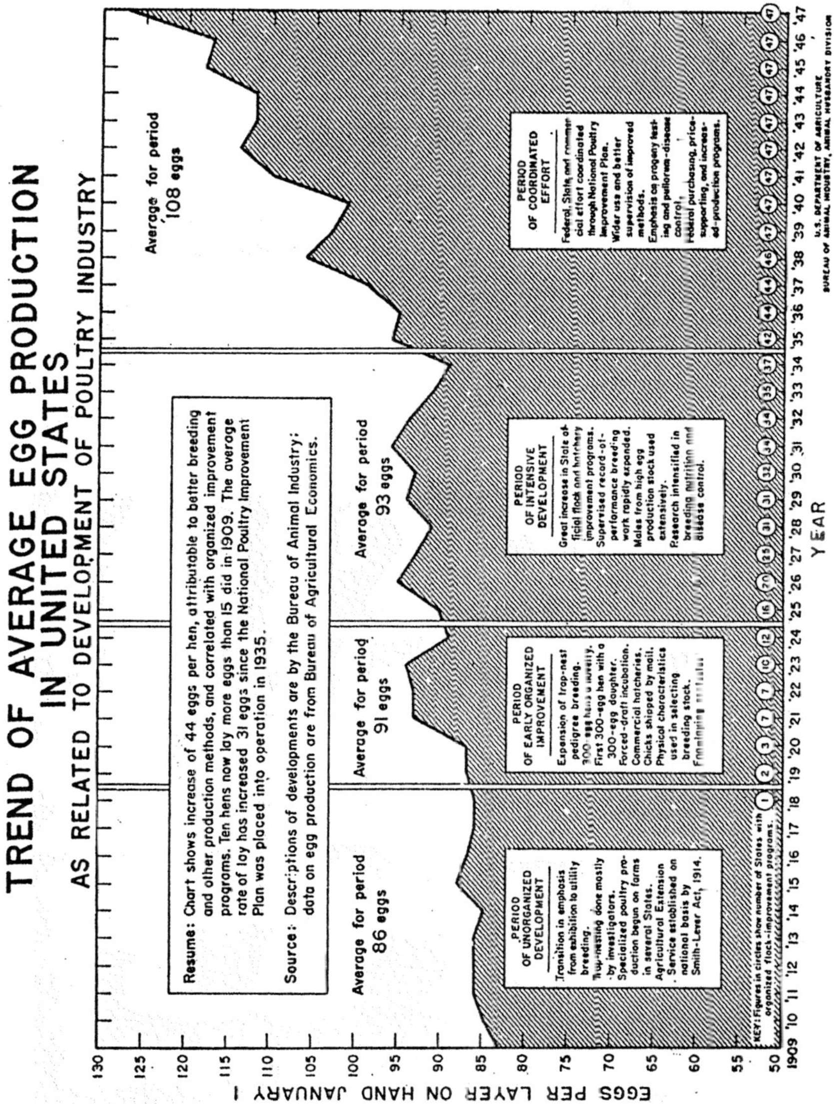
TREND OF AVERAGE EGG PRODUCTION
IN UNITED STATES
AS RELATED TO DEVELOPMENT OF POULTRY INDUSTRY
Resume: Chart shows increase of 44 eggs per hen, attributable to better breeding and other production methods, and correlated with organized improvement programs. Ten hens now lay more eggs than 15 did in 1909. The average rate of lay has increased 31 eggs since the National Poultry Improvement Plan was placed into operation in 1935.
Source: Descriptions of developments are by the Bureau of Animal Industry; data on egg production are from Bureau of Agricultural Economics.
Average for period 86 eggs
Average for period 91 eggs
Average for period 93 eggs
Average for period 108 eggs
PERIOD OF UNORGANIZED DEVELOPMENT
Transition in emphasis from exhibition to utility breeding.
Trap-nesting done mostly by investigators.
Specialized poultry production begun on farms in several States.
Agricultural Extension Service established on national basis by Smith-Lever Act, 1914.
PERIOD OF EARLY ORGANIZED IMPROVEMENT
Expansion of trap-nest pedigree breeding.
First 300-egg hen with a 300-egg daughter.
Forced-draft incubation.
Commercial hatcheries.
Chicks shipped by mail.
Physical characteristics used in selecting breeding stock.
PERIOD OF INTENSIVE DEVELOPMENT
Great increase in State official flock and hatchery improvement programs.
Supervised record-of-performance breeding work rapidly expanded.
Males from high egg production stock used extensively.
Research intensified in breeding, nutrition and disease control.
PERIOD OF COORDINATED EFFORT
Federal, State, and commercial effort coordinated through National Poultry Improvement Plan.
Wider use and better supervision of improved methods.
Emphasis on progeny testing and pullorum-disease control.
Federal purchasing, price-supporting, and increased-production programs.
KEY: Figures in circles show number of States with organized flock-improvement programs.
EGGS PER LAYER ON HAND JANUARY 1
YEAR
U.S. DEPARTMENT OF AGRICULTURE
BUREAU OF ANIMAL INDUSTRY, ANIMAL HUSBANDRY DIVISION

To illustrate:

Let Sl = slow feathering; sl = rapid feathering; and — indicate the absence of a sex chromosome

White Leghorn male (slsl) X Jersey Black Giant female Sl—

First generation offspring Slsl = males and all are slow feathering

sl— = females and all are rapid feathering

The degree of accuracy will depend upon the purity of the two breeds for rapid and slow feathering.

Color sexing:

Color sexing within a pure breed has been practiced with fair accuracy in recent years. This method can be used by breeders who have small orders for sexed chicks, since definite males and females can be segregated with a high degree of ac-

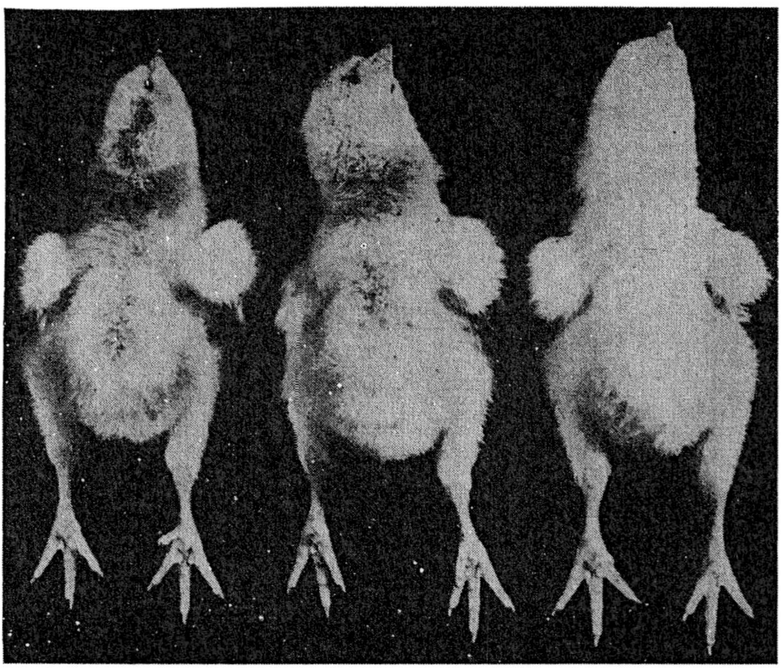

Fig. 107—A series of wings showing the range of flight feather development in day-old male and female chicks from the cross of White Leghorn male by Rhode Island females. Note the well-developed flights in the females (right) and the very short flight feathers in the males (left).

[From Kansas Agr. Exp. Sta. Bulletin 252, by special permission D. C. Warren.]

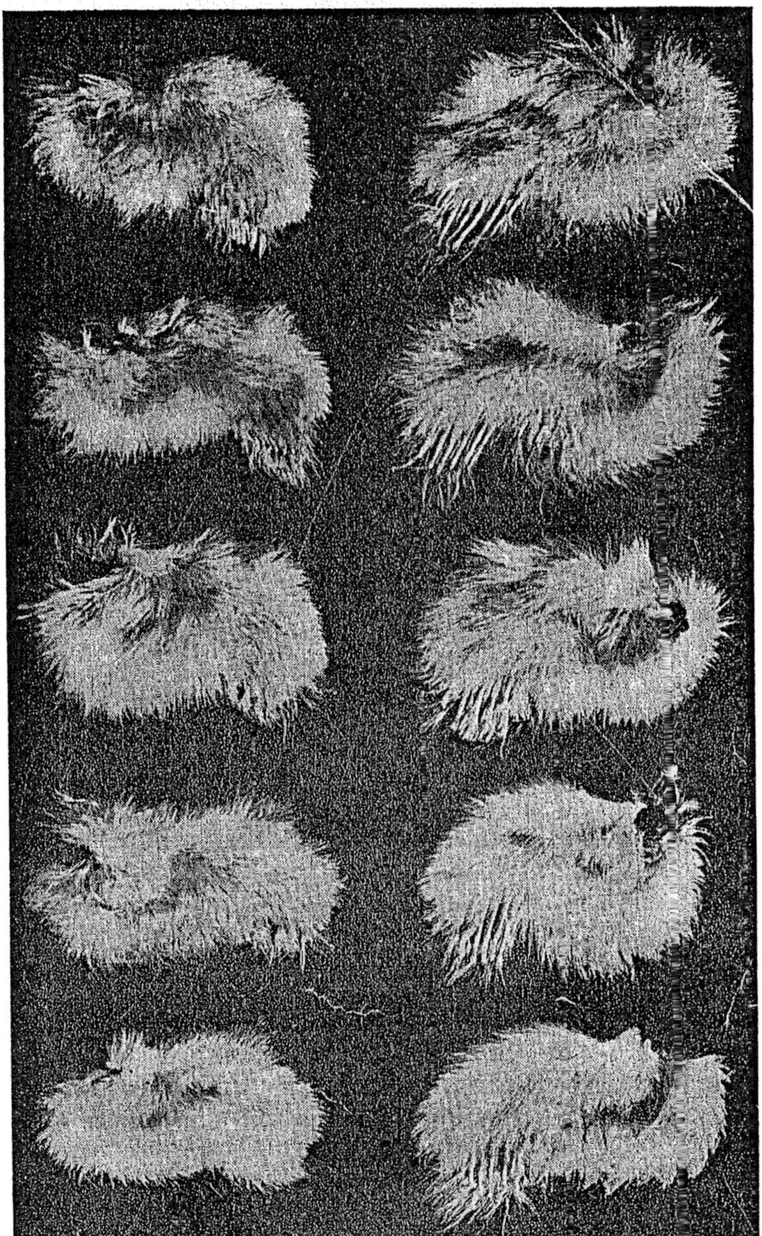

Fig. 108—Male chicks, result of cross of Rhode Island Red male by White Wyandotte female. White or yellowish in color but never showing buff or red. Some males show smokiness in the down, especially on the neck or head.

[From Kansas Agr. Exp. Sta. Bulletin 252, by special permission D. C. Warren.]

curacy. The questionable chicks can be placed with the straight run chicks.

In Rhode Island Reds and in New Hampshires the females are very uniform in color, while the males have a light area of down in the web of the wing and a light area on the throat. The

Fig. 109—Female chicks showing colors ranging from a definitely brown-striped chick to one showing only a light tint of buff. Result of cross of Rhode Island Red male by White Wyandotte female.

[From Kansas Agr. Exp. Sta. Bulletin 252, by special permission D. C. Warren.]

darker red the down color, the easier it usually is to sex by this method.

In Barred Plymouth Rocks, the males are somewhat lighter in pigmentation and they have a larger and often more blurred head spot than the females. The females are more intense in color and the head spot is smaller and more definite.

Sex Identification by Physical Examination:

This method was discovered by the Chinese and exploited by the Japanese. In the male a small white rudimentary copulatory organ is found, while in its place in females, only a small depres

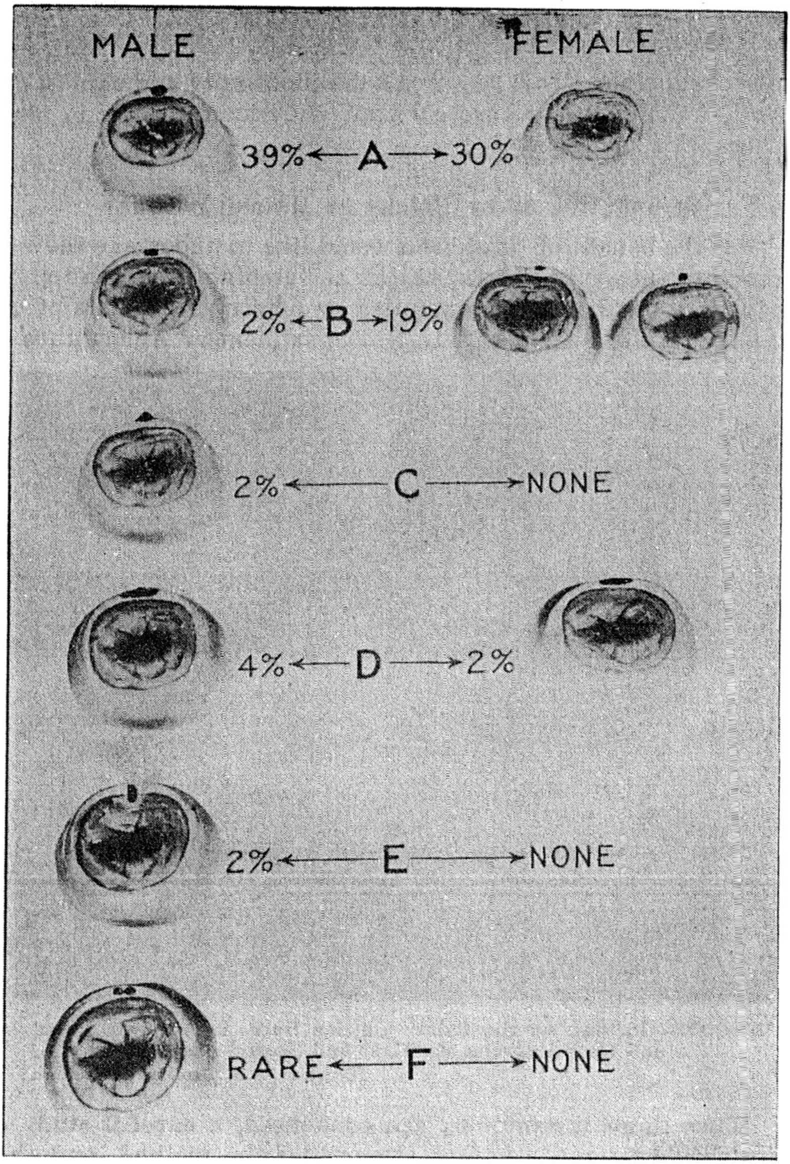

Fig. 110—Photographs of Wax Models Showing Different Types of Copulatory Organs Found in Male and Female Chicks, and Approximate Percentage of Each Type. The Different Types of Male Organs are: A, large, Cone-shaped; B, Medium, Cone-shaped; C, Tongue-shaped; D, Thick, Elongated Fold, with Raised Center; E, Elongated, Crosswise Fold; F, Divided or Forked. The Different Types of Female Organs are: A, No Organ; B, Small, Cone-shaped; D, Thick, Elongated Fold. [Courtesy U. S. Dept. of Agric.]

sion is observable. This requires a detailed study and skill in observation and can be accurately applied by many people in identifying males and females.

Segregating Baby Chicks at Hatching Time*

For the benefit of those who would like to undertake the segregation of the sexes of baby chicks at hatching time, the following instructions have been compiled by poultry specialists of the Bureau of Animal Industry, U. S. Department of Agriculture.

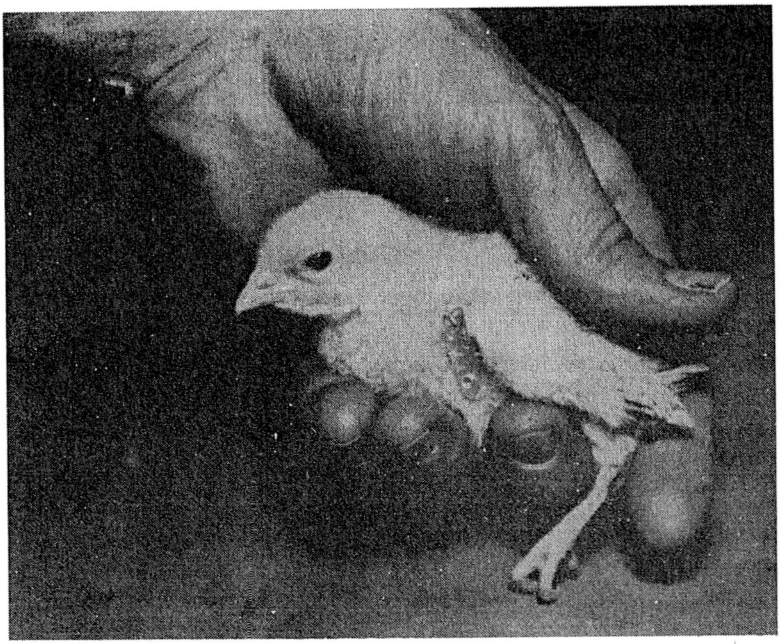

Fig. 111—Picking up the Chick; Chick's Back Against the Palm, and Legs Between the First and Second Fingers.
[Courtesy U. S. Dept. of Agriculture.]

Since these instructions are condensed, a careful study of them should be made before attempting the method, and they should be available for reference during the first few trials. Much delay in progress may be avoided by doing this.

The chick is held in the palm of the left hand with its legs between the first two fingers and its head hanging down. If the

*United States Department of Agriculture Bureau of Animal Industry, Animal Husbandry Division. (Mimeograph No. 2.)

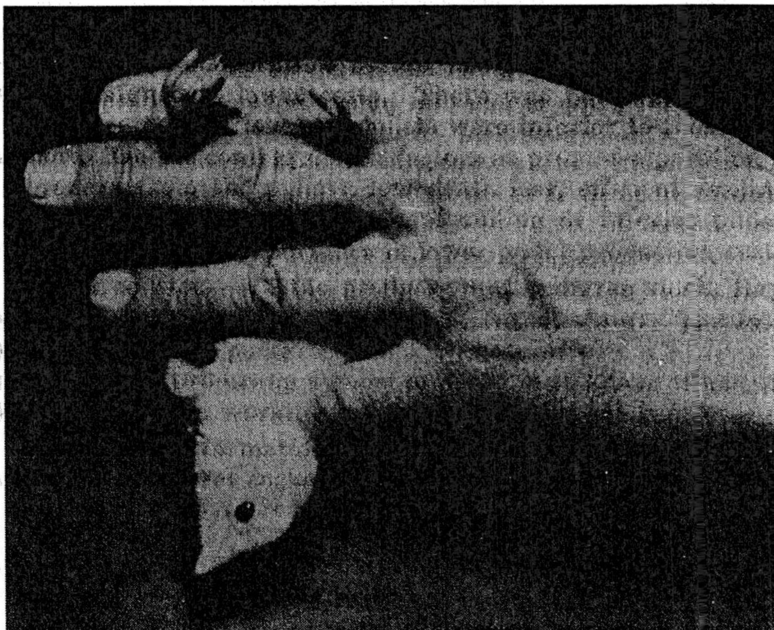

Fig. 112—Holding the Chick with the Little Finger Behind the Neck for Support.
[Courtesy U. S. Dept. of Agriculture.]

little finger is now brought forward, allowing the head to drop between the little finger and the one next to it, the operator will find that he has his two middle fingers over the chick's breast, his forefinger above its legs, his little finger against its back and his thumb will be free.

The free thumb of the left hand is now placed near the lower (ventral) side of the anus (vent) to assist the thumb and forefinger of the right hand, which are placed on each side of the anus a little toward the back. With the two thumbs and one finger, spaced about equally around, the eversion of the anus is easy. However, experience has shown that in some cases the fear of injuring the chicks leads to a tendency to evert the anus only partly. It is essential that the anus be well everted and with a little practice this can be accomplished without injury to the chick.

The genital organ is found near the left thumb on the ventral fold of the anus near the rim. It appears as a tiny tit or pimple which is not easily seen at first. However, with a little practice the eye soon learns to detect it quickly. The size and shape of the organ vary considerably. Most of the males have a relatively prominent organ, while in a majority of the females it is completely absent. Chicks with such characteristics are easily identified, even by a novice.

A small percentage of both males and females have relatively small organs. It is the class of chicks with these small organs that present the real problem in making rapid progress in learning to sex chicks. However, there are differences in the appearance of these small organs, and with persistent practice the op-

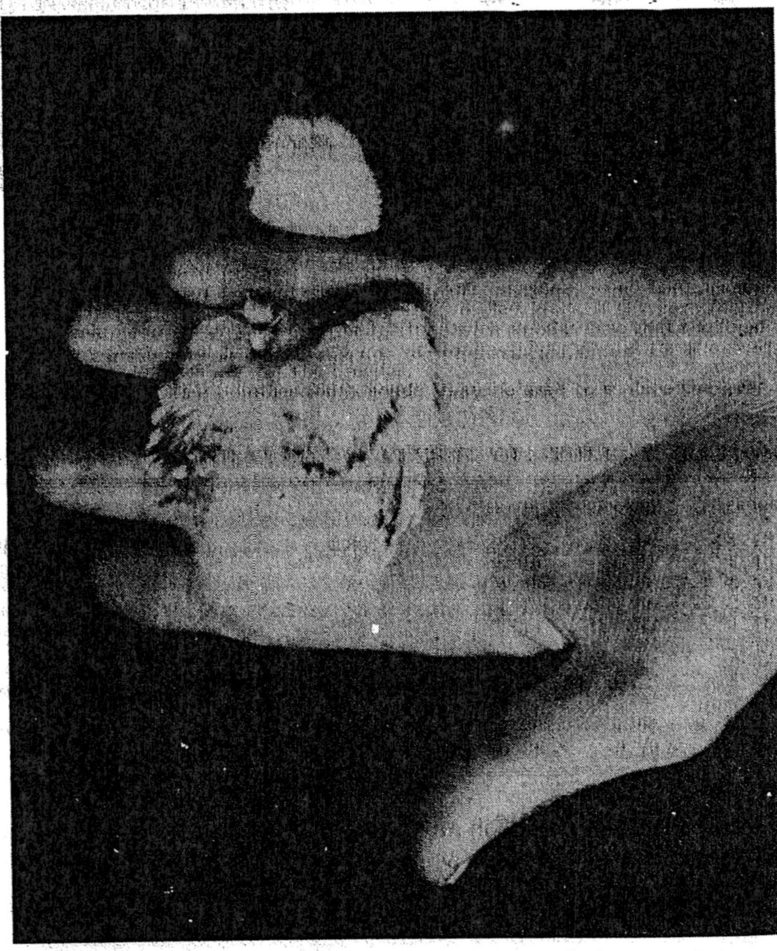

Fig. 113—Rear View of Hand Holding the Chick, Showing Legs and Head in Position Before Cupping the Hand.

[Courtesy U. S. Dept. of Agriculture.]

erator will eventually learn to recognize them for what they are. Until such time, it is safer to assume that the very small organ means that the chick is a female, since the majority of them are.

An accuracy of 70 to 80 per cent should be attained soon after the observer is able to recognize the genital organ. The greatest difficulty comes between 80 and 90 per cent. This is where one learns to differentiate between the males and females of the small-organ class.

In the regular practice of sexing it is not unusual to find that a small run of the first chicks picked up are females, and this is one of the problems that presents itself to the beginner because there is often nothing to see in the female chick. Since he cannot know exactly what he is looking for, there is always the question as to whether he is looking at a male and failing to see what is there or whether the chick is a female. Where "sex-linked" chicks are available, known males can be used until the organ has been identified. A popular sex-lined cross is a Rhode Island Red male mated to Barred Plymouth Rock females, the male chicks produced by this cross being black with a white spot on the back of the head, and the beak and shanks being yellow, whereas the female chicks are solid black in down color and the beak and shanks are also black. Where this is not possible the operator should handle a large enough number of chicks to insure the handling of some males. After the genital organ has been surely recognized, better progress should result from working with chicks the sex of which is unknown to the operator.

It is necessary for the beginner—in the absence of an experienced instructor—to dissect the chicks after segregation in order to check his observations and correct erroneous decisions. After this the number of chicks that will need to be dissected will be considerably smaller, consisting only of those in which the operator doubts his accuracy of segregation. Cull chicks and even those that need to be helped out of the shell can be used for practice.

The chicks to be dissected should first be killed and then cut open along the sides of the abdomen from the anus toward the wing joints. The intestines are then gently removed from the abdominal cavity. In the male chick the testes appear as a pair of small white or yellowish, oblong bodies, regular in shape and lying along the backbone at the level of the lower ribs. In the female chick the ovary is a small yellowish, flat, irregular body, in the same position as the testes, but on the left side only. After a few chicks have been dissected, it should be very easy to distinguish between testes and ovary.

The above procedure should be followed without regard to speed until the operator has reached a degree of accuracy suitable to his needs. For commercial purposes this must be well over 90 per cent. Some increase in speed will occur naturally during the period of learning, but real speed must be the result of consistent practice with many chicks.

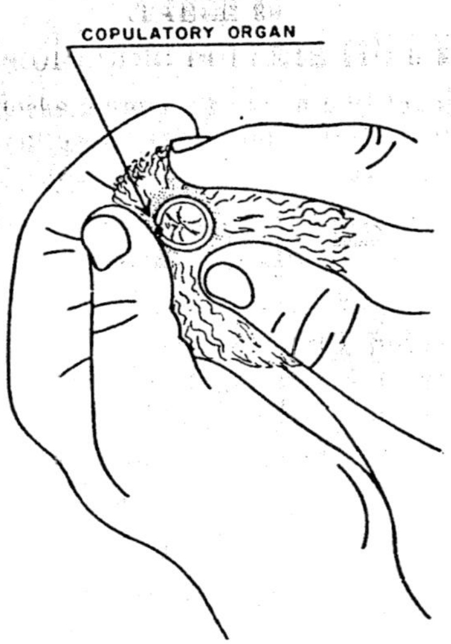

Fig. 114—Drawing Showing the Position of the Fingers in Everting the Anus, and the Location of the Copulatory Organ.
[Courtesy U. S. Dept. of Agriculture.]

A good light is essential. Bright daylight is best, but a 200-watt electric daylight bulb may be used. The operator will find that fewer mistakes will be made if the chick is rotated several ways so that observations can be made from several angles. It is especially true that in some chicks where the genital organ is pale in color, it is difficult to see when looking directly down upon it, but is easily seen as a silhouette against the light.

Last but not least is the fact that newly hatched chicks are much the easiest to sex. The beginner is urged, for the sake of his own progress, to use only chicks that are newly hatched and dried off, about half a day old. After chicks are one day old they become progressively more difficult to sex on each succeeding day. For easier identification, magnifying glasses may be used.

TABLE 87

DOMINANT AND RECESSIVE CHARACTERISTICS IN CHICKENS*

Characteristic	Dominant or Recessive	Autosomal or Sex-linked
White plumage	In White Leghorns, almost completely dominant to color.	Autosomal
" "	In White Dorkings, Langshans, Minorcas, Plymouth Rocks, Wyandottes, recessive to color.	"
" "	In some strains of Rose Comb White Bantams and White Silkies, recessive to color.	"
Black plumage	Dominant to recessive white.	"
Blue plumage	Due to heterozygous condition of color genes.	"
Buff plumage	A multiple gene characteristic dominant to recessive white and recessive to black, and recessive to silver.	Autosomal and sex-linked
Silver plumage	"Silver" in Columbian, silver-laced, and silver-penciled varieties dominant to "gold" in red, buff, golden-laced and golden-penciled varieties.	Sex-linked
Barred plumage	In Plymouth Rocks dominant to non-barring.	"
" "	In Campines, dominant to non-barring.	Autosomal
Red-splashed white plumage	Recessive to solid color and to color pattern.	"
Albino	Recessive to normal.	"
White skin and shank color	Dominant to yellow skin and shank color.	Sex-linked
Dark skin and shank color	Recessive to non-dark skin and shank color.	Autosomal
Rose Comb	Dominant to Single comb.	"
Pea Comb	" " " "	"
Walnut Comb	Dominant to rose, pea, and single comb.	"
Side sprigs	Dominant to normal comb	"
Cerebral Hernia	Recessive to normal cranium.	"
Crest	Dominant to absence of crest.	"
Muff and Beard	Dominant to absence of muff and beard	"
Naked neck	Dominant to normal neck feathering.	"
Feathered shanks	Dominant to non-feathered shanks.	"
Vulture hock	Recessive to normal hock.	"
Rumplessness	Dominant to normal condition.	"
Long tail	Incompletely dominant to normal tail.	"
Hen feathering	Dominant to normal feathering.	"
Frizzle plumage	Incompletely dominant to normal feathering.	"
Silkie plumage	Recessive to normal plumage.	"
"Frayed" feathers	Recessive to normal feather development.	"
Close feathering	Dominant to loose feathering.	"
Slow feathering	Dominant to rapid feathering.	Sex-linked
Flightless	Dominant to normal wing.	Autosomal
Crooked neck	Recessive to normal neck.	"
Creeper condition	Dominant to normal condition; lethal when homozygous.	"
An embryo lethal	Recessive to normal embryo development; lethal when homozygous.	"
"Sticky" embryos	Recessive t normal; lethal when homozygous.	"
Parietal baldness	Dominant to normal.	"
Congenital blindness	" " "	"
Congenital loco	" " "	"
Congenital palsy	" " "	"
Polydactyl	Dominant to four-toed condition.	"
Early sexual maturity	Dominant to late sexual maturity	Autosomal and sex linked
Broodiness	Dominant to non-broodiness.	"

*U.S.D.A. Yearbook, 1936.

Fig. 115—Position of Holding Chicks as Advocated by the Japanese.
[Courtesy New England Poultryman.]

Questions on Chapter XII

1. Does "like beget like"?
2. Distinguish between in-breeding, line-breeding, cross-breeding, and out-crossing.
3. What is meant by the term pedigree breeding?
4. Upon what five heritable factors does annual egg production depend?
5. How could one systematically proceed to improve egg size in subsequent generations?
6. Sex distinction offers some advantages to the market egg and broiler producer. What are these advantages?
7. How can you distinguish sex in day old chicks from matings of Rhode Island Red males with Barred Plymouth Rock females?
8. Is it possible to identify sex in day old chicks on the basis of physical examination?
9. Are fertility and hatchability inherited?
10. What factors are valuable in color sexing?
11. What observations will assist in selection breeders for improvement in rate of feathering?
12. How can you distinguish sex in day old chicks from matings of Rhode Island Red males with Columbian Plymouth Rock females?
13. How can rate of feathering be used for sex distinction?

CHAPTER XIII

TRAPNESTING AND PEDIGREEING

Subject Matter

Advantages of Trapnesting and Pedigreeing
Disadvantages
Who Should Trapnest and Pedigree
Pedigree Breeding Practice
  Selection
  Trapnesting
  Pedigree Hatching
  Progeny Testing
Records
  Aids in Record Keeping
Simplified Record Keeping

Trapnesting consists of securing production records on individual birds and using the information in all the phases of the breeding program. Pedigreeing is the securing and applying of information dealing with individuals and their ancestry and progeny, and includes records on production, hatchability, livability, and many other factors.

Trapnesting and pedigreeing are followed in order to make the most rapid progress in the breeding operations and for the sake of a systematic breeding plan.

Advantages of Trapnesting and Pedigreeing

Their advantages are that they provide definite information:

1. As to production qualities, such as color, shape, texture, and number of eggs.
2. As to hatching qualities.
3. As to livability of chicks produced.
4. As to prepotency (or the progeny test).

Disadvantages

The disadvantages consist of:

1. Labor and expense of keeping detailed records.
2. Too much emphasis on records; not sufficient on the physical make-up of the bird and its conformation to production standard requirements.

3. Inability of many people to analyze and utilize records that are obtained.
4. The need for many small breeding units to secure accurate information as to parentage.
5. The length of time necessary to develop a reputation as a pedigree breeder and to secure monetary rewards.
6. The failure of many buyers to appreciate the expense involved in producing quality pedigreed stock.
7. Lack of official recognition of private records. This affords the unscrupulous producer an opportunity to falsify them.
8. Apparent simplicity, misleading the average beginner, who soon finds pedigreeing a complex problem and his labors in vain.

Who Should Trapnest and Pedigree

Pedigree breeding is particularly adapted to the specialized poultry breeder who is equipped to carry on this type of work and has the knowledge and inclination. It is thus specifically limited to a small group who may serve as sources of foundation and "seed" stock for the commercial poultryman. To succeed with it, the individual must have enthusiasm, "stick-to-itiveness", and the ability to observe. He must not be easily discouraged, must be systematic, and must be ruthless in discarding birds which fail to measure up to a high standard, regardless of what some of their qualifications may be.

The adaptability of this type of breeding for the commercial poultryman lies in the selection of a small percentage of birds representing only the very best individuals contained in the flock. Obviously, if the entire flock were to be trapped, labor and expense would multiply in compiling many mediocre records. Operations must be restricted to a relatively small group of the poultryman's best birds. These will serve as a specialized unit for line breeding to improve production of subsequent groups and of the flock. In addition, they will provide for high quality flock matings. This breeding program clearly has no place on the commercial poultry farm where the flock is renewed annually, either through the purchase of hatching eggs or baby chicks. Progress is very slow or negligible if trapping is restricted to too small a number of birds.

Pedigree Breeding Practice

Selection:

The principles of selection discussed in Chapter XI should be used as the basis for choosing individuals to be used in pedigree breeding. This applies to both the beginner and the experienced breeder. It is best to select first on the basis of constitutional vigor and physical characteristics and secondly on pedigree, or the individual record, if available. This will tend to prevent placing too much emphasis on the record or pedigree.

Trapnesting:

Trapnesting is the procedure of trapping birds in the nests so that each egg may be identified and the right hen credited with its production. Many types of trapnests are on the market, including trapnest fronts which can be installed on the regular nests if they are of the correct type and dimensions. In the

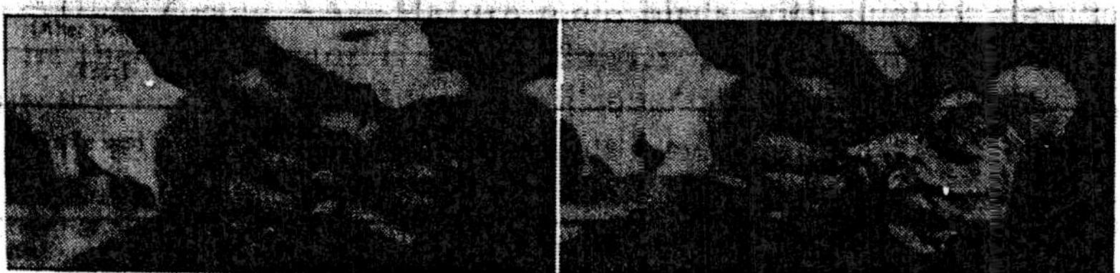

Fig. 116—Wing Banding Pedigreed Chicks at Hatching Time. [Courtesy N. H. Agr. Exp. Sta.]

purchase or construction of trapnests see that they are sufficiently ventilated so that the birds will not become overheated in them during hot weather. In estimating the number necessary, four birds to a nest may be used as a guide. It is preferable to plan for fewer birds to a nest during high peaks of production. The higher the production level of the flock, the larger the number of birds that will lay in the morning and, therefore, the greater the number of nests required.

In trapnesting pullets move them from the range to the laying house before or as soon as production begins, so that you will get information as to age at first egg, or as to sexual maturity. As they go into laying quarters the birds ought to be banded with leg bands that are easy to read, to facilitate the trapping operations. It will prove helpful to double band the birds and record both leg- and wing-band numbers so that if any bands are lost the birds can still be identified. To have a valuable bird lose its single band and thus its identity is extremely exasperating. Double and even triple banding makes this practically impossible.

In attending the traps to release the birds and record the eggs, a definite routine needs to be established. In following this, make as many trips as the level of production requires. In any case, minimum attendance should constitute releasing from traps at least hourly in the morning and at somewhat lengthier periods in the afternoon.

In recording individual egg production records of egg size, shape, color and texture should be taken at regular intervals. Recording the live weight of the birds at the beginning and end of the year, and more frequently if possible, will give an additional indication of how the birds are "standing up" under their rate of production. This double check will give one demonstration of vitality.

Marking individual eggs with the hen's number during the hatching season makes pedigree hatching possible because it offers the possibility of chick identification.

Short interval trapping is resorted to by a number of poultrymen in order to give them information of the relative production qualities of their birds. This reduces labor requirements ordinarily necessitated in a complete trapping program and at the same time furnishes some basic information as a guide to breeding operations.

Pedigree Hatching:

Pedigree hatching consists of maintaining the identity of parentage from egg to chick. In setting eggs from trapnested individuals, it is easy to segregate the eggs from each bird, as they are already marked by a pencil with the dam's number. They can be grouped in the incubator trays so that on the 18th day they may easily be placed in either pedigree baskets or bags in the hatching trays. The more quickly this transfer can be made the less the retardation of embryonic development.

It is to the advantage of the operator to secure all the information possible as he proceeds. This means recording the number of infertile eggs and dead germs at the 18th day when the eggs are candled and transferred to pedigree containers and after the chicks have hatched it is advisable to list the number, indicating the cripples and culls and those dead in the shell. With the number of eggs set from an individual bird at each incubation, in addition to the record of results, the bird's season average may be compiled as one measurement of her value.

In removing chicks from the pedigree containers they should be banded for identification, either by piercing the web of one wing and inserting the band and clinching, or by wrapping the wing band around a shank. In the latter case the band must later be removed to the wing. The identity of the chicks should be placed on the records.

Progeny Testing:

Progeny testing is the maintenance of complete records so that the true breeding value of a mating may be ascertained by the performance of its progeny. The testing of the progeny offers a means of developing favorable family lines and discarding or segregating unpromising families. The rapidity of progress in pedigree breeding will largely be determined by the se-

Fig. 117—Bird with visible wing badge simplifies identification. (Courtesy of Dryden Poultry Breeding Farm.)

lection and utilization in the breeding program of families whose records show a rounded and uniformly high performance, and the rejection of those whose records indicate inconsistent or unsatisfactory activity.

Records

Various types of records are applicable to a pedigree-breeding program. Keeping a few or a complete set will largely be dependent on the object in view. The record sheets as illustrated in these pages represent but one type and are part of a complete set. In many states sample record forms can be secured from the state college or university and adapted to the needs of the individual breeder.

The following represent the types that can be maintained:

1. Monthly egg record.
2. Individual yearly production record.
3. Individual mating record.
4. Individual hatching record.
5. Index sheet.
6. Record of progeny of sire.
7. Summary record of individual hen.

Aids in Record Keeping:

In recording trapnest production, operations will be facilitated by keeping the monthly egg record sheets in the pens to which they belong, close to the nests. This will tend to reduce the labor requirements and also the chances for error. Colored leg bands with large plain numerals will be easier to read. These should be upside down on the birds' left or right legs depending on whether the attendant is left or right handed.

To reduce unproductive record keeping, unworthy individuals should be removed from trapnest pens as soon as they have indicated their mediocrity. Little is to be gained by the continuance of their records; they only add to the burden of pedigree breeding.

In saving trapnested eggs for pedigree hatching, a storage cabinet or table where those from individual hens can be segregated and stored until needed for setting will offer considerable advantage, as the eggs can then be grouped readily in the hatching trays. Similarly many resort to a traying table—an ordinary table through which holes of sufficient size have been bored to hold eggs in place. These holes are in line and each set can be designated with the hen's number. Eggs on such a table can easily be sorted and placed in the incubator trays. The table also comes in handy when sorting out infertiles, dead germs, and chicks dead in the shell, for recording on the pedigree-hatching sheets.

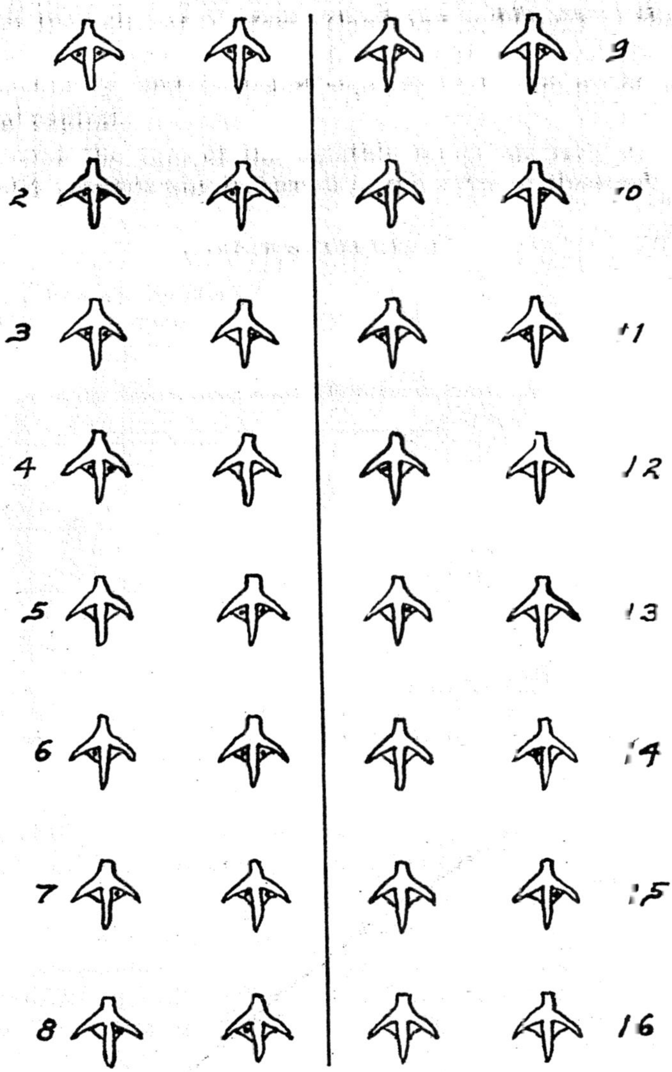

Fig. 118—Chick Toe-Punch System. (16 Possible Combinations.)

Bill of Material for Three Trap Nests Itemized

1—7/8″ x 2″ –12′	corner and partition posts
1—7/8″ x 1 1/2″–54″	top and bottom cleats
1—7/8″ x 1 3/4″–29 1/4″	entrance floor supports, front
1—7/8″ x 1 3/4″–34 1/2″	entrance floor supports, rear
1—7/8″ x 1 3/4″–27 3/4″	upper door strip
1—7/8″ x 2″ –27 3/4″	lower door strip
1—3/8″ x 1 3/4″– 7′	door slats
5—3/8″ x 2″ – 3′	top
3—1/2″ x 2″ – 3′	top
12—1/2″ x 2″ –20″	partitions and ends
3—1/2″ x 2″ – 3′	back
4—1/2″ x 3 1/2″–20″	partitions and ends
1—1/2″ x 3 1/2″– 3′	back
1—1/2″ x 2 1/4″– 3′	front bottom piece
1—1/2″ x 2 1/2″– 7′	lighting boards and cleats
1—1/2″ x 2 1/2″–54″	bottom cleats
4 boards—1/2″ x 4″–3′	bottom.
1—1/2″ x 4 1/2″ –2′	trip door
1—1/2″ x 1/2″ –2′	trip door

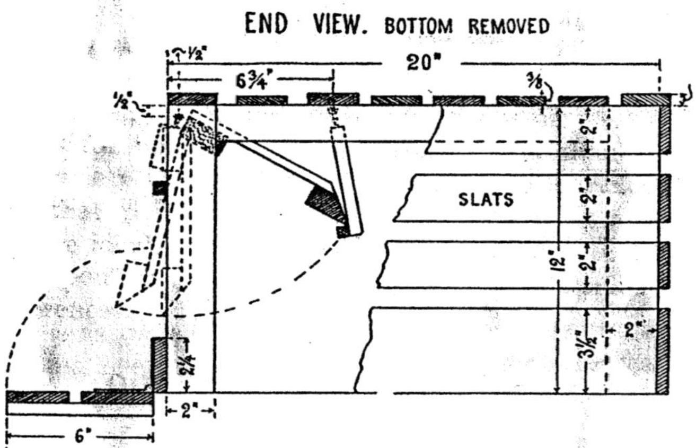

Fig. 119—End View Cornell Trapnest.

With sufficient labor at hatching time, it is comparatively simple to have one person record the results of hatching and the wing band numbers while the others remove the chicks from the pedigree containers and band them.

Simplicity of records is valuable. If they are complex, the task of keeping and using them grows more burdensome and difficult, with the result that in most cases they are neither maintained nor analyzed. Monthly record sheets ruled in weekly intervals aid in the transfer of results to permanent records similarly ruled. The size of the record sheets should suit the equipment available or the equipment the operator intends purchasing.

Loose-leaf books are satisfactory and often used for filing the record forms so that they are readily accessible. If the sheets to be used are large, a regular office letter-filing cabinet is helpful. The Kardex filing case offers many advantages for maintaining permanent record forms. Other filing sytsems may be adapted to other types of records.

N. H. R. O. P. Association and Poultry Dept., U. of N. H.

FAMILY RECORD SHEET

PRODUCTION RECORD OF PROGENY

Progeny of Mating No. ——
Dam's No. —— Sire's No.

SIRE
sire dam
sire dam sire dam
sire dam sire dam sire dam sire dam

BIRD NO.	TOTAL	AUG.	SEPT.	OCT.	NOV.	DEC.	JAN.	FEB.	MAR.	APR.	MAY	JUNE	JULY	AUG.	SEPT.	OCT.	NOV.	

FAMILY RECORD SHEET

MISCELLANEOUS DATA OF PROGENY

Mating No. ——
Dam's No. —— Sire's No. ——

DAM
sire dam
sire dam sire dam
sire dam sire dam sire dam sire dam

BIRD NO.		Date Hatched	Annual Record	Egg Size	Body Wgt.	Hatchability								

Simplified Record Keeping

A great variety of record blanks has been developed during the past years. Most of these systems, however, have been inclined to complicate the process of record keeping. Furthermore, they have made it next to impossible to utilize these records after they have been accumulated.

The blanks herewith included are the basis of R. O. P. statistics in New Hampshire and are an example of simplified records. In addition to the pen trapnest record sheet, the three prin-

Association Record of Performance New Hampshire

PEDIGREE HATCH RECORD AND CHICK INDEX

Poultry Dept. University of New Hampshire

OF PEN No.________

HEADED BY MALE ________

sire ________ dam

sire ________ dam   sire ________ dam

sire ________ dam   sire ________ dam   sire ________ dam   sire ________ dam

SUMMARY OF MATING

No. females in pen________	No. chicks culled________	% Hatch all set________
No. females removed________	No. eggs set________	% Fertility________
________	No. eggs fertile________	% Hatch fertiles________
________	No. chicks hatched________	% Chicks culled________

DATE OF HATCH	SET	FRT	HTH	CLS	CHICK NOS. INDEX	DATE OF HATCH	SET	FRT	HTH	CLS	CHIX NOS. INDEX	DATE OF HATCH	SET	FRT	HTH	CLS	CHIX NOS. INDEX
Hen No.						Hen No.						Hen No.					
Total						Total						Total					
Hen No.						Hen No.						Hen No.					
Total						Total						Total					

cipal reports needed are (1) Family record sheet, (2) Pedigree hatching record and chick index, and (3) a Pullet census.

Those poultrymen operating under the National Poultry Improvement Plan should use records forms furnished under the plan.

The **Family Record Sheet** lists all of the daughters of a single male and female. At the end of each month, the records are brought up to date for all families, so that at the end of the year, the records are complete and can be readily analyzed.

The **Pedigree Hatch Record** and **Chick Index** give a complete story of hatchability of each mating. This sheet has space for 6 females on the front and 9 females on the back (which are not shown)—a total of fifteen.

New Hampshire Record of Performance Association

PULLET CENSUS

Season ....

Poultry Department University of New Hampshire

DAM	SIRE	wg.	leg	wg.	leg	wg.	leg	wg.	leg	wg.	leg	wg.	leg	wg.	leg	wg.	leg

The **Pullet Census** sheet is designed to record the different families of pullets as they are placed on range. This serves two purposes, (1) an accurate check-up on every family, (2) a complete list of pullets for which leg bands must be ordered. Thus when these birds are housed, they are properly banded in consecutive order on the trap sheet, as well as on the Family Record Sheet.

Questions on Chapter XIII

1. What are the advantages of trapnesting and pedigreeing?
2. What are the disadvantages?
3. What type of individual should undertake this kind of work?
4. How would you go about the job of trapnesting?
5. How would you go about the job of pedigree hatching eggs from trapnested birds?
6. List the type of records maintained in a trapnesting-pedigree-breeding program?
7. In what ways may the task of record keeping be facilitated?

CHAPTER XIV

IMPROVEMENT PROGRAMS

Subject Matter

The Development of a National Plan
National Poultry Improvement Plan
National Turkey Improvement Plan
Standard Egg Laying Contests

Programs for the improvement of poultry and poultry products have been numerous. Over a period of years, a number of different plans have been advanced by their proponents as suggested and desirable national standardization programs. The result has been an extensive confusion in the minds of hatching egg and baby chick buyers and somewhat of a misunderstanding between the various sections of the country.

After years of controversial discussion between the interests of the various sections, the federal government inaugurated a national breed improvement plan. Under this plan, the individual states may have designated grades higher than the standard called for but yet conform to the plan in terminology. Furthermore, individual states may participate in as many or as few phases as they desire.

United States Department of Agriculture

(Bureau Animal Industry, Animal Husbandry Division)

THE NATIONAL POULTRY IMPROVEMENT PLAN

The National Poultry Improvement Plan became operative July 1, 1935, under authority of an appropriation by Congress (Public No. 62-74th Congress—H. R. 6718) for the Bureau of Animal Industry, United States Department of Agriculture, to be used in cooperation with State authorities in the administration of regulations for the improvement of poultry, poultry products, and hatcheries.

The acceptance of the plan is optional on the part of States and individual members of the industry within States. It is administered in each State by an official State agency cooperating with the Bureau of Animal Industry of the United States Department of Agriculture.

A list of the official State agencies engaged in the work of the plan may be obtained from this Bureau upon request. The names of members of the industry who are participating may be obtained from the official State agencies.

Article 1. General Regulations

REGULATION 1.—For the purposes of this plan the following words, phrases, names, and terms shall be construed, respectively, to mean—

Paragraph 1. *Plan.*—The National Poultry Improvement Plan.

Par. 2. *Turkey plan.*—The National Turkey Improvement Plan.

Par. 3. *Official State agency.*—The agency recognized by the Bureau of Animal Industry, United States Department of Agriculture, to cooperate in the administration of the plan within the State. It may be the State department of agriculture, State college of agriculture, State poultry-improvement board or association, or other agency recognized officially by the State government.

Par. 4. *State inspector.*—An employee or authorized representative of the official State agency whose duties in connection with the plan comprise the official inspection of flocks, eggs, chicks, hatcheries, poultry premises, records, and methods employed in carrying on the work of the plan.

Par. 5. *R. O. P. supervisor.*—An employee or authorized representative of the official State agency whose duties in connection with the plan are to represent the official State agency in its supervision of U. S. R. O. P. (U. S. Record of Performance) and U. S. R. O. M. (U. S. Register of Merit) work and to vouch for the accuracy of all U. S. R. O. P. and U. S. R. O. M. records made by the U. S. R. O.P. breeders under his jurisdiction.

Par 6. *R. O. P. inspector.*—An employee or authorized representative of the official State agency whose duties in connection with the plan are the inspection of flocks and records of U. S. R. O. P. breeders.

Par. 7. *Flock-selecting agent.*—A person who has taken a course of training in and passed an examination on (1) the selection of breeding stock for standard-bred and production qualities as prescribed by the State college of agriculture or other properly constituted State agency and (2) the plan provisions as applied in the State, and who has been authorized by the official State agency to do the flock-selecting work as provided for in the U. S. Approved and U. S. Certified breeding stages. As an emergency and to take care of individual cases only, the official State agency may, at its discretion, authorize flock-selecting agents who have not taken the specified course of training. Only those persons who have demonstrated to the official State agency through work in the field that they are capable of doing satisfactory selecting work may be so authorized.

Fig. 120—Special egg cooling room, with humidity control.
[Courtesy Guy A. Leader, Leader's Leghorn Farms.]

Par. 8. *State Pullorum tester.*—An employee or authorized representative of the official State agency, who has training in pullorum-testing technique, a working knowledge of the plan provisions as applied in the State, and other qualifications necessary to do the pullorum-testing work as provided for in the plan; however, no person can qualify as a State pullorum tester for work in a hatchery in which he has a financial interest, either entire ownership, partial ownership, or employment.

Par. 9. *Pullorum-testing agent.*—A person who has taken a course of training in and passed an examination on *(1)* pullorum testing as prescribed by the State livestock sanitary authorities or officials of the State college of agriculture and *(2)* the plan provisions as applied in the State, and who has been authorized by the official State agency to do the pullorum-testing work. As an emergency and to take care of individual cases only, the official State agency may, at its discretion, authorize pullorum-testing agents who have not taken the specified course of training. Only those persons who have demonstrated to the official State agency through work in the field that they are capable of doing satisfactory testing work may be so authorized.

Par. 10. *Blood collector.*—A person who has been trained in the proper method of collecting blood samples for the pullorum disease test, who has passed an examination on the plan provisions as applied in the State and who has been authorized by the official State agency to do the blood-collecting work.

Par. 11. *Eggs and hatching eggs.*—Chicken eggs and chicken hatching eggs.

Par. 12. *Baby chicks.*—Chicks that have not been fed or watered.

Par. 13. *Started chicks.*—Chicks that have been fed or watered.

Par. 14. *Breeding stock.*—Chickens to be used for breeding purposes.

Par. 15. *Flock.*—All chickens of one breed or variety on the same premises, except that cross-mated birds as provided in regulation 11 shall be considered as one flock.

Par. 16. *Flock owner.*—A person, firm, or corporation maintaining a breeding flock of chickens.

Par. 17. *Hatchery.*—Hatchery equipment on one premises operated or controlled by any person, firm, or corporation producing chicks.

Par. 18. *U. S. R. O. P.*—U. S. Record of Performance.

Par. 19. *U. S. R. O. M.*—U. S. Register of Merit.

Par. 20. *Official label.*—A label, obtained through or approved by the official State agency, on which are indicated the product, the breeding stage, and the pullorum class.

Par. 21. *Crossbred hatching eggs and chicks.*—Hatching eggs and chicks produced: (a) By crossing two different breeds or varieties; or

(b) By crossing first generation poultry resulting from the combination of two different breeds or varieties with another and different breed or variety; or

(c) By crossing the first generation poultry resulting from the combination of two different breeds or varieties with the first generation poultry resulting from the combination of two other different breeds or varieties.

Par. 22. *Inbred.*—The first generation poultry, chicks, or eggs, produced by a mating of poultry of known relationship, related to the degree of first cousins or closer.

Par. 23. *Inbred line.*—A group of inbred poultry, or the chicks or eggs, resulting from at least four generations of inbreeding. The poultry constituting the line must be individually identified as to origin and so interrelated that the mating of any pair within the line would result in progeny with an amount of inbreeding exceeding that resulting from three successive generations of full brother-sister matings (coefficient of more than 50 per cent).

Par. 24. *Incrossbred.*—The first generation poultry, chicks, or eggs produced:

(a) By crossing two inbred lines of different breeds or varieties; or

(b) By crossing first-generation poultry resulting from the combination of two inbred lines of different breeds or varieties with another inbred line of different breed or variety; or

(c) By crossing the first-generation poultry resulting from the combination of two inbred lines with the first-generation poultry resulting from the combination of two other inbred lines when each line used is of different breed or variety than any other used.

First-generation poultry produced by crossing two inbred lines of the same breed or variety may be used in lieu of an inbred line of that breed or variety.

REGULATION 2.—The official State agency shall conduct efficiently the inspection work called for in the plan and shall direct, administer, supervise, and be responsible for all the work done relative to flock selection in the various breeding stages and to testing for pullorum disease as provided for in the plan.

The Bureau of Animal Industry shall have the right to terminate, after opportunity for informal hearing, in accordance with its memorandum of understanding with any official State agency, its cooperation with such agency if in the opinion of the Bureau the agency is not properly administering the National Poultry Improvement Plan.

REGULATION 3.—Poultry laying and breeding houses shall be kept clean and in good condition. The land in the immediate vicinity of the poultry laying breeding houses shall be kept in sanitary condition.

REGULATION 4.—Hatcheries and all equipment shall be kept in strictly sanitary condition. All incubators must be thoroughly cleaned and disinfected by spraying or fumigating after each hatch by means of an officially recognized method.

REGULATION 5.—All eggs used in any breeding stage of the plan shall be reasonably uniform in size and shape and sound in shell. In the case of all white-egg breeds, U. S. Approved hatching eggs shall be reasonably free from tints, and eggs of other U. S. breeding stages shall be free from tints.

REGULATION 6.—The management of a hatchery participating in the plan shall maintain the identity of hatching eggs being incubated, including those which are being custom-hatched, and shall keep accurate records concerning the name and address of each flock owner; the number of eggs received from each flock owner; the name and address of each purchaser; the number, hatchery grade, if any, breed and variety, and date of shipment or delivery of all chicks and hatching eggs. These records shall be open for official inspection at all times. Eggs received and chicks hatched shall be subject to inspection at all times.

REGULATION 7.—All chicks sold or offered for sale under the plan terminology shall be normal and representative of the breed and variety.

REGULATION 8.—Shipment of chicks shall be made in approved types of chick boxes properly ventilated and securely tied, and chicks shall be shipped to reach the ultimate purchaser within the time limit prescribed by the United States Postal Regulations.

REGULATION 9.—Chicks and hatching eggs produced under the rules and regulations of the plan shall lose their identity as of the plan terminology when they are purchased for resale by nonparticipants.

REGULATION 10.—Chicks and hatching eggs of identical U. S. breeding stages and pullorum classes may be exchanged or bought for resale by members of the industry who are participating in the same phases of the plan. Participants may purchase hatching eggs or chicks of a higher stage or class than those which they produce, but such products may not be sold as of the higher stage or class.

REGULATION 11.—Only one variety of chickens shall be kept on a farm unless, where more than one variety is kept, ample provision has been made to keep these varieties and their eggs separate in a manner satisfactory to the official State agency. With the approval of the official State agency, crossbred or incrossbred hatching eggs and chicks, as defined in regulation 1, paragraphs 21 and 24, may be produced and sold by participants in the plan, provided the parent stock is well selected for good vitality, growth, market, quality egg production, feathering,

Fig. 121—White Plymouth Rocks.
[Courtesy Poultry Tribune.]

and egg weight. The parent stock must also be leg-banded and the flock inspected as required in the U. S. Approved breeding stage. These flocks shall be designated as U. S. Approved crossmated or U. S. Certified crossmated, dependent upon the breeding of the males, or U. S. Approved incrossmated, and hatching eggs or chicks from each flocks shall be designated as U. S. Approved crossbred, U. S. Certified crossbred, and U. S. Approved incrossbred, respectively. Backcross and other matings of crossbred stock not covered by these specific provisions may be made with the permission of the official State agency, but the progeny produced from such matings shall not be advertised or sold as crossbred or incrossbred.

REGULATION 12.—When more than one hatchery located with a State is operated under the same name, ownership, or management, one or more of these hatcheries may not participate in the plan unless all of them participate in the same breeding stage and pullorum class. In the event that any person, firm, or corporation has non-participating hatcheries in other States, no chicks or hatching eggs shall be shipped or delivered to the participating hatcheries from the nonparticipating hatcheries.

REGULATION 13.—The use of the official terminology or any portion thereof of any particular breeding stage or pullorum class of the plan in which members of the industry are participating shall be limited to hatching eggs, chicks, and breeding stock that meet at least the minimum requirements of that particular stage or class when produced by those under the jurisdiction of the official State agency.

REGULATION 14.—Breeding stock and each bundle or package of hatching eggs and chicks sold or offered for sale by participants shall be designed by official labels or tags specifying the breeding stage and pullorum class in accord with the regulations of the plan, except that chicks that are seconds, culls, rejects, or abnormal for breed and variety characteristics shall not be sold or offered for sale under official labels. (See regulations 7 and 9.) With the approval of the official State agency, hatching-egg supply flock owners of a participating hatchery need not officially label each consignment of hatching eggs delivered to the hatchery.

REGULATION 15.—All advertising must be in accordance with the rules and regulations of the plan and the Trade Practice rules of the Federal Trade Commission for the baby-chick industry. A participant using in his advertising the terms "R. O. P. sired," "R. O. P. pedigreed sired," "R. O. P. mated," or any other term or letters of similar meaning or implication shall clearly show the breeds and the approximate percentage or proportion of the birds in his matings (flocks supplying him with hatching eggs or chicks) of each breed that are of that quality. Participants who are not R. O. P. breeders shall not use advertising or other sales efforts which imply that they are R. O. P. breeders. Copies of proposed folders, catalogs, circulars, or other printed literature shall be submitted to, and approved by, the official State agency before being printed. Copies of such literature, after being printed, shall be filed with the official State agency. Copies of all publication advertising shall be mailed to the official State agency at or before the time these advertisements are submitted to the publisher or publishers.

The advertising literature of hatcheries which have places of business in more than one State shall be approved by the official State agency in which the main or home office is located after being checked by the Bureau of Animal Industry, except that local advertising literature for each State branch, or branches shall be approved by the official State agency of each such State in which branches are located. Refer also to regulation 22 for proper use of terms.

REGULATION 16.—Participation in one of the pullorum classes shall be a prerequisite to participation in any of the breeding stages of the plan, except that an R. O. P. breeder who fails to qualify for one of the pullorum classes may continue to participate in the R. O. P. breeding stage, so long as he satisfies the official State agency that he is proceeding as rapidly as possible to qualify for one of the pullorum classes. During this period, each sale of hatching eggs, baby chicks, and breeding stock must first be approved by the official State agency.

Beginning with the season of 1950-51, participation in one of the breeding stages of the plan shall be a prerequisite to participation in any of the pullorum classes of the plan.

REGULATION 17.—The use of the word "accredited" to indicate breeding improvement or disease eradication or control or its use in any other way is prohibited.

REGULATION 18.—The National Poultry Improvement Plan emblem (see cover page), the official terms and various designs illustrative of the breeding stages and the pullorum classes (figs. 1 to 8) may be used only by participants who have met the specific requirements and in each case where the National Poultry Improvement Plan emblem is used in advertising, the breeding stage and pullorum classification shall be shown.

REGULATION 19.—The selection of birds in a flock to qualify as U. S. Approved or U. S. Certified by a State inspector shall be regarded as one official inspection, provided a complete inpection is made and so reported.

REGULATION 20.—Males that are candidates for U. S. R. O. P. qualification may be dubbed, provided each bird has been officially examined by a State inspector and passed in respect to comb disqualifications. Such birds must otherwise comply with all regulations pertaining to the qualifications of U. S. R. O. P. males not earlier than 6 months of age. A record of the number of males passed for dubbing in each flock, the breeder from whom the males were procured, and the complete wing-band numbers of those disqualified shall be made and sent to the official State agency within 10 days after such examination and dubbing. With the approval of the official State agency, flock-selecting agents may officially examine and approve cockerels in respect to comb disqualifications.

REGULATION 21.—Any person, firm, or corporation desiring to participate in the plan except flockowners who are affiliated with participating hatcheries shall sign an agreement with the official State agency and shall not purchase, receive, offer for sale, or be a dealer in or agent for breeding stock, hatch eggs, or chicks other than those produced under the plan except with permission from the official State agency. In certain cases special permission may be granted by the official State agency, in writing, to purchase hatching eggs, chicks, or breeding stock not produced under the plan for use in breeding or hatching-egg supply flocks. In such cases a report shall be made to the official State agency on each shipment received and to whom delivered. With the permission of the official State agency, market eggs may be bought and sold.

REGULATION 22.—Flocks, hatching eggs, hatcheries, and chicks that have met the requirements of both a breeding stage and a pullorum class of the plan shall be designated in accordance with the stage and class attained as follows: U. S. Approved, Pullorum-Tested; U. S. Approved, Pullorum-Controlled; U. S. Approved, Pullorum-Passed; U. S. Approved, Pullorum-Clean; U. S. Certified, Pullorum-Tested; U. S. Certified, Pullorum-Controlled; U. S. Certified, Pullorum-Passed; or U. S. Certified, Pullorum-Clean.

Matings, hatching eggs, chicks, and birds that have met the requirements of the U. S. R. O. P. breeding stage and a pullorum class of the plan shall be designated in accordance with the class attained as follows: U. S. R. O. P., Pullorum-Tested; U. S. R. O. P., Pullorum-Controlled; U. S. R. O. P., Pullorum-Passed; or U. S. R. O. P., Pullorum-Clean.

Where descriptive terms are used in advertising in addition to the official terminology of the plan, such terms shall be clearly explained.

(E.g., if more than one hatchery grade of U. S. Approved chicks is offered for sale and the chicks in one of such grades are sired entirely by U. S. R. O. P. males, it is sufficient and proper to describe such a grade as "U. S. Approved chicks sired by U. S. R. O. P. males.")

Participants in the plan may use the terms "R. O. P. sired," "R. O. P. pedigreed sired," "R. O. P. mated," or any other term or letters of similar meaning or implication to describe chicks, only when the males siring the chicks so described have been officially leg-banded as U. S. R. O. P. males.

REGULATION 23.—The minimum hatching egg weight for the U. S. Approved and U. S. Certified stages for the months December through June shall be 1 11/12 ounces each. The minimum for the months July through November shall be 1 10/12 ounces each except that during these months, eggs weighing as low as 1 8/12 ounces each may be set for the production of chicks designated as "broiler chicks." These minimum egg-weight requirements shall not apply to breeds and varieties for which the standard weight of the hen is less than 4 pounds.

REGULATION 24.—Started chicks shall be labeled as such, and all the chicks in each lot sold or offered for sale shall be of the same age (hatching date) unless otherwise specified.

REGULATION 25.—With the approval of the official State agency, hatching-egg and chick dealers may participate in the plan and enjoy all the privileges of participating hatcheries in the handling of hatching eggs and chicks, provided all the applicable requirements for participating hatcheries are fully and completely complied with.

REGULATION 26.—Noncompliance with the provisions of the plan which is not corrected within a reasonable time shall bar any participant from further participation for a period of 2 years. Such action shall not be taken until a thorough investigation and hearing have been conducted by the official State agency.

REGULATION 27.—Participation in any phase of the plan shall be construed as maintaining supervision and compliance throughout the 12 months of the year. (It is understood that a participant may withdraw from participation or that his agreement may be terminated by the official State agency for cause.)

REGULATION 28.—A participant in any phase of the plan may not incubate or hatch turkey eggs in the same room as chicken eggs unless the turkeys are of the same or higher pullorum classification.

REGULATION 29.—A new or nonstandard breed or variety may only be accepted in the U. S. Approved and U. S. Certified breeding stages of the plan provided a suitable standard for the breed or variety has been approved by a joint committee of the National Poultry Improvement Plan and the American Poultry Association. With the permission of the official State agency, a limited number of chicks may be distributed to selected patrons for the purpose of obtaining large scale field trials in the developments of new breeds.

REGULATION 30.—Effective July 1, 1949, a hatchery to be eligible for participation in the National Poultry Improvement Plan may not brood or keep started chicks or poultry of any age in the same room in which the incubators are located.

REGULATION 31.—All applicants for participation in the National Poultry Improvement Plan shall demonstrate to the satisfaction of their official State agency, prior to their admission to such participation by the agency, that their facilities, personnel, and practices are adequate for carrying out the applicable provisions of the National Poultry Improvement Plan.

Article 2. Breeding Stages

SECTION 1. U. S. APPROVED BREEDING STAGES

Paragraph 1. *U. S. Approved flocks.*—These flocks shall meet the following requirements:

(a) Females shall be rigidly and thoroughly selected once each year for constitutional vigor and for egg production, such selected females to combine standardbred and production qualities to a reasonably high degree; (b) males shall be selected especially for constitutional vigor and standardbred qualities; (c) the selection of the flocks shall be approved by a State inspector.

Par. 2. The method of selecting poultry breeding stock prescribed by the State college of agriculture or other properly constituted State agency shall serve as the basis for the selection of males and females. All birds to be used in U. S. Approved flocks shall be selected by a State inspector or by a flock-selecting agent.

Par. 3. All birds selected for U. S. Approved flocks shall be banded at the time of selection with sealed and numbered leg bands obtained through the official State agency. With the permission of the official State agency, individually wing-banded females on an R. O. P. breeder's own farm need not be leg-banded for the production of U. S. Approved and U. S. Certified chicks, provided (1) all the females on the premises are individually wing-banded, and (2) that a complete record of the wing-band numbers of the females in each pen is made at the time of selecting and testing. Birds rejected by the State inspector or by the flock-selecting agent shall have their tails clipped, have their leg bands removed, and be kept separate from the breeding flocks. The selection of birds for a U. S. Approved flock shall be completed at least 10 days before any eggs are saved for hatching purposes, except that this 10-day interval shall not be required if all males found in the flock are approved.

Par. 4. Before any flock other than one affiliated with a participating hatchery may be recognized as a U. S. Approved flock, it must be inspected and approved by a State inspector. Each year at least 15 per cent of the hatchery flocks supplying eggs to a U. S. Approved hatchery and at least 15 per cent of the flocks selected by each flock-selecting agent shall be inspected and approved by a State inspector. The State inspector shall determine which flocks shall be inspected, and the inspection shall consist in the physical examination of a sufficient number of males and females to satisfy him that the flock is worthy of approval.

Reg. U. S. Pat. Off.

(GREEN)

Fig. 1.—Design authorized to identify U. S. Approved hatching eggs, chicks, flocks, and hatcheries.

Par. 5. No birds except those with sealed and numbered official leg bands or wing bands as provided in paragraph 3, shall be allowed in the breeding flocks, and no male birds shall be allowed in any other mature flock on the same premises except in U. S. Certified or U. S. R. O. P. flocks. This shall not be construed as permitting a U. S. Approved and a U. S. Certified flock of the same breed and variety on one farm.

Par. 6. *U. S. Approved hatching eggs.*—Such eggs shall come only from U. S. Approved flocks or a higher U. S. breeding stage and shall weigh at least 1 11/12 ounces each, except as provided for in regulation 23.

Par. 7. *U. S. Approved chicks.*—They shall be hatched only from U. S. Approved eggs and in U. S. Approved hatcheries.

Par 8. *U S. Approved hatcheries.*—These hatcheries may sell only hatching eggs and chicks produced under the breeding stages of the plan, except as provided for in regulation 11. Hatching eggs and chicks of the U. S. Certified stage of varieties other than those sold as U. S. Approved may be hatched or sold in a U. S. Approved hatchery. A U. S. Approved hatchery operated by a U. S. R. O. P. breeder may sell U. S. R. O. P. hatching eggs and chicks. U. S. Approved hatcheries shall be inspected and approved by a State inspector at least once during the hatching season

SECTION 2. U. S. CERTIFIED BREEDING STAGE

Reg. U. S. Pat. Off.
(YELLOW)
Fig. 2.—Design authorized to identify U. S. Certified hatching eggs, chicks, flocks, and hatcheries.

Paragraph 1. *U. S. Certified flocks.*—These flocks shall fulfill all the requirements of U. S. Approved flocks and shall be mated to U. S. R. O. P. males. For the first 2 years of U. S. R. O. P. work carried on by any breeder, the males for his own flock need not be U. S. R. O. P. males but must be individually pedigreed and of breeding which meets the qualifications for U. S. R. O. P. males. All such males may be used in subsequent breeding seasons if reinspected and certified by the official State agency. The males and females shall be of at least qualifying standard weight as designated by the official State agency in cooperation with the State college of agriculture or other properly constituted State agency, which gives the prescribed training in selecting poultry breeding stock. The males and females shall be selected by either a State inspector or a flock-selecting agent once each year, and if selected by a flock-selecting agent, at least 50 per cent of the flocks selected by each agent shall be inspected by a State inspector. This flock inspection shall consist in the examination of a sufficient number of males and females and the checking of the wing bands on a sufficient number of males to satisfy the State inspector that the flock is worthy of certification.

Par. 2. *U. S. Certified hatching eggs.*—Such eggs shall come only from U. S. Certified flocks and shall weigh at least 1 11/12 ounces each and average at least 24 ounces to the dozen, except as provided for in regulation 23.

Par. 3. *U. S. Certified chicks.*—They shall be hatched only from U. S. Certified eggs and in U. S. Certified hatcheries, except that a hatchery may produce and sell U. S. Certified chicks of one breed or variety and U. S. Approved of another breed or variety, in which case the hatchery shall be recognized as a U. S. Approved hatchery.

Par. 4. *U. S. Certified hatcheries.*—These hatcheries may sell only U. S. Certified hatching eggs and chicks except as provided for in regulation 11. A U. S. Certified hatchery operated by a U. S. R. O. P. breeder may sell U. S. R. O. P. hatching eggs and chicks. U. S. Certified hatcheries shall be inspected and approved by the State inspector at least twice during the hatching season.

Par. 5. With the approval of the official State agency, U. S. Certified flocks may be mated to Canadian R. O. P. males, provided such males meet U. S. R. O. P. qualifications.

Par. 6. *U. S. Certified crossmated flocks.*—The flocks shall fulfill the requirements of U. S. Approved crossmated flocks and shall be mated to U. S. R. O. P. males.

Par. 7. *U. S. Certified crossbred hatching eggs.*—Such eggs shall come only from U. S. Certified crossmated flocks and shall weigh at least 1 11/12 ounces each and average at least 24 ounces to the dozen except as provided for in regulation 23.

Par. 8. *U. S. Certified crossbred chicks.*—These chicks shall be hatched only from U. S. Certified crossbred hatching eggs. A hatchery may not sell U. S. Certified crossbred chicks and U. S. Approved crossbred chicks of the same cross.

SECTION 3. U. S. RECORD OF PERFORMANCE BREEDING STAGE

Paragraph 1. U. S. Record of Performance embraces records of egg production, egg weight, and body weight made on the breeder's premises under official supervision and similar records made at officially conducted egg-laying tests, State colleges of agriculture, and State and Federal experiment stations, when such records are passed on by the official State agency and when the individual birds meet other U. S. R. O. P. requirements, except that the birds produced by State colleges of agriculture and State and Federal experiment stations may be recognized as U. S. R. O. P. only for the use of U. S. R. O. P. breeders in their breeding work. With the approval of the official State agency, Canadian R. O. P. males may be used by U. S. R. O. P. breeders, provided such males meet U. S. R. O. P. qualifications.

Par. 2. The trap nests used in U. S. R. O. P. work shall be of an approved type and shall be kept in perfect working order at all times. The poultry plant and all egg-production records shall be subject at all times to unannounced inspection by a State inspector. At the time of each inspection, the inspector shall have sole charge of the trap nests and of eggs laid. Each U. S. R. O. P. candidate shall be banded with a sealed and numbered official leg band. No females other than those entered in U. S. R. O. P. shall be allowed in the pens where U. S. R. O. P. candidates are being trap-nested, except with the approval of the official State agency.

Reg. U. S. Pat. Off.

(BLUE)
Fig. 3—Design authorized to identify U. S. R. O. P. hatching eggs, chicks, and breeding stock.

Par. 3. U. S. R. O. P. eggs and chicks may be produced only on a U. S. R. O. P. breeder's premises except that with the approval of the official State agency (a) a U. S. R. O. P. breeder may have eggs hatched by any hatchery operating under the plan provided the U. S. R. O. P. breeder does the pedigreeing of the chicks; and (b) a U. S. Approved or U. S. Certified hatchery owner may produce U. S. R. O. P. chicks from purchased U. S. R. O. P. eggs for the use of his hatchery-flock owners only, provided proper records of purchases are sent to the official State agency and a State inspector does the wing-banding of these chicks.

Par. 4. When U. S. R. O. P. hatching eggs are transferred from one U. S. R. O. P. breeder to another U. S. R. O. P. breeder or hatcheryman for the purpose of hatching U. S. R. O. P. chicks, the conditions that shall govern such a transaction include the following: (a) The U. S. R. O. P. breeder who sells U. S. R. O. P. hatching eggs must send a report in triplicate at the time the eggs are shipped consisting of a list of the eggs sold, the pen or sire number and dam number of each egg, and the egg production, average egg weight, and body weight of each dam, to the official State agency of this State. If the eggs are shipped outside the State, these records shall be attested to by the R. O. P. supervisor, who shall retain one copy and mail one copy to the purchaser and one to the official State agency of the State to which the eggs are shipped. (b) Within 5 days after hatching time the U. S. R. O. P. breeder or hatcheryman who purchased the eggs must send to his official State agency a list of the chicks hatched from each dam and their respective wing-band numbers.

Par. 5. When U. S. R. O. P. chicks, cockerels, or mature birds are sold or transferred by members of the industry participating in the plan, the individual making the sale must send a report in triplicate to the official State agency at the time these products are shipped giving the number of birds sold; the number, egg production, average egg weight, and body weight of each dam; and the pen or leg-band number of the sire of each chick, cockerel, or mature bird. If the chicks, cockerels, or mature birds are shipped outside the State, these records shall be attested to by the R. O. P. supervisor, who shall retain one copy and mail one copy to the purchaser and one to the official State agency of the State to which these products are shipped.

Par. 6. To be eligible for U. S. R. O. P. entry, all females shall have been raised from U. S. R. O. P. or U. S. Certified flocks or other flocks of breeding satisfactory to the official State agency. Eligibility shall be confined to any standard breed or variety or any other breed or variety recognized by the United States Record of Performance Federation.

Par. 7. All birds trap-nested on the premises of a U .S. R. O. P. breeder, shall be considered as entered under U. S. R. O. P. supervision with the following exceptions: (a) Preliminary trap-nesting of all birds may be permitted during the first 30-day period. (b) An additional preliminary trap-nesting period of pedigreed candidates for 3 calendar months may be permitted to allow more time to select and enter candidates on a family basis; however, if a pedigreed candidate is continued in U. S. R. O. P. all her full sisters trap-nested beyond the 30-day period as candidates for U. S. R. O. P. must be considered as having been entered. (c) Candidates that react to the test for pullorum disease when officially tested within 4 months after trap-nesting has begun will not be considered as entered. (d) All qualified U. S. R. O. P. hens may be trap-nested. (e) With the written approval of the official State agency, a U. S. R. O. P. breeder may keep birds in individual laying cages or trap-nest inbred or other birds of experimental breeding, provided the trap-nest records of such birds are not referred to in advertising. (f) He may, with the approval of the official State agency, trap-nest any other breed or variety which he is developing as a basis for future U. S. R. O. P. work. (g) With the approval of the official State agency, hens in their second or subsequent production years may be privately trapped to complete lifetime egg records that were begun before U. S. R. O. P. work was undertaken.

Par. 8. All flocks shall be inspected at least five times a year. One or more consecutive days of inspection by the R. O. P. inspector shall represent a single inspection. In the event of an outbreak of contagious disease in a poultry plant where U. S. R. O. P. work is carried on, inspection may be temporarily suspended.

Par. 9. *Duties of R. O. P. supervisor.*—The R. O. P. supervisor shall represent the official State agency in its supervision of U. S. R. O. P. work, and he shall vouch for the accuracy of all U. S. R. O. P. and U. S. R. O. M. records made by U. S. R. O. P. breeders under his jurisdiction. He shall keep on file in his office for at least 5 years, (a) a record of all qualified males, with at least one-generation pedigree showing egg records and egg weights of the female ancestors, and (b) a record of all qualified females with their egg records and egg weights for their first year of production. He shall also keep on file a record of all U. S. R. O. P. matings and all U. S. R. O. P. chicks with at least one-generation pedigree

Par 10. The R. O. P. supervisor shall furnish the Bureau of Animal Industry, for publication, a summary of the U. S. R. O. P. work under his supervision each year, which shall include the following information of each flock.

(a) Total number of pullets of that particular breed on the farm.

(b) Number of pullets originally trapnested for the U. S. R. O. P. entry.

(c) Number of pullets withdrawn during trial period.

(d) Number of pullets entered in U. S. R. O. P.

(e) Number of pullets entered in U. S. R. O. P. which are individually pedigreed.

(f) Number meeting the U. S. R. O. P. requirements.

(g) Percentage of birds entered meeting the U. S. R. O. P. requirements.

(h) Number of birds qualifying which lay from 200 to 249 eggs.

(i) Number of birds qualifying which lay 250 to 299 eggs.

(j) Number of birds qualifying which lay 300 or more eggs.

(k) Average egg production of all birds entered.

(l) Average egg weight of all birds entered.

(m) Average body weight of all birds entered.

Par. 11. The R. O. P. supervisor may bring charges to the official State agency of his State against any U. S. R. O. P. breeder for falsifying records, for unethical conduct in his business methods, or for any other cause which he may consider detrimental to U. S. R. O. P. work. On such charges being proved in a fair hearing, the offending breeder must be reprimanded or suspended by the official State agency according to the nature of the offense.

Par. 12. *Duties of R. O. P. inspector.*—The R. O. P. inspector shall work under the direction of the R. O. P. supervisor. It shall be his duty to visit and inspect the work of each U. S. R. O. P. breeder at least five times each year, his visits to be so timed that each season of the year shall be represented in his reports, and at least three of these visits shall be unannounced. On each visit he shall do the trap-nesting for the day and shall record all eggs laid, and during the official egg-weighing period for the breeder shall weigh and record the weight of the egg laid by each individual bird entered in U. S. R. O. P. He shall examine hens apparently out of production to see that they are not being credited with eggs. During the breeding season he shall examine all birds in U. S. R. O. P. matings to see that the birds which constitute these matings are properly listed with the supervisor. He shall compare the number of eggs being incubated from each hen with the number she is credited with having laid during the corresponding period, and shall have authority to test these eggs for fertility. When the hatch is reported he shall compare his findings with the number of chicks reported hatched and wing-banded. He shall be courteous to breeders at all times but shall be on the alert to discover errors in records or violations of rules, intentional or otherwise. He shall immediately report to the supervisor any suspicious circumstances or appearance of fraud in the records or business dealings of a U. S. R. O. P. breeder.

Par. 13. *U. S. R. O. P. breeder.*—A U. S. R. O. P. breeder is a person, firm, or corporation maintaining a flock of chickens in accordance with the provisions of the U. S. R. O. P. breeding stage. In additional to the U. S. R. O. P. matings, all other mated birds supplying hatching eggs to such person, firm, or corporation or to any hatchery owned or operated by such person, firm, or corporation shall be under the supervision of the plan in one of the breeding stages.

Par. 14. The candidate flock or flocks of a U. S. R. O. P. breeder must be trap-nested at regular intervals every day throughout the year, except that the official State agency may approve as a substitute for daily trap-nesting of U. S. R. O. P. candidates (a) 6-day-a-week trap-nesting (the official egg record to be computed by multiplying the trap-nest record by 7/6), (b) 5-day-a-week trap-nesting (the official egg record to be computed by multiplying the trap-nest record by 7/5), or (c) 3-day-a-week trap-nesting after a breeding flock has been established for a period of 2 years (the official egg record to be computed by multiplying the trap-nest record by 7/3). The breeder shall send to his supervisor, not later than the tenth day of each month, a complete record of all U. S. R. O. P. egg-production records and egg weights for the previous month. The egg production, pedigree, and sales records shall at all times be open to examination by the supervisor or inspector.

Par. 15. *U. S. R. O. P. qualification through egg-laying tests.*—The U. S. R. O. P. breeder who desires to have his laying-test entry considered as entered in U. S R. O. P. shall make application in writing to the manager of the test not later than January 1. The manager of the laying test will be expected to keep a special record on such entries on N. P. I. P. Form 17, Report of Birds Entered in Egg-Laying Tests by U. S. R. O. P. breeders. At the end of the laying-test year the birds shall be examined for final qualification and officially leg-banded by the R. O. P. inspector or supervisor in the State to which the birds are sent from the test. Only those entries of U. S. R. O. P. breeders who are doing home trap-nesting of pullets of the same breed and variety under R. O. P. supervision are eligible for U. S. R. O. P. entry and qualification. Necessary records will be forwarded to the breeder and the supervisor by the laying-test manager.

Egg weights for laying-test entries will be based on the average weight of eggs laid on three consecutive days during the first full week of each month from January to May, inclusive. Additional eggs may be weighed after the 5-month period to meet the minimum number of eight (8) egg weights required for qualification in R. O. P.

Par. 16. *U. S. R. O. P. females.*—Pullets shall have laid 200 or more eggs during the first laying year and yearling and older hens shall have laid 200 or more eggs during the trap-nest year. The first laying year and the trap-nest year, in the case of birds on the breeder's premises shall mean 365 days from the date of laying the first egg in a trap-nest. In the case of egg-laying-test pullets and hens, the first laying year or the trap-nest year shall mean the duration of the official laying-test year. To facilitate the preparation of the annual R. O. P. summary the R. O. P. supervisor may close all records by November 30, but birds that have not completed their first laying year may be permitted to do so and be included in the following year's summary

Par. 17. The qualifying average egg weight for pullets shall be 24 ounces to the dozen and for yearling or older hens shall be 25 ounces to the dozen. Official egg weights shall be based on the average of not less than eight individual egg weights. Eggs shall be weighed either four consecutive days each month, or on the same day each week for any four consecutive months between January 1 and July 1 previously indicated by the breeder. When the 3-day-a-week trap-nesting system is used, eggs shall be weighed 3 consecutive days each month for a period of 5 months between January 1 and July 1, previously indicated by the breeder. The inspectors' weights taken during the official weigh period shall be included in computing the average and the number of weights. Additional eggs may be weighed after the official weigh period to meet the minimum number of 8 egg weights required for qualification.

Par. 18. The U. S. R. O. P. breeder and the R. O. P. inspector shall record the leg-band numbers of all U. S. R. O. P. candidates showing a standard disqualification, laying eggs of undesirable shape or shell texture, or laying tinted eggs in the case of white-egg breeds. Such birds shall not qualify as U. S. R. O. P. females. The R. O. P. inspector may remove the U. S. R. O. P. leg band from any bird which develops these defects subsequent to qualification.

Par. 19. All females shall be of at least qualifying body weight for hens and be reasonably good representatives for each breed or variety as designated in the American Standard of

Perfection, allowance being made for the standard weight tolerance except that 3 pounds' tolerance shall be allowed for Plymouth Rocks, Jersey Giants, and breeds in the English and Asiatic classes. The following table gives the specific minimum qualifying female body weights for breeds and varieties participating in the U. S. R. O. P. breeding stage in 1946-47:

	Pounds
Australorp	3.5
Jersey White Giant	7.0
Langshan	4.5
Leghorn	3.5
Minorcas: White and Buff	4.5
New Hampshire	4.5
Orpington	5.0
Plymouth Rock	4.5
Rhode Island Red	4.5
Rhode Island White	4.5
Wyandotte	4.5

Each female shall be weighed between January 1 and June 1 by the R. O. P. inspector and this weight shall be used for the purpose of qualifying the bird for the U. S. R. O. P. At the option of the breeder and the official State agency, one additional weight taken by the breeder during the same period may be used in determining the official body weight.

Par. 20. All birds meeting the above requirements with respect to egg production, egg weight, eggshell color and texture, egg shape, body weight, and freedom from standard disqualifications, and that are reasonably good representatives of the breed or variety as designated in the American Standard of Perfection, shall be considered as having met the requirements for U. S. R. O. P. females. Before such birds may be sold as U. S. R. O. P. females or used in U. S. R. O. P. mating or flock matings of U. S. R. O. P. males and U. S. R. O. P. females, they shall be examined and banded by the R. O. P. inspector or supervisor with a U. S. R. O. P. sealed, official leg band.

Par. 21. *U. S. R. O. P. males.*—Males produced from U. S. R. O. P. chicks may be qualified as U. S. R. O. P. males for participants in the plan only. To qualify, such males shall have passed the final individual inspection of a State inspector not earlier than at 6 months of age, except that a flock-selecting agent may be permitted to select and leg-band U. S. R. O. P. males for use in U. S. Approved and U. S. Certified flocks. The State inspector shall examine a sufficient number of the males selected and leg-banded by each flock-selecting agent to determine that the selection was satisfactory. Each male passed by an inspector or flock-selecting agent shall possess strong constitutional vigor and be a good representative of the breed and variety as designated in the American Standard of Perfection, shall be free from standard disqualifications, and shall be banded with a U. S. R. O. P. sealed and numbered official leg band when selected. At the time of selection and banding the inspector or flock-selecting agent shall make a complete record of the wing- and leg-band numbers of each male banded. Privately pedigreed males referred to in the next paragraph and the non-R. O. P. males referred to in paragraph 29 that have met the progeny-test requirements for U. S. R. O. M. males shall be considered as having met the requirements for U. S. R. O. M. and U. S. R. O. P. males and may be officially leg-banded as such.

Par. 22. *U. S. R. O. P. matings.*—These matings shall be comprised of only U. S. R. O. P. females mated to a U. S. R. O. P. male. The breeder may place in the U. S. R. O. P. mating females that are not U. S. R. O. P., provided he does not sell pedigreed eggs, chicks, or stock from such females as U. S. R. O. P. products. During the first year of U. S. R. O. P. breeding work carried on by any breeder, the male used in U. S. R. O. P. matings need not be a U. S. R. O. P. male, provided he is of equal pedigree. Such males may be used in similar matings during the second year of U. S. R. O. P. breeding work, if reinspected and passed by the inspector, but not thereafter unless they qualify as U. S. R. O. M. males. No eggs or progeny from such a mating shall be sold as U. S. R. O. P. products, except that male offspring produced during the first breeding year may be permitted to head U. S. Certified flocks in the State where produced or to head U. S. R. O. P. matings on the breeder's premises.

Par. 23. Only one male shall be allowed in a single U. S. R. O. P. mating at any one time, and the date of entry and removal of males shall be properly recorded.

Par. 24. All breeding hens in U. S. R. O. P. matings shall be trap-nested during the breeding season and all egg-production records shall be submitted to the official State agency. An annual inspection of pedigreed hatching facilities, method, and chick-banding techniques shall be made on each R. O. P. farm at time of banding.

Par. 25. *U. S. R. O. P. hatching eggs.*—U. S. R. O. P. hatching eggs shall come only from U. S. R. O. P. matings. Such eggs offered for sale shall weigh at least 2 ounces each and 25 ounces or more to the dozen except that no minimum weight requirements shall apply to eggs from U. S. R. O. M. dams. Each egg shall bear the U. S. R. O. P. number of the hen that laid it and the number of the mating.

Par. 26. *U. S. R. O. P. chicks.*—They shall come only from U. S. R. O. P. eggs. The chicks from each female shall be hatched separately in pedigree baskets or sacks and shall be banded individually with sealed and numbered official wing bands at the time of removal from pedigree baskets or sacks. Such bands shall be stamped with the letters "U. S. R. O. P." and shall be so numbered as to permit no two chicks on the same breeder's farm being identically numbered. When eggs are incubated to produce U. S. R. O. P. chicks, an accurate record of the number of eggs from each hen with her pen number and a record of all chicks banded shall be furnished the R. O. P. supervisor on N. P. I. P. Form 8, or a substitute form, at the end of each four hatches. The chicks shall be subject to inspection at any time. The bands from dead chicks shall be retained for the information of the inspector and shall be furnished to him on request.

Par. 27. *U. S. R. O. P. candidate matings.*—These matings shall be comprised of the U. S. R. O. P. candidate females mated to U. S. R. O. P. males. During the first year of U. S. R. O. P. breeding work carried on by any breeder, the males in U. S. R. O. P. candidate matings need not be U. S. R. O. P. males, provided they are of equal pedigree. Such males may be used in similar matings during the second year of U. S. R. O. P. breeding work, if reinspected and passed by the R. O. P. inspector. The breeder shall send the R. O. P. supervisor at the end of each month, a complete report on the birds comprised in these matings, the number of eggs incubated and sold, and the number of chicks hatched and sold. Hatching eggs sold as coming from such matings may be properly identified or stamped with the letters "Cand." Chicks sold as coming from such matings may properly be identified or wing-banded with official wing-bands stamped "Cand."

Par. 28. *U. S. R. O. P. single-male candidate matings.*—Females entered in U. S. R. O. P. may be mated to a U. S. R. O. P. male for the purpose of obtaining progeny records a year earlier than would otherwise be possible. During the first year of U. S. R. O. P. breeding work carried on by any breeder, the male used in a U. S. R. O. P. candidate mating need not be a U. S. R. O. P. male, provided he is of equal pedigree. Such males may be used in similar matings during the second year of U. S. R. O. P. breeding work, if reinspected and passed by the inspector, but not thereafter unless they qualify as U. S. R. O. M. males. The breeder shall send the R. O. P. supervisor a complete report on the birds comprised in this mating, the number of eggs incubated, and the wing-band numbers of the chicks hatched and banded. Sealed and numbered official wing bands marked "Cand." shall be used to differentiate such chicks from U. S. R. O. P. chicks. The male progeny of the females which qualify as U. S. R. O. P. females may qualify as U. S. R. O. P. males for use on the breeder's own premises and to head U. S. Certified flocks in the State where produced, provided they meet all the requirements of U. S. R. O. P. males except that of being produced from U. S. R. O. P. chicks.

Par. 29. *Trial matings.*—With the approval of the official State agency a U. S. R. O. P. breeder may make trial matings in which a non-U. S. R. O. P. male may be mated to either U. S. R. O. P. or non-U. S. R. O. P. females. No hatching eggs, baby chicks, or breeding stock may be sold from such matings. The females from such matings may be entered as U. S. R. O. P. candidates, provided they are reasonably good representatives of the breed or variety as designated in the American Standard of Perfection. Such males and females, including those females referred to in the second sentence of Paragraph 22, page 16 that meet the progeny-test requirement for U. S. R. O. M. shall be considered as having met the requirements for U. S. R. O. M. The sires meeting U. S. R. O. M. requirements may be officially leg-banded as U S. R. O. P. males provided they are free from standard disqualifications. The females meeting U. S. R. O. M. requirements may be leg-banded as U. S. R. O. P. females provided (1) they are free from standard disqualifications and (2) they have a family average egg production and egg weight equal to or better than the family average for all the daughters of the sire mate.

The sons from such matings may be eligible to qualify as U. S. R. O. P. males after their sires and dams qualify for U. S. R. O. M., provided they are reasonably good representatives of the breed and variety as designated in the American Standard of Perfection.

Par. 30. *Flock matings of U. S. R. O. P. males and U. S. R. O. P. females.*—Such matings may properly be maintained by U. S. R. O. P. breeders provided the hatching eggs and chicks produced therefrom are not represented, sold or identified as U. S. R. O. P. products.

Par. 31. *Wing and leg-banding provisions.*—(a) All bands shall be obtained through or approved by the official State agency and all purchases of such shall be authorized by and recorded with the official State agency.

(b) The use of the letters R. O. P. alone or in combination with any other words or letters on wing bands or leg bands is limited to U. S. R. O. P. chicks, qualified males, and qualified females.

(c) Wing bands used on U. S. R. O. P. chicks shall be stamped with the letters U. S. R. O. P. and shall be so numbered as to permit no two chicks on the same breeder's farm being identically numbered.

(d) Chicks from matings of U. S. R. O. P. candidates with U. S. R. O. P. males may properly be identified with wing bands stamped with the letters "Cand."

(e) Chicks from flock matings of U. S. R. O. P. males and U. S. R. O. P. females may properly be identified with wing bands stamped with the words "Flock Mating" or abbreviations thereof.

SECTION 4. REGISTER OF MERIT BREEDING STAGE

Paragraph 1. For the purpose of qualifying for the U. S. Register of Merit stage, the progeny records for one breeding season only shall be considered.

Par. 2. *U. S. R. O. M. sire.*—A U. S. R. O. M. sire is a male which, when used in a supervised single-male mating, has at least one-half and a minimum of 20 of his daughters that are entered in U. S. R. O. P. qualify.

Par. 3. *U. S. R. O. M. dam.*—A U. S. R. O. M. dam is a female which, when used in a supervised single-male mating, has at least one-half and a minimum of 4 of her daughters entered in U. S. R. O. P. qualify.

Par. 4. *U. S. R. O. M. honor roll sire.*—A U. S. R. O. M. sire whose daughters had a family average egg weight of 24 or more ounces per dozen, and a hen-housed family average egg production as follows:

Number of daughters:	Egg production	Number of daughters:	Egg production
40 or more	200	32	219
39	202	31	222
38	204	30	225
37	206	29	230
36	208	28	235
35	210	27	240
34	213	26	245
33	216	25	250

Par. 5. *U. S. R. O. M. honor-roll dam.*—A U. S. R. O. M. dam whose daughters had a family average egg weight of 24 or more ounces per dozen, and a hen-housed family average egg production as follows:

Number of daughters:	Egg production
8 or more	200
7	210
6	225
5	250

Reg. U. S. Pat. Off.

(PURPLE)

Fig. 4—Design authorized to identify U. S. R. O. M. sires, dams, and matings.

Par. 6. *U. S. R. O. M. matings.*—These matings shall be comprised of U. S. R. O. M. females, or U. S. R. O. P. females from U. S. R. O. M. families (i.e., either their sire or a dam qualified for U. S. R. O. M.) mated to a U. S. R. O. M. male or a son of a U. S. R. O. M. male.

Article 3. Pullorum-Control and Eradication Classes

SECTION 1. GENERAL PROVISIONS

Paragraph 1. The official State agency shall assume responsibility for the accuracy of the pullorum-testing work done.

Par. 2. All tests with pullorum, typhoid, or polyvalent antigens of flocks in any pullorum class or candidates for any pullorum class must be reported to the official State agency.

Par. 3. In order for a flock to qualify as U. S. Pullorum-Tested, U. S. Pullorum-Controlled, or U. S. Pullorum-Passed, the testing of the birds for pullorum disease shall be done by a State pullorum tester or a pullorum-testing agent.

Par. 4. In order for a flock to qualify as U. S. Pullorum-Clean, the testing of the birds for pullorum disease shall be done by a State pullorum tester, or pullorum testing agent, satisfactory to the official State agency, provided that when the testing is done by a pullorum testing agent, a represenative of the official State agency shall check-test 15 per cent or more and a minimum of 50 birds in each of 35 per cent or more of the flocks tested by the agent or agents for each hatchery to satisfy the official State agency the testing work was satisfactory.

Par. 5. Chickens more than 5 months of age may not be added to U. S. Pullorum-Passed or U. S. Pullorum-Clean flocks except after the approval of the official State agency. Such birds from only U. S. Pullorum-Passed or U. S. Pullorum-Clean flocks may be added to U. S. Pullorum-Passed flocks, and only those birds from U. S. Pullorum-Clean flocks may be added to U. S. Pullorum-Clean flocks. The number of chickens more than 5 months of age and the name and address of the person from whom the purchase is to be made shall be furnished the official State agency. In moving these birds from one farm to another, due precautions should be taken to use clean, sanitary coops.

Par. 6. The official test for pullorum disease in the plan shall be one of the following agglutination tests:

(a) The standard tube agglutination test, as described in the Proceedings of the United States Live Stock Sanitary Association, November 30 to December 2, 1932, pages 487 to 491;

(b) The stained-antigen, rapid, whole-blood test[2] as described by Schaffer MacDonald, Hall, and Bunyea, in the Journal of the American Veterinary Medical Association, volume 79 (n. s. 32), No. 2, pages 236-240, 1931; or other such rapid, whole-blood test as may be approved by the Bureau of Animal Industry, United States Department of Agriculture;

(c) The rapid serum test, as described by Runnells, Coon, Farley, and Thorp Journal of the American Veterinary Medical Association, volume 70 (n. s. 23), No. 5, pages 660-662, 1927.

(For convenience of reference the essential features of the three tests are reprinted in the appendix of this plan.) Modifications from these procedures may be made operative by an official State agency upon approval of the Bureau of Animal Industry, United States Department of Agriculture.

Par. 7. All chickens to be used as breeders must be tested for pullorum disease when more than five months of age by means of one of the three official tests. Reactors must be removed from the premises on the completion of the test, and the premises must be immediately carefully cleaned and disinfected to the satisfaction of the official State agency.

Since only birds to be used as breeders are required to be tested, when other birds are kept on the same premises, ample provision must be made to keep these birds and their eggs separate.

---

[2]Complete information on this method of testing is contained in Miscellaneous Publication No. 349, The Use of the Rapid, Whole-Blood Test.

Par. 8. Each lot of antigen used in the whole-blood testing (see par. 6b) for pullorum disease shall be approved by the Bureau of Animal Industry, United States Department of Agriculture.

Par. 9. All domesticated fowl such as turkeys, ducks, geese, pheasants, guineas, pea fowl, bantams, etc., shall either be properly tested to meet the same pullorum tolerance as the breeding flock of chickens, or be segregated from the breeding flock of chickens, at all stages of development and during all seasons of the year.

Par. 10. In the event that tests to qualify or requalify flocks as U. S. Pullorum-Passed or U. S. Pullorum-Clean reveal any reactors, the owner shall have the right to submit, or the official State agency the right to request the submission of all reactors if five or less, or 20 per cent of all reactors with a minimum of five and a maximum of 20, to a laboratory designated by the official State agency for autopsy and bacteriological examination. In case such bacteriological examination fails to reveal any conclusive evidence of pullorum infection, the flock shall be construed as having had no pullorum. All reactors not slaughtered for bacteriological examination shall be removed and disposed of in a manner satisfactory to the official State agency.

Par. 11. In the event of a thorough investigation or the presentation of other evidence which indicates that pullorum disease or fowl typhoid outbreaks are occurring in chicks, the official State agency may, at its discretion, require additional retesting of the flock or flocks, or make further autopsies and bacteriological examinations.

Par. 12. The custom hatching of eggs from non-U. S. Pullorum-Tested flocks is permitted in U. S. Pullorum-Tested and U. S. Pullorum-Controlled hatcheries, provided the incubation and hatching of such eggs are done in a separate room. The custom hatching of eggs from flocks of a lower pullorum class than U. S. Pullorum-Passed is not permitted in a U. S. Pullorum-Passed hatchery. The custom hatching of eggs from flocks of a lower pullorum class than U. S. Pullorum-Clean is not permitted in a U. S. Pullorum-Clean hatchery.

Par. 13. In separating a room for the use of two or more incubators, to provide for custom hatching, a tight partition is absolutely necessary and must be provided. Outside ventilation and an outside entry for each room are required. Any door or opening between the two rooms must be sealed, as by weather stripping and locking, or by battening the cracks around the room where eggs from nonqualified flocks are being incubated or hatched.

Par. 14. In the case of U. S. Pullorum-Controlled, U. S. Pullorum-Passed, and U. S. Pullorum-Clean flocks, all birds which are returned to the premises from poultry exhibits or egg-laying tests or which have been removed from the premises for any purpose whatsoever, shall be quarantined for at least 30 days and officially tested at the time of their return and again at least 30 days later, and any reactors shall be removed immediately after each test. In the event reactors are found such birds shall not be added to a U. S. Pullorum-Clean flock until they have passed two negative tests at least 30 days apart, and shall not be added to a U. S. Pullorum-Passed flock until they have passed one negative test for pullorum disease, each retest being at least 30 days apart.

Par. 15. Hatcheries participating in the pullorum classes only shall be inspected and approved by a State inspector at least once during the hatching season.

Par. 16. Each year at least 15 per cent of the flocks participating in the pullorum classes only shall be inspected and approved by a State inspector. Field testing or blood collecting by a State pullorum tester or a State inspector shall be regarded as an official flock inspection provided a complete inspection is made and so reported.

Par. 17. After a flock has been classified as U. S. Pullorum-Tested, U. S. Pullorum-Controlled, or U. S. Pullorum-Passed by the official State agency, the results of any retesting of such flock during the current breeding and hatching season shall not adversely affect its pullorum classification for that season, except that if in the opinion of the official State agency the amount of reaction found in any flock or flocks is such as to be dangerous or detrimental to the operation of the plan, the official State agency shall require that such flock or flocks be retested after an interval of at least 30 days or that such flocks be immediately discontinued as hatching-egg flocks. Furthermore, in such cases the official State agency may require that the hatching eggs from such flocks be removed form the incubator and destroyed prior to hatching.

Par. 18. When more than one breeding flock of chickens is maintained on the same premises, all such flocks shall be given an official pullorum classification no higher than the lowest classification attained by any one of the flocks.

Par. 19. In the event that eggs other than chicken or turkey eggs are to be incubated or hatched by a hatchery, such eggs shall be either produced by flocks that have been properly tested to meet at least the minimum pullorum tolerance of the hatchery or be hatched in separate rooms in accordance with paragraphs 12 and 13 of this section.

Par. 20. Ducks meeting the requirements for a pullorum classification may be given the same official recognition that is given chickens.

SECTION 2. U. S. PULLORUM-TESTED CLASS

Paragraph. 1. *U. S. Pullorum-Tested flocks.*—These flocks shall meet the following requirements: (a) All chickens to be used as breeders shall be tested for pullorum disease when more than 5 months of age, under the supervision of an official State agency and shall contain fewer than 3 per cent of reactors, the last test being made within 12 months immediately preceding the date of sale of hatching eggs or chicks from such flocks. At the end of the 1948-49 hatching season, the U. S. Pullorum-Tested class shall be deleted from the plan; (b) all indicated carriers of pullorum disease shall be removed from the premises on completion of the test and disposed of in a manner satisfactory to the official State agency; (c) all birds remaining in the flock shall be properly leg-banded; (d) individual birds introduced into U. S. Pullorum-Tested flocks shall have passed, within 12 months, a negative official test for pullorum disease.

Par. 2. A flock containing 3 per cent or more of reactors on the first test on being retested at intervals of not less than 30 days, and all reactors being removed after each test until the percentage of reactors is less than 3, may qualify as a U. S. Pullorum-Tested flock.

Reg. U. S. Pat. Off.

(GREEN)

Fig. 5.—Design authorized to identify U. S. Pullorum-Tested hatching eggs, chicks, flocks and hatcheries.

Par. 3. *U. S. Pullorum-Tested hatching eggs.*—Such eggs are from officially tested birds in U. S. Pullorum-Tested flocks.

Par. 4. *U. S. Pullorum-Tested chicks.*——These are hatched from eggs produced by U. S. Pullorum-Tested, U. S. Pullorum-Controlled, U. S. Pullorum-Passed, or U. S. Pullorum-Clean flocks, in a U. S. Pullorum-Tested hatchery.

Par. 5. *U. S. Pullorum-Tested hatcheries.*—These hatcheries operate under the supervision of an official State agency and (1) buy or sell chicken or turkey hatching eggs; (2) incubate and hatch chicken or turkey eggs; (3) brood or sell chicks or poults; (4) be dealers in or agents for such products from only U. S. Pullorum-Tested, U. S. Pullorum-Controlled, U. S. Pullorum Passed, or U. S. Pullorum-Clean flocks, except that custom hatching may be carried on under the conditions prescribed in article 3, section 1, paragraphs 12 and 13.
key eggs; (3) brood or sell chicks or poults; or (4) be dealers in or agents for such products from only U. S. Pullorum-Tested, U. S. Pullorum-Controlled, U. S. Pullorum Passed, or U. S. Pullorum-Clean flocks, except that custom hatching may be carried on under the conditions prescribed in article 3, section 1, paragraphs 12 and 13.

SECTION 3. U. S. PULLORUM-CONTROLLED CLASS

Paragraph 1. *U. S. Pullorum-Controlled flocks.* — These flocks shall meet the following requirements: (a) All chickens to be used as breeders shall be tested for pullorum disease when more than 5 months of age, under the supervision of an official State agency, and shall contain fewer than 2 per cent of reactors, the last test being made within 12 months immediately preceding the date of sale of hatching eggs or chicks from such flocks; (b) all indicated carriers of pullorum disease shall be removed from the premises on completion of the test and disposed of in a manner satisfactory to the official State agency; (c) all birds remaining in the flocks shall be properly leg-banded; (d) individual birds introduced into U. S. Pullorum-Controlled flocks shall have passed, within 12 months, a negative official test for pullorum disease.

Reg. U. S. Pat. Off.

(YELLOW)

Fig. 6.—Design authorized to identify U. S. Pullorum-Controlled hatching eggs, chicks, flocks, and hatcheries.

Par. 2. A flock containing 2 per cent or more of reactors on the first test on being retested at intervals of not less than 30 days, and all reactors removed after each test until the percentage of reactors is less than 2, may qualify as a U. S. Pullorum-Controlled flock.

Par. 3. *U. S. Pullorum-Controlled hatching eggs.*—Such eggs are from officially tested birds in U. S. Pullorum-Controlled flocks.

Par. 4. *U. S. Pullorum-Controlled chicks.*—These are hatched from eggs produced by U. S. Pullorum-Controlled, U. S. Pullorum-Passed, or U. S. Pullorum-Clean flocks, in a U. S. Pullorum-Controlled hatchery.

Par. 5. *U. S. Pullorum-Controlled hatcheries.*—These hatcheries operate under the supervision of an official State agency and (1) buy or sell chicken or turkey hatching eggs, (2) incubate and hatch chicken or turkey eggs, (3) brood or sell chicks or poults, or (4) be dealers in or agents for such products from only U. S. Pullorum-Controlled, U. S. Pullorum-Passed, or U. S. Pullorum-Clean flocks and hatcheries, except that custom hatching may be carried on under the conditions prescribed in article 3, section 1, paragraphs 12 and 13.

SECTION 4. U. S. PULLORUM-PASSED CLASS

Paragraph 1. *U. S. Pullorum-Passed flocks.*—These flocks shall meet the following requirements: All chickens to be used as breeders shall be tested for pullorum disease when more than 5 months of age under the supervision of an official State agency, shall be officially leg-banded, and shall contain no reactors, the last test being made within the testing year immediately preceding the date of sale of hatching eggs or chicks from such flocks.

Reg. U. S. Pat. Off.

(RED)

Fig. 7.—Design authorized to identify U. S. Pullorum-Passed hatching eggs, chicks, flocks and hatcheries.

Par. 2. *U. S. Pollorum-Passed hatching egg.*—Such eggs are from officially tested birds in U. S. Pullorum-Passed flocks.

Par. 3. *U. S. Pullorum-Passed chicks.*—These are hatched in a U. S. Pullorum-Passed hatchery from eggs produced by U. S. Pullorum-Passed or U. S. Pullorum-Clean flocks.

Par. 4. *U. S. Pullorum-Passed hatcheries.*—These hatcheries operate under the supervision of an official State agency and (1) buy or sell chicken- or turkey-hatching eggs, (2) incubate and hatch chicken or turkey eggs, (3) brood or sell chicks or poults, or (4) be dealers in or agents for such products from only U. S. Pullorum-Passed or U. S. Pullorum-Clean flocks and hatcheries.

SECTION 5. U. S. PULLORUM-CLEAN CLASS

Paragraph 1. *U. S. Pullorum-Clean flocks.*—These flocks shall meet the following requirements: (a) All chickens to be used as breeders shall be tested for pullorum disease when more than 5 months of age, under the supervision of an official State agency, shall be officially leg-banded, and shall contain no reactors either in two consecutive tests not less than 6 months apart, or in three consecutive tests no less than 30 days apart, the last being made within the testing year immediately preceding the date of sale of hatching eggs or chicks from such flocks; (b) once a flock is established as U. S. Pullorum-Clean it remains so as long as no reactors are found in the official annual test of birds used as breeders; (c) flocks developed exclusively from U. S. Pullorum-Passed or U. S. Pullorum-Clean chicks may be recognized as U. S. Pullorum-Clean flocks on one annual test conducted under supervision of the official State agency if no reactors are found; (d) in order to retain a U. S. Pullorum-Clean classification all purchases of baby chicks must be from U. S. Pullorum-Passed or U. S. Pullorum-Clean sources. This requirement applies to chicks to be raised for meat production purposes as well as for breeding stock.

Par. 2. When not more than one-half of 1 per cent, but in no case not more than four reactors are found in a flock having a U. S. Pullorum-Clean rating, one-retest with no reactors of the unit or units involved followed by a retest with no reactors of the entire breeding flock may be used at the discretion of the official State agency in requalifying a flock for the U. S. Pullorum-Clean class. The retest shall be made at intervals of not less than 30 days, nor more than 6 weeks apart. Such a flock shall temporarily lose its status as U. S. Pullorum-Clean and, until requalification, chicks or hatching eggs from the flock shall not be represented to be U. S. Pullorum-Clean.

Reg. U. S. Pat. Off.

(BLUE)

Fig. 8.—Design authorized to identify U. S. Pullorum-Clean hatching eggs, chicks, flocks, and hatcheries.

Par. 3. *U. S. Pullorum-Clean hatching eggs.*—Such eggs are from officially tested birds in U. S. Pullorum-Clean flocks.

Par. 4. *U. S. Pullorum-Clean chicks.*—These are hatched in a U. S. Pullorum-Clean hatchery from eggs produced by U. S. Pullorum-Clean flocks.

Par. 5. *U. S. Pullorum-Clean hatcheries.*

—These hatcheries operate under the supervision of an official State agency and (1) buy or sell chicken- or turkey-hatching eggs, (2) incubate and hatch chicken or turkey eggs, (3) brood or sell chicks or poults, or (4) be dealers in or agents for such products from only U. S. Pullorum-Clean flocks or hatcheries.

Fig. 122—Buff Orpingtons.
[Courtesy Poultry Tribune.]

Summary of the National Poultry Improvement Plan

OBJECTIVES OF THE PLAN

To improve the production and breeding qualities of poultry.

To reduce the mortality of chicks from pullorum disease.

To identify, authoritatively, breeding stock, hatching eggs, and chicks, with respect to quality, by describing them in terms uniformily accepted in all parts of the United States.

To serve as a means through which scientific research may be applied for the improvement of poultry and poultry products.

The National Turkey Improvement Plan[1]

In 1937 a tentative turkey-improvement plan was developed by a National Turkey Improvement Committee. That plan was intended as a guide for State turkey-improvement work until

[1]The turkey improvement plan was originally described in a mimeographed statement issued in October 1943. It has been revised annually at meetings of official delegates, representatives of the National Turkey Federation, and representatives of the Bureau of Animal Industry.

such time as there was sufficient demand to justify the formulation of a National Turkey Improvement Plan.

In 1941 the National Turkey Federation appointed a special committee to study State programs and to develop a turkey-improvement plan that could be applied throughout the country. The plan developed by the committee was presented at the 1943 annual meeting of the Federation and unanimously adopted by the general session.

The National Turkey Improvement Plan became operative September 25, 1943, with the approval of the Secretary of Agriculture and under authority of an appropriation made by Congress for the Bureau of Animal Industry, United States Department of Agriculture, to be used in cooperation with State authorities in the administration of regulations for the improvement of poultry, poultry products, and hatcheries. Authority for the administration of the plan is now contained in the Department of Agriculture Organic Act 1944. (Public 425, 78th Congress, H. R. 4278.)

Acceptance of the turkey plan is optional with States and individual members of the industry within States. It is administered in each State by an official State agency cooperating with the Bureau of Animal Industry, United States Department of Agriculture.

The turkey plan will require revision from time to time to conform with the development of the industry and with new information as it becomes available. Revisions will be considered at meetings of authorized representatives of the turkey breeders and hatcheries with representatives of the Bureau of Animal Industry.

The primary objectives of the National Turkey Improvement Plan are to improve the production and market quality of turkeys and to reduce losses from disease. Through it turkey breeding stock, hatching eggs, and poults may be identified, authoritatively, with respect to breeding quality and degree of freedom from pullorum disease. The application of the provisions of the plan should cause turkey production to be more efficient and thus aid in making the enterprise more profitable.

Article 1. General Regulations

REGULATION 1.—For the purpose of this plan the following words, phrases, names, and terms shall be construed, respectively, to mean—

Paragraph 1. *Turkey plan.*—The National Turkey Improvement Plan.

Par. 2. *Chicken plan.*—The National Poultry Improvement Plan.

Par. 3. *Official State agency.*—The agency recognized by the Bureau of Animal Industry, United States Department of Agriculture, to cooperate in the administration of the turkey plan within the State. It may be the State department of agriculture, State college of agriculture, State poultry improvement board or association, or other agency recognized officially by the State government.

Par. 4. *State inspector.*—An employee or authorized representative of the official State agency whose duties in connection

Fig. 123—Australorps.
[Courtesy Poultry Tribune.]

with the turkey plan comprise the official inspection of flocks, eggs, poults, hatcheries, turkey premises, records, and methods employed in carrying on the work of the turkey plan

Par. 5. *R. O. P. supervisor.*—An employee or authorized representative of the official State agency whose duties in connection with the turkey plan are to represent the official State agency in its supervision of U. S. R. O. P. (U. S. Record of Performance) and U. S. R. O. M. (U. S. Register of Merit) work and to vouch for the accuracy of all U. S. R O. P. and U. S.

R. O. M. records made by the U. S. R. O. P. breeders under his jurisdiction.

Par 6. *R. O. P. inspector.*—An employee or authorized representative of the official State agency whose duties in connection with the turkey plan are the inspection of flocks and records of U. S. R. O. P. turkey breeders.

Par. 7. *Flock-selecting agent.*—A person who has taken a course of training in and passed an examination on (1) the selection of turkey breeding stock according to standardbred and meat-production qualities as prescribed by the State college of agriculture or other properly constituted State agency and

Fig. 124—Barred Plymouth Rocks.
[Courtesy Poultry Tribune.]

(2) the turkey-plan provisions as applied in the State, and who has been authorized by the official State agency to do the flock-selecting work as provided for in the breeding stages of the turkey plan. As an emergency and to take care of individual cases only, the official State agency may, at its discretion, authorize flock-selecting agents who have not taken the specified course of training. Only those persons who have demonstrated to the official State agency through work in the field that they are capable of doing satisfactory selecting work may be so authorized.

Par. 8. *State pullorum tester.*—An employee or authorized representative of the official State agency, who has training in pullorum-testing technique, a working knowledge of the plan provisions as applied in the State, and other qualifications necessary to do the pullorum-testing work as provided for in the plan; however, no person can qualify as a State pullorum tester for work in a hatchery in which he has a financial interest, either entire ownership, partial ownership, or employment.

Par. 9. *Pullorum-testing agent.*—A person who has taken a course of training in and passed an examination on (1) pullorum testing as prescribed by the State livestock sanitary authorities or officials of the State college of agriculture and (2) the plan provisions as applied in the State, and who has been authorized by the official State agency to do the pullorum-testing work as provided for in the U. S. Pullorum-Controlled and U. S. Pullorum-Passed classes. As an emergency and to take care of individual cases only, the official State agency may, at its discretion, authorize pullorum-testing agents who have not taken the specified course of training. Only those persons who have demonstrated to the official State agency through work in the field that they are capable of doing satisfactory testing work may be so authorized.

Par. 10. *Blood collector.*—A person who has been trained in the proper method of collecting blood samples for the pullorum disease test, who has passed an examination on the turkey plan provisions as applied in the State, and who has been authorized by the official State agency to do the blood-collecting work.

Par. 11. *Eggs and hatching eggs.*—Turkey eggs and turkey hatching eggs.

Par. 12. *Baby poults.*—Poults that have not been fed or watered.

Par. 13. *Started poults.*—Poults that have been fed or watered.

Par. 14. *Breeding stock.*—Turkeys that are to be used for breeding purposes.

Par. 15. *Flock.*—All turkeys of one variety on the same premises except that crossmated birds as provided in Regulation 9 shall be construed as one flock.

Par. 16. *Flock owner.*—A person, firm, or corporation maintaining a breeding flock of turkeys.

Par. 17. *Hatchery.*—Hatchery equipment on one premises operated or controlled by any person, firm, or corporation producing poults.

Par. 18. *U. S. R. O. P.*—U. S. Record of Performance.

Par. 19. *U. S. R. O. M.*—U. S. Register of Merit.

Par. 20. *Official label.*—A label, obtained through or approved by the official State agency, on which are indicated the product, the breeding stage, and the pullorum class.

Par. 21. *Standard varieties.*—Varieties of turkeys recognized by the American Poultry Association.

Par. 22. *Small varieties.*—Varieties of turkeys with weights below those recognized by the American Poultry Association for standard varieties and developed primarily to provide for small-family trade.

Par. 23. *Broad-Breasted Bronze.*—A variety of turkey developed especially for large size and extremely wide breast. In order to qualify as a Broad Breasted Bronze in the turkey plan, the width of the breast at a point 1¾ inches above the keel shall be at least 3½ inches for both toms and hens prior to the time the birds are 30 weeks of age.

Par. 24. *Crossbred poults and hatching eggs.*—Poults and hatching eggs produced from a mating of toms of one variety with hens of another variety.

Par. 25.—*Hatchability.*—A percentage figure indicating the proportion of poults hatched of the total (not fertile only) eggs set.

REGULATION 2.—The official State agency shall conduct efficiently the inspection work called for in the turkey plan and shall direct, administer, supervise, and be responsible for all the work done relative to flock selection in the various breeding stages and to testing for pullorum disease as provided for in the turkey plan. The Bureau of Animal Industry shall have the right to terminate, after opportunity for informal hearing, in accordance with its memorandum of understanding with any official State agency, its cooperation with such agency if in the opinion of the Bureau the agency is not properly administering the National Turkey Improvement Plan.

REGULATION 3.—Turkey-breeding houses and pens shall be kept clean and in good condition. The land in the immediate vicinity of the turkey-breeding houses and pens shall be kept in sanitary condition.

REGULATION 4.—Hatcheries and all equipment shall be kept in strictly sanitary condition. All incubators must be thoroughly cleaned and disinfected by spraying or fumigating after each hatch, by means of an officially recognized method.

REGULATION 5.— All eggs used in any breeding stage of the turkey plan shall be reasonably uniform in size and shape, sound in shell, and shall weigh at least 2½ ounces each, except that a minimum of 2 1/6 ounces each or 26 ounces to the dozen shall apply to the Beltsville Small White and other small varieties. Nonpedigreed hatching eggs shall be trayed and the poults boxed with a view to uniformity of size.

REGULATION 6.—The management of a hatchery participating in the turkey plan shall maintain the identity of hatching eggs being incubated, including those which are being custom-hatched, and shall keep accurate records concerning the name and address of each flock owner; the number of eggs received from each flock owner; the name and address of each purchaser; the number, hatchery grade, if any, variety, and date of shipment or delivery of all poults and hatching eggs. These records shall be open for official inspection at all times. Eggs received and poults hatched shall be subject to inspection at all times.

REGULATION 7.—All poults sold or offered for sale under the turkey-plan terminology shall be normal and representative of the variety.

REGULATION 8.—Shipment of poults shall be made in approved types of poult boxes properly ventilated and securely tied, and poults shall be shipped to reach the ultimate purchaser within the time limit prescribed by the United States Postal Regulations.

REGULATION 9.—Only one variety of turkeys shall be kept on a farm unless, where more than one variety is kept, ample provision has been made to keep these varieties and their eggs separate, in a manner satisfactory to the official State agency. With the approval of the official State agency, crossbred hatching eggs and poults may be produced and sold by participants in the turkey plan, provided the parent stock is selected and banded, and the flock is inspected as required in the U. S. Approved breeding stage. These flocks shall be designated as U. S. Approved Crossmated flocks. Hatching eggs or poults from such flocks shall be designated as U. S. Approved Crossbred hatching eggs or poults, with a descriptive phase designating the variety of the toms and the variety of the hens used in the cross. (The variety of the tom shall be designated first in all crosses.)

REGULATION 10.—When more than one hatchery located within a State is operated under the same name, ownership, or management, one or more of these hatcheries may not participate in the turkey plan unless all of them participate in the same breeding stage and pullorum class. In the event that any person, firm, or corporation has nonparticipating hatcheries in other States no poults or hatching eggs shall be shipped to the participating hatcheries from the non-participating hatcheries.

REGULATION 11.—Breeding stock and each bundle or package of hatching eggs and poults sold or offered for sale by participants shall be designated by official labels or tags specifying the breeding stage and pullorum class, except that poults that are seconds, culls, rejects, or abnormal for variety characteristics shall not be sold or offered for sale under official labels. (See regulations 7 and 21.) With the approval of the official State agency, hatching egg supply flock owners of a participating hatchery need not officially label each consignment of hatching eggs delivered to the hatchery.

REGULATION 12.—All advertising must be in accord with the rules and regulations of the turkey plan and the Trade Practice Rules of the Federal Trade Commission for the baby chick industry. Copies of proposed folders, catalogs, circulars, or other literature shall be submitted to and approved by the official State agency before being printed. Copies of such literature, after being printed, shall be filed with the official State agency. Copies of all publication advertising shall be mailed to the official State agency at or before the time these advertisements are submitted to the publisher or publishers. The advertising literature of hatcheries which have places of business in more than one State shall be approved by the official State agency in which the main or home office is located after being checked by the Bureau of Animal Industry, except that local advertising literature for each State, branch or branches, shall be approved by the official State agency of each such State in which branches are located. Since U. S. R. O. P. and U. S. R. O. M. are generally understood to apply to chickens, a participant in the turkey plan who uses these terms shall prefix them with the word "turkey" in letters of the same size of type (Refer also to regulation 15 for proper use of terms.)

REGULATION 13.—The National Turkey Improvement Plan emblem (see cover page) and the official terms and various designs illustrative of the breeding stages and pullorum classes (figs. 1-7) may be used only by participants who have met the specific requirements. Whenever the National Turkey Improvement Plan emblem or other reference to participation in the plan is used in advertising, the breeding stage and pullorum classification shall be shown.

REGULATION 14.—Any person, firm, or corporation desiring to participate in the turkey plan, except flock owners who are affiliated with participating hatcheries, shall sign an agreement with the official State agency and shall not purchase, receive, offer for sale, or be a dealer in or agent for breeding stock, hatching eggs, or poults other than those produced under the turkey plan, except with permission from the official State agency. In certain cases special permission may be granted by the official State agency, in writing, to purchase hatching eggs, poults, or breeding stock not produced under the turkey plan for use in breeding or hatching-egg supply flocks. In such cases a report shall be made to the official State agency on each shipment received and to whom delivered.

REGULATION 15.—Flocks, hatching eggs, hatcheries, and poults that have met the requirements of both a breeding stage and pullorum class of the turkey plan shall be designated in accordance with the stage and class attained as follows: U. S. Approved Pullorum-Controlled; U. S. Approved, Pullorum-Passed; U. S. Approved, Pullorum-Clean; U. S. Certified, Pullorum-Controlled; U. S. Certified, Pullorum-Passed; and U. S. Certified, Pullorum-Clean.

Matings, hatching eggs, and poults that have met the requirements of one of the higher breeding stages and a pullorum class of the turkey plan shall be designated in accordance with the class attained as follows: U. S. R. O. P., Pullorum-Controlled; U. S. R. O. P., Pullorum-Passed; U. S. R. O. P., Pullorum-Clean; U. S. R. O. M., Pullorum-Controlled; U. S. R. O. M., Pullorum-Passed; and U. S. R. O. M., Pullorum-Clean.

Where descriptive terms are used in advertising, in addition to the official terminology of the turkey plan, such terms shall be clearly explained.

REGULATION 16.—Standard poults shall be labeled as such, and all the poults in each lot sold or offered for sale shall be of the same age (hatching date) unless otherwise specified.

REGULATION 17.—With the approval of the official State agency, hatching-egg and poult dealers may participate in the turkey plan and enjoy all the privileges of participating hatcheries in the handling of hatching eggs and poults, provided all the applicable requirements for participating hatcheries are fully and completely complied with.

REGULATION 18.—Noncompliance with the provisions of the plan which is not corrected within a reasonable time shall bar any participant from further participation for a period of 2 years. Such action shall not be taken until a thorough investigation and hearing have been conducted by the official State agency.

REGULATION 19.—Participation in one of the pullorum classes shall be a prerequisite to participation in any of the breeding stages of the turkey plan, except that an R. O. P. breeder who fails to qualify for one of the pullorum classes may continue to participate in the R. O. P. breeding stage so long as he satisfies the official State agency that he is proceeding as rapidly as possible to qualify for one of the pullorum classes. During this period each sale of hatching eggs, poults, and breeding stock must first be approved by the official State agency.

REGULATION 20.—Participation in any phase of the turkey plan shall be construed as maintaining supervision and compliance throughout the 12 months of the year. (It is understood that a participant may withdraw from participation or that his agreement may be terminated by the official State agency for cause.)

REGULATION 21.—Poults and hatching eggs produced under the rules and regulations of the turkey plan shall lose their identity as of the turkey-plan terminology when they are purchased for resale by non-participants.

REGULATION 22.—The following body weights by ages for various varieties should be considered as suggested minimums or maximums in all breeding stages (table 1):

REGULATION 23.—The selection of birds in a flock to qualify as U. S. Approved or U. S. Certified by a State inspector shall be regarded as one official inspection, provided a complete inspection is made and so reported.

REGULATION 24.—The term, "broad-breasted" shall not be used in describing any variety of turkey unless the width of the breast at a point 1¾ inches above the keel is at least 3½ inches for both toms and hens prior to the time the birds are 30 weeks of age.

REGULATION 25.—All applicants for participation in the National Turkey Improvement Plan shall demonstrate to the satisfaction of their official State agency, prior to their admission to such participation by the agency, that their facilities, personnel, and practices are adequate for carrying out the applicable provisions of the National Turkey Improvement Plan.

Article 2. Breeding Stages

SECTION 1. U. S. APPROVED BREEDING STAGE

Paragraph 1. *U. S. Approved flocks.*—These flocks shall meet the following requirements: Toms and hens shall be carefully selected once each year for constitutional vigor and standardbred qualities to a reasonably high degree. The following specifications should be emphasized:

TABLE I

SUGGESTED MAXIMUM BODY WEIGHTS FOR STANDARD AND BELTSVILLE SMALL WHITE VARIETIES AND SUGGESTED MAXIMUM BODY WEIGHTS FOR SMALL VARIETIES OF TURKEYS

Sex and age	Minimum body weight				Maximum body weight	
	Broad-Breasted Bronze	Standard varieties other than Bronze	Standard Bronze	Beltsville Small White	Beltsville Small White	Other small varieties
	Pounds	Pounds	Pounds	Pounds	Pounds	Pounds
Toms:						
24 weeks	18	14	15	12	16	17.5
26 weeks	20	15.5	16.5	13.5	18	19.5
28 weeks	22	17	18	15	20	21.5
30 weeks	23	18	19	16	22	23.5
32 weeks	24	19	20	17	23	24.5
Hens:						
24 weeks	12	8.5	9.5	7	10	11
26 weeks	12.5	9	10	7.5	10.5	11.5
28 weeks	13	9.5	10.5	8	11	12
30 weeks	13.5	10	11	8.5	12	13
32 weeks	14	10.5	11.5	9	12.5	13.5

(a) Conformation: Body should be deep and relatively rectangular in shape, flesh of breast well developed, giving this region a rounded fullness, in contrast to sharp-keeled birds lacking in muscular development. The front point of the keel should be well fleshed and free from coarseness. The back should be flat and wide over its entire length. Leg bones should be sturdy, of medium length, and without coarseness. Normal turkey carriage, or posture (body carried at an angle of 35° to 45° from the horizontal) is important as are sturdy, strong legs placed squarely under the body.

(GREEN)

Fig. 1—Design used to identify U. S. Approved flocks, eggs, poults, and hatcheries.

(b) Shape of breastbone: Breastbone should be straight, relatively long, and parallel with the back. It should extend well up between the lower thighs (drumsticks) as the bird is suspended by the legs.

(c) Rate of maturity: Birds showing definite indications of late maturity should not be approved.

Par. 2. The method of selecting turkey breeding stock for standardbred and meat-production qualities prescribed by the State college of agriculture or other properly constituted State agency shall serve as a basis for the selection of toms and hens. Recommendations of special committees of the National Turkey Federation and the American Poultry Association on turkey type, weight, and other variety characteristics may properly be used as guides in developing and improving standards of selection. Emphasis at all times should be placed upon uniformity as regards conformation, size, color, carriage, and appearance of the birds remaining in the breeding flock after selection is completed. All birds to be used in U. S. Approved breeding flocks shall be selected by a State inspector or by a flock-selecting agent.

Par. 3. All birds selected for U. S. Approved breeding flocks shall be banded at the time of selection, with sealed and numbered leg or wing bands obtained from the official State agency. Birds rejected by the State inspector or the flock-selecting agent shall have leg and wing bands removed, and shall be kept separate from the breeding flocks. The selection of birds for U. S. Approved flocks shall be completed prior to March 1, and at least 10 days before any eggs are saved for hatching purposes, except that this 10-day interval shall not be required if all the toms found in the breeding flock are approved. No birds shall be selected and approved earlier than at 22 weeks of age.

Par. 4. Before any flock other than one affiliated with a participating hatchery may be recognized as a U. S. Approved flock, it must be inspected and approved by a State inspector. Each year at least one-third of the hatchery flocks supplying eggs to a U. S. Approved hatchery and at least one-third of the flocks selected by each flock-selecting agent shall be inspected and approved by a State inspector. The inspection of a flock by a State inspector shall consist in the physical examination, by handling, of a sufficient number of toms and hens to satisfy him that the flock-selection work performed meets the requirements of the U. S. Approved breeding stage.

Par. 5. No birds except those with official sealed and numbered leg or wing bands shall be allowed in the breeding flocks, and no toms shall be allowed with any other hens of breeding age on the same premises except in U. S. R. O. P., U. S. R. O. M., or candidate matings

Par. 6. *U. S. Approved hatching eggs.*—Such eggs shall come from U. S. Approved flocks.

Par. 7. *U. S. Approved poults.*—They shall be hatched from U. S. Approved hatching eggs and in U. S. Approved hatcheries.

Par. 8. *U. S. Approved hatcheries.*—These hatcheries shall be inspected and approved by a State inspector at least once during the hatching season. They may not incubate, hatch for sale, or sell hatching eggs and poults other than those produced under the breeding stages of the turkey plan. U. S. Approved hatcheries may hatch for sale and sell both U. S. Approved and U. S. Certified poults of the same variety, provided adequate segregation of eggs and poults is practiced and ample records are kept to satisfy the official State agency that there is no mixing or interchanging from one breeding stage to another.

SECTION 2. U. S. CERTIFIED BREEDING STAGE

Paragraph 1. *U. S. Certified flocks.*—These flocks shall fulfill all the requirements of U. S. Approved flocks and shall be free from disqualifications. The hens shall be mated to U. S. R. O. P. toms or to sons of U. S. R. O. M. matings. The toms shall be selected by a State inspector and the hens by a State inspector or a specially authorized flock-selecting agent. All flocks shall be inspected and approved by a State inspector at least once during the breeding season. After the second year of U. S. Certified breeding work on any farm, hens to be used as breeders shall be from U. S. Certified flocks or from U. S. R. O. P. or U. S. R. O. M. matings.

Par. 2. *U. S. Certified hatching eggs.*—Such eggs shall come from U. S. Certified flocks.

Par. 3. *U. S. Certified poults.*—They shall be hatched from U. S. Certified eggs in U. S. Certified hatcheries or in U. S. Approved hatcheries. When hatched in U. S. Approved hatcheries adequate segregation of eggs and poults must be provided and ample records must be kept to satisfy the official State agency that there is no mixing or interchanging from one breeding stage to another.

(YELLOW)

Fig. 2—Design used to identify U. S. Certified flocks, eggs, poults, and hatcheries.

Par. 4. *U. S. Certified hatcheries.*—These hatcheries may hatch for sale and sell U. S. Certified eggs or poults and eggs and poults from U. S. R. O. P. and U. S. R. O. M. matings only. They shall fulfill all the requirements of U. S. Approved hatcheries.

SECTION 3. U. S. RECORD OF PERFORMANCE BREEDING STAGE

Paragraph 1. U. S. Record of Performance embraces records of egg production, hatchability, and body weight made on the breeder's premises under official supervision and similar records made at State colleges of agriculture, and State and Federal experiment stations, when such records are passed on by the official State agency and when the individual birds meet other U. S. R. O. P. requirements. The birds produced by State colleges of agriculture and State and Federal experiment stations may recognized as U. S. R. O. P. only for the use of U. S. R. O. P. breeders in their breeding work. The primary objective of this and the U. S. R. O. M. breeding stage is to produce young toms to head U. S. Certified flocks. When any disastrous event occurs that affects the pedigree breeding and progeny-testing work of any advanced turkey breeder, and when such event is immediately brought to the attention of his official State agency, this agency may, with the consent of the Bureau of Animal Industry, make an equitable revision of the qualifying requirements.

Par. 2. The trap-nest used in U. S. R. O. P. work shall be of an approved type and shall be kept in perfect working order at all times. The breeding farm and all egg-production records shall be subject at all times to unannounced inspections by a State inspector. At the time of each inspection, the inspector shall have sole charge of the trap nests and of eggs laid. Each U. S. R. O. P. candidate shall be banded with a sealed and numbered official leg band. No hens other than those entered in U. S. R. O. P. shall be allowed in the pens where U. S. R. O. P. candidates are being trap-nested, except with the approval of the official State agency.

Par. 3. U. S. R. O. P. eggs and poults may be produced only on a U. S. R. O. P. breeder's premises except that with the approval of the official State agency (a) a U. S. R. O. P. breeder may have eggs hatched by any hatchery operating under the turkey plan provided the U. S. R. O. P. breeder does the pedigreeing of the poults; and (b) a U. S. Approved or U. S. Certified hatchery owner may produce U. S. R. O. P. poults from purchased U. S. R. O. P. eggs for the use of his hatchery-flock owners only, provided proper records of purchases are sent to the official State agency and a State inspector does the wing-banding of these poults.

Par. 4. When U. S. R. O. P. hatching eggs are transferred from one U. S. R. O. P. breeder to another U. S. R. O. P. breeder or hatcheryman for the purpose of hatching U. S. R. O. P. poults, the conditions that shall govern such a transaction include the following: (a) The U. S. R. O. P. breeder who sells the U. S. R. O. P. hatching eggs must send a report in triplicate at the time the eggs are shipped consisting of a list of the eggs sold, the pen or sire number and dam number of each egg, and the egg production, hatchability, and body weight of each dam, to his R. O. P. supervisor. If the eggs are shipped outside the State, these records shall be attested to by the R. O. P. supervisor, who shall retain one copy and mail one copy to the purchaser and one to the official State agency of the State to which the eggs are shipped. (b) Within 5 days after incubation has begun, the purchasing U. S. R. O. P. breeder of hatcheryman must submit

(BLUE)

Fig. 3—Design used to identify U. S. R. O. P. hatching eggs, poults, and breeding stock.

to his official State agency a list of the U. S. R. O. P. eggs bought. (c) Within 5 days after hatching time the U. S. R. O. P. breeder or hatcheryman who purchased the eggs must send to his official State agency a list of the poults hatched from each dam and their respective wing-band numbers.

Par. 5. When U. S. R. O. P. poults, young toms, or mature birds are sold or transferred by members of the industry participating in the turkey plan, the individual making the sale must send a report in triplicate to his R. O. P. supervisor at the time these products are shipped giving the number of birds sold; the leg- or wing-band number, egg production, hatchability, and body weight of each dam, and the pen or leg-band number of the sire of each poult, young tom, or mature bird. If the poults, young toms, or mature birds are shipped outside the State, these records shall be attested to the R. O. P. supervisor, who shall retain one copy and mail one copy to the purchaser and one to the official State agency of the State to which these products are shipped.

Par. 6. To be eligible for U. S. R. O. P. entry, all hens shall have been raised from U. S. R. O. M. matings, U. S. R. O. P. matings, or U. S. Certified flocks or other flocks of breeding satisfactory to the official State agency and the individual birds must meet the body type requirements prescribed by the official State agency. Eligibility shall be confined to any standard variety or other variety recognized by the official State agency.

Par. 7. All hens trap-nested on the premises of a U. S. R. O. P. breeder shall be considered as entered under U. S. R. O. P. supervision with the following exceptions: (a) All qualified U. S. R. O. P. hens may be trap-nested. (b) He may, with the approval of the official State agency, trap-nest any other variety which he is developing as a basis for future U. S. R. O. P. work. (c) With the approval of the official State agency, hens in their second or subsequent production years may be privately trapped to complete records that were begun before U. S. R. O. P. work was undertaken.

Par. 8. All matings shall be inspected at least two times each year. One or more consecutive days of inspection by the R. O. P. inspector shall represent a single inspection. In the event of an outbreak of contagious disease on a turkey farm where U. S. R. O. P. work is carried on, inspection may be temporarily suspended.

Par. 9. *Duties of R. O. P. supervisor.*—The R. O. P. supervisor shall represent the official State agency in its supervision of U. S. R. O. P. work, and he shall vouch for the accuracy of all U. S. R. O. P. records made by authorized U. S. R. O. P. breeders under his jurisdiction. He shall keep on file in his office at least five years, (a) a record of all qualified toms, with at least one-generation pedigree showing U. S. R. O. P. records of the female ancestors, and (b) a record of all qualified hens with their U. S. R. O. P. records. He shall also keep on file a record of all U. S. R. O. P. matings and all U. S. R. O. P. poults with at least one-generation pedigree.

Par.10. The R.O.P. supervisor shall furnish the Bureau of Animal Industry, for publication, a summary of the U. S. R. O. P. and U. S. R. O. M. work under his supervision each year, which shall include the following information for each flock:

For U. S. R. O. P. candidates:

(a) Total number of young hens of that particular variety on the farm.

(b) Number of U. S. R. O. P. candidate matings.

(c) Number of young hens entered in U. S. R. O. P.

(d) Number of young hens meeting the U. S. R. O. P. requirements.

(e) Percentage of young hens entered meeting the U. S. R. O. P. requirements.

(f) Average rate of (percentage) egg production of all candidate hens.

(g) Average body weight of all the candidate hens.

(h) Average hatchability of all eggs set from all the candidate hens.

For U. S. R. O. M. candidates:

(a) Number of U. S. R. O. M. candidate matings.

(b) Number of young hens entered in U. S. R. O. M.

(c) Number and percentage of young toms entered meeting U. S. R. O. M. requirements.

(d) Number and percentage of young hens entered meeting the U. S. R. O. M. requirements.

(e) Average rate of (percentage) egg production of all candidate hens.

(f) Average body weight of all the candidate hens and toms.

(g) Average hatchability of all eggs set from all the candidate hens.

(h) Average viability of the progeny from all candidate toms and hens.

(i) Percentage of Grade A progeny from all candidate toms and hens.

Par. 11. The R. O. P. supervisor may bring charges to the official State agency of his State against any U. S. R. O. P. breeder for falsifying records, for unethical conduct in his business methods, or for any other cause which he may consider detrimental to U S. R. O. P. work. On such charges being proved in a fair hearing the offending breeder must be reprimanded or suspended by the official State agency according to the nature of the offense. (See Regulation 18.)

Par. 12. *Duties of R. O. P. inspector.*—The R. O. P. inspector shall work under direction of the R. O P. supervisor. It shall be his duty to visit and inspect the work of each U. S. R. O. P. breeder at least three times each year, his visits to be unannounced, and two of which shall be during the trap-nesting period. On each visit during the trap-nesting period he shall do the trap-nesting for the day and shall record all eggs laid by each individual bird entered in U. S. R. O. P. He shall examine hens apparently out of production to see that they are not being credited with eggs. On at least one visit he shall examine all birds in U. S. R. O. P. and U. S. R. O. P. candidate matings to see that the birds which constitute these matings are properly listed with the supervisor. He shall compare the number of eggs being incubated from each hen with the number she is credited with having laid during the corresponding period, and shall have authority to test those eggs for fertility. When the hatch is reported he shall compare his findings with the number of poults reported hatched and wing-banded. He shall be courteous to breeders at all times, but shall be on the alert to discover errors in records or violations of rules, intentional or otherwise. He shall immediately report to the supervisor any suspicious circumstances or appearance of fraud in the records or business dealings of a U. S. R. O. P. breeder.

Par. 13. *U. S. R. O. P. turkey breeder.*—A U. S. R. O. P. turkey breeder is a person, firm, or corporation maintaining turkey matings in accordance with the provisions of the U. S. R. O. P. breeding stage. In addition to the U. S. R. O. P. and U. S. R. O. P.

candidate matings, all other mated hens supplying hatching eggs to such turkey breeder or to any hatchery owned or operated by him shall qualify for one of the breeding stages of the turkey plan.

Par. 14. Each candidate hen of a U. S. R. O. P. breeder shall be trap-nested at regular intervals every day for a period of at least 8 consecutive weeks from the date the first normal egg is laid in a trap-nest. The period for measuring hatchability shall be a minimum of 8 consecutive weeks by individual hens, pens, or farms agreed upon during the hatching season by the supervisor and the breeder.

Par. 15. The breeder shall send to his supervisor, not later than the tenth day of each month, a complete report of all trap-nesting for the previous month. He shall also send to his supervisor a complete record of all pedigreed eggs set, poults hatched, and poults banded on N. P. I. P. Form 8 or a substitute form at the end of each four hatches. The egg production, pedigree, and sales records shall at all times be open to examination by the R. O. P. supervisor or inspector.

Par. 16. *U. S. R. O. P. hens.*—In order to qualify for U. S. R. O. P., hens shall meet the following requirements:

(a) Be trap-nested in a single-tom mating under official supervision as specified in paragraph 14 of this section.

(b) Produce eggs at the rate of at least 50 per cent during the first 8 weeks of trap-nesting.

(c) Produce eggs that hatch at the rate of 70 per cent or more of all eggs set, provided that the official percentage hatchability shall be based on all the normal eggs produced by the candidate during the period for measuring hatchability and a minimum of 20 poults. At altitudes of 3,000 to 3,499 feet the minimum qualifying hatchability percentage shall be 65 and at altitudes of 3,500 feet or more, the minimum qualifying hatchability percentage shall be 60. All poults hatched shall be counted. The official percentage of hatchability shall be expressed as a whole number.

To arrive at a whole number with which to express the average, round down on all decimals less than 0.50 and round up on all decimals of 0.50 or more; e.g., where the calculated average is 69.49 per cent, the official average shall be reported as 69 and where the average is 69.50 or 69.51, the official average shall be reported as 70 per cent.

(d) Be a reasonably good representative of the variety in the judgment of the R. O. P. inspector and be of at least qualifying body weight for young hens (due allowance being given to the 6-pound tolerance), as set forth in the latest revision of the American Standard of Perfection. For varieties not included in the Standard of Perfection, the type and qualifying weight shall be established by the official State agency. The official body weight shall be expressed as a whole number and shall be taken by the breeder or the R. O. P. inspector at or about the time the candidate begins egg production, provided that when the weights are taken by the breeder they shall be checked by the R. O. P. inspector at the time of his next inspection visit to the flock. In checking weights a sufficient number of candidates shall be weighed by the R. O. P. inspector to satisfy him that the weights taken by the breeder are correct.

Par. 17. All hens meeting the above requirements shall be considered U. S. R. O. P. hens. Before such hens may be sold as U. S. R O. P. hens or used in U. S. R. O. P. matings they shall be examined and banded by the R. O. P. supervisor or inspector with a U. S. R. O. P. sealed official leg band.

Par. 18. The U. S. R. O. P. breeder and the R. O. P. inspector shall record the leg-band numbers of all U. S. R. O. P. candidates showing a disqualification or laying eggs of undesirable shape or shell texture. Such hens shall not qualify as U. S. R. O. P. hens. The R. O. P. inspector may remove the U. S. R. O. P. leg band from any hen which develops these disqualifications or defects subsequent to qualification.

Par. 19. *U. S. R. O. P. toms.*—Toms produced from U. S. R. O. P. poults shall qualify as U. S. R. O. P. toms provided they shall have passed the final individual inspection of a State inspector not earlier than at 22 weeks of age. Young toms from U. S. R. O. P. candidate matings are eligible for qualifications as U. S. R. O. P. toms after their dams qualify as U. S. R. O. P. hens. Each tom passed by an inspector shall possess strong constitutional vigor and be a good representative of the variety, shall be free from disqualifications, and shall be banded with a U. S. R. O. P. sealed and numbered official leg band when passed by the inspector. At the time of inspection and banding, the inspector shall make a complete record of the wing- and leg-band numbers and body weight of each tom banded. Privately pedigreed toms referred to in the next paragraph and that have met the progeny-test requirements for U. S. R. O. M. toms shall be considered as having met the requirements for U. S. R. O. M. and U. S. R. O. P. toms and may be officially leg-banded as such.

Par. 20. *U. S. R. O. P. candidate matings.*—These matings shall be comprised of U. S. R. O. P. candidate hens mated to a U. S. R. O. P. tom to which the following provisions apply:

(a) Only one tom shall be allowed in a mating at any one time and the date of entry and removal of toms shall be properly recorded.

(b) The hens shall be trap-nested for a minimum period of 8 consecutive weeks as specified in paragraph 14.

(c) Each egg shall be accurately identified as to the hen and the tom or pen from which it came.

(d) For each hen an accurate record shall be kept of all eggs laid during the trapping period, the number of eggs sold, the number of eggs incubated, the number of poults hatched, and the number banded.

(e) The poults from such matings shall be banded with sealed and numbered official wing bands marked "Cand." to differentiate them from U. S. R. O. P. poults.

(f) The young tom progeny of the hens which qualify as U. S. R. O. P. hens may qualify as U. S. R. O. P. toms as set forth in paragraph 19 of this section provided they meet all the requirements of U. S. R. O. P. toms except that of being produced from U. S. R. O. P. poults.

(g) To make it possible for a breeder to get started in U. S. R. O. P. breeding work within his own strain during the first year of such work on his farm, the tom mated with U. S. R. O. P. candidate hens need not be a U. S. R. O. P. tom provided he is of equal pedigree or is a U. S. Approved tom of outstanding quality. Such toms may be used in similar matings during the second year of U. S. R. O. P. breeding work, if reinspected and passed by the inspector, but not thereafter unless they qualify as U. S. R. O. M. toms. No eggs or progeny from such a mating shall be sold as U. S. R. O. P. products, except that young toms whose dams qualify for U. S. R. O. P. may be permitted to head U. S. Certified flocks owned or controlled by the breeder and to head U. S. R. O. P. matings on the breeder's premises.

Par. 21. *U. S. R. O. P. matings.*—These matings shall be comprised of only U. S. R. O. P. hens mated to a U. S. R. O. P. tom. The breeder may place in the U. S. R. O. P. mating, hens that are not U. S. R. O. P. provided he does not sell pedigreed eggs, poults, or stock from such hens as U. S. R. O. P. products. The provisions of paragraph 20 shall apply to U. S. R. O. P. matings.

Par. 22. *U. S. R. O. P. hatching eggs.*—Such eggs shall come only from U. S. R. O. P. matings. Each egg shall bear the U. S. R. O. P. number of the hen that laid it and the pen or tom number of the mating.

Par. 23. *U. S. R. O. P. poults.*—They shall come only from U. S. R. O. P. eggs. The poults from each hen shall be hatched separately in pedigree baskets or sacks and shall be banded individually with sealed and numbered official wing bands at the time of removal from pedigree baskets or sacks. The poults shall be subject to inspection at any time.

Par. 24. *Wing- and leg-banding provisions:*

(a) All bands shall be obtained through or approved by the official State agency and all purchases of such shall be authorized by and recorded with the official State agency.

(b) The use of the letters R. O. P. alone or in combination with any other words or letters on wing or leg bands is limited to U. S. R. O. P. poults, qualified toms, and qualified hens.

(c) Wing bands used on U. S. R. O. P. poults shall be stamped with the letters U. S .R. O. P. and shall be so numbered as to permit no two poults on the same breeder's farm being identically numbered.

(d) Poults from matings of U. S. R. O. P. candidates with U. S. R. O. P. toms may properly be identified with wing bands stamped with the letters, "Cand"

SECTION 4. U. S. REGISTER OF MERIT BREEDING STAGE

Paragraph 1. This breeding stage incorporates all the requirements of the U. S. R. O. P. breeding stage and in addition includes progeny test records of viability and market quality of the progeny of qualified U. S. R. O. P. toms and hens. U. S. R. O. M. qualification is further limited to only those individual U. S. R. O. P. toms and hens which, on initial inspection, indicated strong vigor, excellent body conformation and fleshing condition, and were excellent representatives of the variety in the judgment of the R. O. P. supervisor or inspector. When any disastrous event occurs that affects the pedigree breeding and progeny testing work of any advanced turkey breeder, and when such event is immediately brought to the attention of his official State agency, this agency may, with the consent of the Bureau of Animal Industry, make an equitable revision of the qualifying requirements.

Par. 2. The U. S. R. O. P. breeder who desires to qualify birds for U. S. R. O. M. shall rear on his premises or have reared under his supervision poults hatched as set forth in paragraphs 4 or 5 of this section in order that viability and market quality requirements may be determined.

Par. 3. *U. S. R. O. M. candidate matings.*—These matings are the same as U. S. R. O. P. candidate matings except that the toms and hens must be of excellent quality as provided in paragraph 1 of this section. The toms and hens in such matings shall be considered U. S. R. O. M. candidates until they qualify as provided in paragraphs 4 and 5, respectively, of this section. After this qualification, the progeny of either the qualified tom or hen shall be considered as coming from a U. S. R. O. M. mating.

(PURPLE)

Fig. 4—Design used to identify U. S. R. O. M. males, females, and matings.

Par. 4. ***U. S. R. O. M. toms.***—U. S. R. O. M. toms when mated as in paragraph 20, section 3 of this article, shall qualify for U. S. R. O. M. when the following requirements are met:

(a) All hatching eggs produced from the mating during the 8-week period for measuring hatchability and a minimum average of 20 eggs per hen and 200 eggs per mating shall be set.

(b) The average hatchability of all eggs set shall be 65 per cent or more. At altitudes of 3,000 to 3,499 feet, the minimum qualifying hatchability percentage shall be 60, and at altitudes of 3,500 feet or more the minimum qualifying hatchability percentage shall be 55.

(c) The minimum number of poults raised to 8 weeks of age shall be 75 per cent of all poults started, provided that at least 100 poults from each pen shall be started.

(d) Each young turkey reared, of those referred to in (c) above, shall be carefully examined not earlier than at 22 weeks of age by a State inspector or authorized agent of the official State agency. At the time of this examination, at least 60 per cent of all young toms and at least 80 per cent of all young hens shall be of U. S. Grade A market quality, due consideration being given to the age of the progeny and the climatic environment as related to their fattening and pinfeather condition, provided a minimum of 70 turkeys from each mating meet these requirements.

(e) Having met these requirements the tom may be banded by the R. O. P. supervisor or inspector with a sealed official leg band lettered "U. S. R. O. M."

(f) Before any progeny shall be classed as meeting the requirements of U. S. Grade A, it shall fulfill the requirements for U. S. Approved as set forth in article 2, paragraphs 1 and 2, except for the last sentence.

Par. 5. *Mass matings of U. S. R. O. M. hens with sons of U. S. R. O. M. matings.*—A breeder may make mass matings of U. S. R. O. M. hens with sons of U. S. R. O. M. matings for the purpose of producing toms to head U. S. Certified flocks. Such matings need not be trap-nested but eggs produced by such matings shall be segregrated and properly identified at all times. The poults from such matings shall be wing-banded at hatching time with sealed and numbered official wing bands marked "U. S. Cert." The young toms from such matings may be used to head U. S. Certified flocks after having been approved by the State inspector. After such approval they may be banded with official sealed and numbered wing or leg bands marked "U. S. Cert." The breeder shall furnish his supervisor with a complete record of the number of eggs produced by each such mating, the number of poults hatched, and the disposition of such eggs and poults.

Par. 6. *U. S. R. O. M. hens.*—U. S. R. O. P. hens when mated as in paragraph 20, section 3 of this article shall qualify for U. S. R. O. M. when the following requirements are met:

(a) A minimum of 20 poults shall be started of those hatched during the 8-week period for measuring hatchability referred to in paragraph 14, section 3, of this article.

(b) Eighty per cent of all poults started shall be alive at 8 weeks of age.

(c) Each young turkey reared of those referred to in (a) above shall be carefully examined not earlier than at 22 weeks of age by a State inspector or authorized agent of the official State agency. At the time of this examination, at least 60 per cent of all young toms and at least 80 per cent of all young hens shall be of U. S. Grade A market quality, due consideration being given to the age of the progeny and the climatic environment as related to their fattening and pinfeather condition, provided a minimum of 12 turkeys meet these requirements.

(d) Having met these requirements the hen may be banded by the R. O. P. supervisor or inspector with a sealed official leg band lettered "U. S. R. O. M."

(e) Before any progeny shall be classed as meeting the requirements of U. S. Grade A, it shall fulfill the requirements for U. S. Approved as set forth in article 2, paragraphs 1 and 2, except for the last sentence.

Par. 7. *U. S. R. O. M. matings.*—These matings may consist of:

(a) U. S. R. O. M. tom remated to the same hens with which he qualified; or

(b) U. S. R. O. M. hen remated to the same tom with which she qualified; or

(c) U. S. R. O. M. tom remated to the same U. S. R. O. M. hens with which he qualified.

Par. 8. *Poults from U. S. R. O. M. matings.*—These poults shall be hatched as provided in paragraph 23, section 3, of this article and shall be banded with U. S. R. O. M. wing bands.

Par. 9. *Sons of U. S. R. O. M. matings.*—In order to qualify to head U. S. Certified flocks, a son of a U. S. R. O. M. mating shall have passed the final individual inspection of a State inspector not earlier than at 22 weeks of age and shall be banded at the time of inspection with an official sealed and numbered leg band. These leg bands shall be stamped with the letters "U. S. R. O. P."

Par. 10. At the time of inspection the son of a U. S. R. O. M. mating shall indicate strong vigor, shall show, upon handling, excellent body conformation and fleshing, and shall be an excellent representative of the variety in the judgment of the State inspector. The inspector shall make a complete record of the body weight, wing- and leg-band numbers of each tom banded.

Par. 11. The young toms from U. S. R. O. M. candidate matings described in paragraph 3 of this section are eligible for qualification as sons of U. S. R. O. M. matings after their parent mating qualifies as a U. S. R. O. M. mating.

Article 3. Pullorum Classes

The aim or goal in an official pullorum disease testing program for turkeys shall be eradication rather than mere control of the disease. To that end every possible step should be taken not only to eradicate infected birds from the breeding flocks, but also to eliminate all possible or probable avenues of infection for adult breeding stock, poults, and growing turkeys. Complete segregation of turkeys from chickens on the farm and in hatcheries at all stages of development, and during all seasons of the year is urgently recommended.

SECTION 1. GENERAL PROVISIONS

Paragraph 1. The official State agency shall assume responsibility for the accuracy of the pullorum testing work.

Par. 2. The official tests for pullorum disease in turkeys shall be (1) the standard tube agglutination test as described in the Proceedings of the United States Livestock Sanitary Association, November 30 to December 2, 1932, pages 487-491; (2) the rapid serum plate test as described by Runnells, Coon, Farley, and Thorp, Journal of the American Veterinary Medical Association, volume 70 (N.S. 23), No. 5, pages 660-662, 1927; and (3) the rapid, whole-blood test for restricted use when approved by the Official State Agency and when carried out with lots of K antigen or with lots of Turkey Redigen that have been approved for use by the Bureau of Animal Industry. The use of the rapid, whole-blood tests are restricted to testing flocks attempting to qualify for U. S. Pullorum-Controlled and U. S. Pullorum-Passed classes.

DIRECTIONS FOR CONDUCTING RAPID, WHOLE-BLOOD TESTS FOR PULLORUM DISEASE IN TURKEYS

When approved by the official State agency, rapid, whole-blood tests intended to qualify flocks for the U. S. Pullorum-Controlled and U. S. Pullorum-Passed classes shall be carried out as follows:

When approved lots of K antigen are used, the test shall be carried out in a manner similar to that used in the whole-blood test of chickens as described in Miscellaneous Publication No. 349. The same proportions of blood to antigen (1 to 2 or 3) shall be used. However, the reactions with pullorum-infected turkeys are usually less pronounced and develop more slowly than reactions with positive chicken blood. It is, therefore necessary when testing turkeys to allow a longer time for reading and to observe the smear more closely than when testing chickens. For these reasons, the smear shall be observed for at least 3 minutes after mixing, and all birds that give a visible reaction, however slight, shall be considered positive reactors and must be removed.

Unless provisions are made for heating the plate, testing shall not be conducted when the air temperature is below 70° F.

More than one test may be placed upon the plate at one time, but the full 3-minute period shall be allowed for reading each test.

The materials, other than antigen, to be used in the test are those recommended for use in the rapid, whole-blood plate test of chickens, namely, a porcelain or glass plate provided with a white background, a loop that delivers 0.02 to 0.025 cc. of blood, a dropper that delivers 0.05 cc. of antigen, and equipment for cleaning or washing the plates and loops. A thermometer and a timepiece must also be provided.

When approved lots of Turkey Redigen are used, the test shall be carried out in the manner described in directions furnished by the producer of the antigen.

Par. 3. In testing for pullorum disease by means of the standard tube agglutination test or the rapid serum plate test, blood samples shall be drawn by an authorized blood collector and the testing shall be done in a laboratory recognized and approved by the official State agency.

Par. 4. All tests with pullorum, typhoid, or polyvalent antigens of flocks in any pullorum class or candidates for any pullorum class must be reported to the official State agency, and reactors occurring in unofficial tests shall be considered on the same basis as reactors in official tests.

Par. 5. All new breeding stock (except turkeys of the same or a higher pullorum class) and stock returned to the breeder's farm from exhibits, tests, etc., must be quarantined at considerable distance from the other turkeys. Such turkeys must be kept in quarantine for at least 15 days and then officially tested and found to be negative before being added to the flock.

Par. 6. When more than one breeding flock of turkeys are maintained on the same premises, all such flocks shall be given an official pullorum classification no higher than the lowest classification attained by one of the flocks.

Par. 7. When other than breeding birds are kept on the same premises, ample provisions must be made to keep these birds and their eggs separate.

Par. 8. All domesticated fowl such as chickens, ducks, geese, pheasants, guineas, pea fowl, bantams, etc., shall either be properly tested to meet the same pullorum tolerance as the breeding flock of turkeys, or be segregated from the breeding flock of turkeys at all stages of development and during all seasons of the year.

Par. 9. In the event that tests to qualify or requalify flocks as U. S. Pullorum--Passed or U. S. Pullorum-Clean reveal any reactors, the owner shall have the right to submit, or the official State agency the right to request the submission of, all reactors if 5 or less, or 20 per cent of all reactors with a minimum of 5 and a maximum of 20 to a laboratory designated by the official State agency for autopsy and bacteriological examination. In case such bacteriological examination fails to reveal any conclusive evidence of pullorum infection, the flock shall be construed as having had no pullorum. All reactors not slaughtered for bacteriological examination shall be removed and disposed of in a manner satisfactory to the official State agency.

Par. 10. In the event that a thorough investigation or other evidence indicates that pullorum disease or fowl typhoid outbreaks are occurring in poults, the official State agency may, at its discretion, require additional retesting of the flock or flocks, or make further autopsies and bacteriological examinations.

Par. 11. The custom-hatching of turkey eggs or eggs of other poultry from nonqualified flocks is permitted in U. S. Pullorum-Controlled hatcheries, provided the incubation and hatching of such eggs are done in a separate room under the conditions prescribed in paragraph 12 of this section, but such custom-hatching may not be done in U. S. Pullorum-Passed and U. S. Pullorum-Clean hatcheries.

Par. 12. In separating a room for the use of two or more incubators, to provide for custom-hatching, a tight partition is absolutely necessary and must be provided. Outside ventilation and an outside entry for each room are required. Any door or opening between the two rooms must be sealed, as by weather-stripping and locking, or by battening the cracks around the door in the room where eggs from non-U. S. Pullorum-Controlled flocks are being incubated or hatched.

Par. 13. In the event that eggs other than chicken or turkey eggs are to be incubated or hatched by a hatchery, such eggs shall either be produced by flocks that have been properly tested to meet at least the minimum pullorum tolerance of the hatchery or be hatched in accordance with paragraphs 11 and 12 of this section.

Par. 14. Hatcheries participating in the pullorum classes only shall be inspected and approved by a State inspector at least once during the hatching season.

Par. 15. Each year at least one-third of the flocks participating in the pullorum classes only shall be inspected and approved by a State inspector. Field-testing or blood-collecting shall be regarded as an official flock inspection provided a complete inspection is made and so reported.

Par. 16. After a flock has been classified as U. S. Pullorum-Controlled, U S. Pullorum-Passed, or U. S. Pullorum-Clean by the official State agency, the results of any retesting of such flocks during the current breeding and hatching season shall not adversely affect its pullorum classification for the season, except that if in the opinion of the official State agency the amount of reaction found in any flock is such as to be dangerous or detrimental to the operation of the plan, the official State agency shall require that flock or flocks be retested after an interval of at least 21 days or that such flocks be immediately discontinued as hatching egg flocks. Furthermore, the official State agency may require that the hatching eggs from such flocks be removed from the incubator and destroyed prior to hatching.

SECTION 2. U. S. PULLORUM-CONTROLLED CLASS

Paragraph 1. This is a tolerance class. No breeder or hatchery should be content with having qualified for this class. Control and eradication measures should be practiced continuously with a view to reducing the percentage of reactors each year and qualifying for the U. S. Pullorum-Passed class and then the U. S. Pullorum-Clean class as soon as possible.

(YELLOW)

Fig. 5—Design used to identify U. S. Pullorum-Controlled flocks, eggs, poults, and hatcheries.

Par. 2. *U. S. Pullorum-Controlled flocks.*—These flocks shall meet the following requirements:

(a) All turkeys to be used as breeders shall be tested for pullorum disease when more than 4 months of age under the supervision of an official State agency, and shall contain fewer than 2 percent reactors, the last test being made within 6 month simmediately preceding the date of first sale of hatching eggs, poults, or breeding stock from such flocks.

(b) All indicated carriers of pullorum disease shall be removed from the premises on completion of the test and disposed of in a manner satisfactory to the official State agency.

(c) All birds remaining in the breeding flock shall be properly leg- or wing-banded.

(d) Individual birds introduced into U. S. Pullorum-Controlled flocks shall have passed, within 6 months, a negative official test for pullorum disease.

Par. 3. A flock containing 2 per cent or more of reactors on the first test, on being retested at intervals of not less than 21 days, and all reactors disposed of after each test until the percentage of reactors is fewer than 2 per cent, may qualify as a U. S. Pullorum-Controlled flock.

Par. 4. *U. S. Pullorum-Controlled hatching eggs.*—Such eggs are from U. S. Pullorum-Controlled flocks.

Par. 5. *U. S. Pullorum-Controlled poults.* — These are hatched from eggs produced by U. S. Pullorum-Controlled, U. S. Pullorum-Passed, or U. S. Pullorum-Clean flocks in a U. S. Pullorum-Controlled hatchery.

Par. 6. *U. S. Pullorum-Controlled hatcheries.*—These hatcheries operate under the supervision of an official State agency and (1) buy or sell turkey or chicken hatching eggs, (2) incubate and hatch turkey or chicken eggs, (3) brood or sell poults or chicks, or (4) be dealers in or agents for such products from only U. S. Pullorum-Controlled, U. S. Pullorum-Passed, or U. S. Pullorum-Clean flocks and hatcheries, except that custom hatching may be carried on under the conditions prescribed in section 1, paragraph 12 of this article.

SECTION 3. U. S. PULLORUM-PASSED CLASS

Paragraph 1. This class, which embodies the recognition of infected premises, permits no tolerance in the official recognition of the flock. It emphasizes the advisability of retesting at frequent intervals until no reactors are found and, therefore, is a more effective step toward eradication than the tolerance class. No breeder should be content with having qualified his flock for this class. Control measures should be practiced continuously with a view to eliminating all infection and qualifying for the U. S. Pullorum-Clean class as soon as possible.

(RED)

Fig. 6—Design used to identify U. S. Pullorum-Passed flocks, eggs, poults, and hatcheries.

Par. 2. *U. S. Pullorum-Passed flocks.*—These flocks shall meet the following requirements:

(a) All turkeys to be used at breeders shall be tested for pullorum disease when more than 4 months of age under the supervision of an official State agency, and shall contain no reactors, the last test being made within 6 months immediately preceding the date of first sale of hatching eggs, poults, or breeding stock from such flock. When reactors are found the flock may qualify as U. S. Pullorum-Passed, provided it is retested at intervals of not less than 21 days, until one negative test is obtained.

(b) All birds remaining in the breeding flock shall be properly leg- or wing-banded.

(c) Individual birds introduced into U. S. Pullorum-Passed flocks shall come from U. S. Pullorum-Passed or U. S. Pullorum-Clean flocks only, except that individual birds from other sources may be introduced provided they are quarantined for 15 days, then officially tested for pullorum disease and found to be negative.

Par. 3. *U. S. Pullorum-Passed hatching eggs.*—Such eggs are from U. S. Pullorum-Passed flocks.

Par. 4. *U. S. Pullorum-Passed poults.*—These are hatched from eggs produced by U. S. Pullorum-Passed and U. S. Pullorum-Clean flocks in a U. S. Pullorum-Passed hatchery.

Par. 5. *U. S. Pullorum-Passed hatcheries.*—These hatcheries operate under the supervision of an official State agency and (1) buy or sell turkey or chicken hatching eggs, (2) incubate and hatch turkey or chicken eggs, (3) brood or sell poults or chicks, or (4) be dealers in or agents for such products from only U. S. Pullorum-Passed or U. S. Pullorum-Clean flocks and hatcheries.

SECTION 4. U. S. PULLORUM-CLEAN CLASS

Paragraph 1. *U. S. Pullorum-Clean flocks.*—These flocks shall meet the following requirements:

(a) All turkeys to be used as breeders shall be tested for pullorum disease when more than 4 months of age under the supervision of an official State agency and shall contain no reactors on the first or any subsequent test made within 6 months immediately preceding the date of first sale of hatching eggs, poults, or breeding stock from such flock.

(BLUE)

Fig. 7—Design used to identify U. S. Pullorum-Clean flocks, eggs, poults, and hatcheries.

(b) A Pullorum-Passed flock may qualify for the Pullorum-Clean class provided it has two additional consecutive negative tests for pullorum disease at least 21 days apart, the last test being made within 6 months immediately preceding the date of sale of hatching eggs or poults from such flock.

(c) All birds remaining in the breeding flock shall be properly leg- or wing-banded.

Par. 2. Individual turkeys introduced into U. S. Pullorum-Clean flocks shall come from U. S. Pullorum-Clean bocks only, except that individual turkeys from other sources may be introduced provided they are quarantined for 15 days, then officially tested for pullorum disease and found to be negative.

Par. 3. Flocks of this classification shall be designated as having been U. S. Pullorum-Clean for 1, 2, 3, etc., years.

Par. 4. *U. S. Pullorum-Clean hatching eggs.*—Such eggs shall come from U. S. Pullorum-Clean flocks.

Par. 5. *U. S. Pullorum-Clean poults.*—These are hatched in a U. S. Pullorum-Clean hatchery from eggs produced by U. S. Pullorum-Clean flocks.

Par. 6. *U. S. Pullorum-Clean hatcheries.*—These hatcheries operate under the supervision of an official State agency and (1) buy or sell turkey or chicken hatching eggs, (2) incubate and hatch turkey or chicken eggs, (3) brood or sell poults or chicks, or (4) be dealers in or agents for such products from only U. S. Pullorum-Clean flocks or hatcheries.

Fig. 125—Silkie Male.

[Courtesy U. S. D. A.]

Standard Egg Laying Contests:

An understanding of the importance of egg laying contests as an important consideration in poultry breeding projects may be gleaned from the 1930 year book of the Canadian National Record Association in the following statement:

"Canada's National Poultry Registration Program, embracing as it does the standardization of all official egg-laying contests conducted by the Federation Government of Canada and then using these contests as the basis for poultry registration work, the registering of all birds qualifying in these contests and the breeding of these birds under a system of Government inspection, constitutes one of the most advanced national breeding policies ever undertaken by any country."

Considerable attention has been given to the development of egg laying contests in the United States. The objects in the conduction of a contest may be listed briefly as:

1. To improve the average egg production of the standard pure-bred commercial hen by encouraging interest in selection and breeding.

2. To give all poultrymen an opportunity to have their birds officially trapnested.

3. To afford a testing station where poultrymen may compare their strain of birds with other strains and breeds.

4. To make possible better breeding, through the return of the contest trapnested birds to their owners.

5. To demonstrate the value and to furnish the public with records showing the relative value of the various strains and breeds represented in competition.

6. To accumulate valuable data on egg production performance records which, when studied statistically and completely, may aid in the development of a more efficient program of selection and breeding.

For a period of years, awards in egg laying contests were given on the basis of egg production alone. Standardization of egg laying contests brought to the fore the need for a more adequate measure of the performance of the birds, and the point system was inaugurated as the yardstick. Birds which lay standard two-ounce eggs receive one point for each such egg produced. An egg that will average 25 or 26 ounces to the dozen receives additional point compensation while those smaller than standard are cut in proportion. The contests, therefore, are not primarily interested in the number of eggs a bird lays but in the marketability of the eggs produced.

Since then contests have inaugurated a system whereby no alternate birds are permitted in a pen, and consequently the records of all 13 birds in a standard egg laying contest are counted. Previously, only the 10 high constituted the pen record.

This change in procedure was decided upon in order to debit adult mortality more stringently. In such a method, egg laying contests focused the attention of the breeder upon and directed his efforts towards improving the adult livability of his strain. This procedure was first effective in 1937-38 egg laying contests. Therefore egg laying contests are contributing greatly to improvement in breeding programs.

Their contribution to improved egg production is exemplified by the more or less continual improvement in the performances over a period of years, frequently culminating in new world's records. Many new records have held their places for but a few years, with newer records shattering the old.

Such records are not haphazard but represent the utmost in breeding skill. Breeders everywhere should take a more active interest in supporting the tests being conducted.

Standard Contest in Operation During 1948-49 Season

Arizona Egg Laying Test, Tucson, Ariz.
Florida National Egg Laying Test, Chipley, Fla.
Georgia National Egg Laying Test, Athens, Ga.
Hunterdon County Egg Laying Test, Flemington, N. J.
Illinois Egg Laying Test, Quincy, Illinois.
Maine Egg Laying Test, Monmouth, Maine.
Missouri National Egg Laying Test, Mountain Grove, Mo.
New York State Egg Laying Test, Farmingdale, Long Island, N. Y.
Oklahoma Egg Laying Test, Stillwater, Okla.
Pennsylvania Official Egg Laying Test, Harrisburg, Pa.
Rhode Island State Egg Laying Test, Kingston, R. I.
Storrs Egg Laying Test, Storrs, Conn.
Texas Egg Laying Test, Stephenville, Texas.
Western New York Official Egg Laying Test, Stafford, N. Y.

Questions on Chapter XIV

1. List the stages of the National Poultry Improvement Plan that have reference to production qualities?
2. List the stages that have reference to Pullorum Disease control?
3. List the breeding stages and the Pullorum classes of the National Turkey Improvement Plan.
4. What are the objects in the conduction of an egg laying contest?
5. Are winnings in egg laying contests based on number of eggs produced without consideration to egg size?

CHAPTER XV

MARKETING

Subject Matter

- **Methods of Marketing**
  - **Direct to Consumer**
  - **Selling to Retailer**
  - **Commission Mercnants, Wholesalers and Jobbers**
  - **Co-operative Marketing**
  - **Hucksters**
  - **Auction Selling**
  - **Miscellaneous Channels**
- **Preparation of Poultry for Market**
  - **Methods of Killing**
  - **Dressing**
  - **Wax Picking**
  - **Drawing**
  - **Dressing and Drawing Loss**
- **Selling Live Versus Dressed Poultry**
  - **Tentative U. S. Standards for Classes and Grades of Live Poultry**
  - **Definition of Market Terms**
- **Marketing Eggs**
  - **Official U. S. Egg Standards**
  - **Tentative U. S. Wholesale Grades**
  - **Tentative U. S. Consumer Grades**
  - **Cleaning Dirty Eggs**
  - **Grading Eggs and Candling**
  - **Cartoning and Packing**
- **Egg Marketing Programs**
- **Salesmanship**

The poultry industry consists primarily of two phases—production and marketing—making the disposal of products a most important consideration. Marketing of poultry products received little attention from the producers, however, until recently.

Consumer demand is the ultimate factor on which marketing must be based. Thus it becomes necessary for the producer to study this carefully and produce and market his products in a way to meet the approval of the buying public. The industry has been too intently "production minded"; it must become "market minded" if it is to dispose of its goods to the best advantage.

Methods of Marketing

Many channels are open for the disposal of poultry products. The one to be used will be decided by the producer's individual need and attitude, the results secured, the volume to be distributed, and local market conditions.

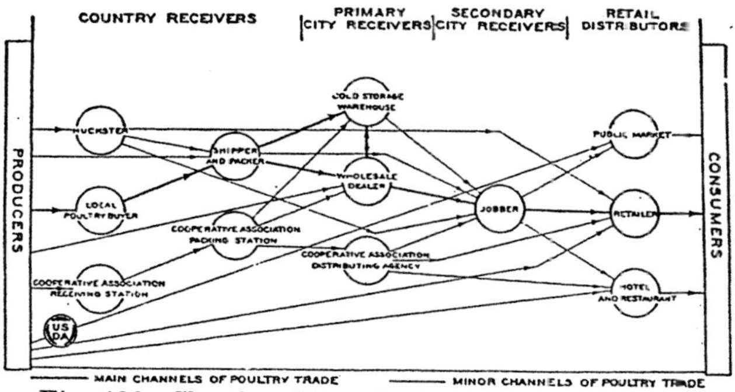

Fig. 126—Showing Channels Through Which Eggs Pass to the Consumer. [U. S. D. A.]

Direct to the Consumer:

Selling direct to the consumer eliminates all intermediate handlers and may be done by delivery to the consumer by truck or parcel post, through curb or farmer's retail markets, or through roadside stands. For the direct seller to satisfactorily cater to this type of trade he must maintain and standardize quality, and this must yield a sufficient price to justify its additional expense. The volume in direct selling will also need to be fairly constant and large enough to sustain operations. The necessity of keeping charge accounts with customers involves the possibility of poor accounts. The cost of distribution is another factor for consideration, covering as it does the requirements for cartoning, grading, packing and delivery.

TABLE 38

WHERE IOWA FARMERS SELL THEIR POULTRY*

	Number	Per cent total dealers
Buying stations	1,597	48.0
Grocery stores	913	27.5
Grocery stores with meat market in conjunction	177	5.3
Hucksters	171	5.2
Meat markets	139	4.2
Carlot shippers, plant equipped with refrigeration	124	3.7
Carlot shippers, plant without refrigeration	55	1.6
Miscellaneous	149	4.5
Total	3,325	100.00

*Iowa State College; eggs sold through similar channels.

The following tables indicate the increased cost of marketing eggs direct to the consumer and the increased price received.

TABLE 39

AVERAGE COSTS OF MARKETING EGGS BY 88 FARMERS IN PENNSYLVANIA*

Method of Marketing	No. of Farmers	Cost of Marketing Per Doz.[1]	Cost of Marketing Per Case[2]
Huckster	21	$.006	$ .17
Country store	20	.019	.56
New York wholesaler	23	.043	1.29
Consumer	24	.047	1.42

*"Egg Marketing by Farmers in Pennsylvania," F. F. Linsinger, Pennsylvania State College, Bulletin 214.

[1]Total costs divided by total number of dozen.

[2]Total costs divided by total number of cases.

TABLE 40

NET PRICES REALIZED BY 88 FARMERS IN PENNSYLVANIA

Method of Marketing	Cents Per Dozen: Wt. Gross Price	Cents Per Dozen: Marketing Cost	Net Price
Country store	35.21	1.86	33.35
New York wholesaler	41.71	4.74	36.97
Huckster	38.41	.58	37.83
Consumer	48.17	4.30	43.87

Fig. 127—Cartoning Eggs for Retail Trade.

[Courtesy N. E. Poultryman.]

Selling to Retailer:

Selling to the retailer combines many of the advantages of sales direct to consumers, since the products are marketed to

*After marketing costs are deducted from gross prices, net prices to producers are highest to those selling direct to consumers and lowest to those selling to country stores. Net prices realized from New York wholesalers and hucksters are not significantly different. Both, however, are lower than direct-to-consumer prices and higher than country-store prices.

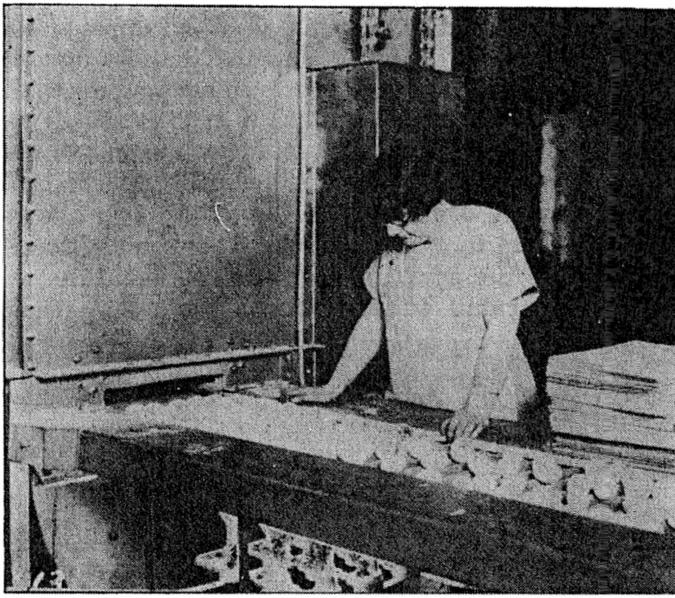

Fig. 128—Inspecting Eggs as They Come from a Sander.
[Courtesy of E. W. Benjamin.]

Fig. 129—Grading and Packing Sanded Eggs.
[Courtesy of E. W. Benjamin.]

restaurants, hotels, soda fountains, grocery stores, milk route operators, and butcher shops. It has the distinct advantage of fewer deliveries because of increased volume, and the probability of fewer losses from poor accounts. It is difficult, however, to make and maintain contact with retailers due to the necessary volume, continuity of supply, and standardization of product.

The chain-store organizations are taking over the retailing of poultry products to a large extent. This competition is gradually eliminating the small independent retailer, thus lessening the chances for retail outlets for the individual poultryman.

Fig. 130—Active Bidding at an Egg Auction at Flemington, N. J.

Commission Merchants, Wholesalers and Jobbers:

Where the producer is so tied up with production activities that only a limited amount of time and effort remain for selling his product, distribution through commission merchants, wholesalers and jobbers will undoubtedly be satisfactory, providing a reliable dealer is secured.

The function of a commission merchant is to dispose of products to the best advantage for a stipulated compensation on a percentage basis.

The wholesaler is ordinarily a cash buyer who purchases outright and resells. This type of buyer is increasing in numbers.

The jobber is the intermediate between the commission merchant and the wholesaler or the poultrymen and the retailer. His primary function is to supply the needs of the retailer. Some dealers serve both as wholesale receiver and jobber.

In making contact with distributors in the market areas, the producer should select an outlet consistent with the product he has to place on the market.

Co-operative Marketing:

Poultry products lend themselves to co-operative marketing as well as other agricultural products. This method of marketing poultry products has been brought about through unsatisfactory price levels and a surplus over local demands, making shipment to more distant areas necessary.

Fig. 131—Buyers Attending a Connecticut Egg Auction.
[Courtesy N. E. Poultryman.]

Co-operative efforts of this type are successful only where there is a real and urgent need, where it has the backing of the producer and where the volume is sufficient to justify the overhead. Even with these requirements, failure may ensue due to inefficiencies in marketing practices, lack of knowledge of merchandising the product, unsatisfactory trade connections, insufficient capital, and expansion covering too many and varied products.

Hucksters:

In many sections local marketing conditions have led hucksters to purchase eggs on more or less regular farm-to-farm routes. Some of the more progressive buyers of this type have developed specialized outlet for good quality products and are thus able to pay the producer a fair market price. This method of selling is convenient for the producer in that the buyer comes directly to his door and saves him the time of hauling his product to market. Table 40 shows that Pennsylvania farmers received a slightly better price where eggs were sold to the huckster than where they were either taken to the country store or shipped direct to New York wholesalers.

Auction Selling:

The selling of poultry produce by auction has been gaining headway in certain sections. The advantage claimed for this method is that sales are made direct to the jobber or retailer at a central point, thus saving on transportation and distribution charges and placing the product in the hands of the retailer in a fresher condition. It also enables the individual jobber or retailer to get the exact volume and quality necessary for the continuance of his business.

Miscellaneous Channels:

Marketing channels under this heading include the country store, car-lot buyers, the storage and packing industries, and egg breaking and drying establishments. All of these mean the lapse of considerable time before the product reaches the consumer, due to the fact that it must go through several hands.

TABLE 41

PRACTICAL EQUIVALENT VALUES, EGGS

1-lb. frozen or liquid eggs is equivalent to 10.2 eggs in shell.
1-lb. dried whole eggs is equivalent to 3.6 lbs., liquid whole eggs.
1-lb. dried yolks is equivalent to 2.25 lbs., liquid yolk.
1-lb. dried albumen is equivalent to 7.3 lbs., liquid albumen.

(Poultry Tribune)

Weight obtained from 30 dozen case, 24 ounce eggs.

1. Dried whole eggs ---------- 9 to 10 pounds.
2. Dried yolks ---------- 7 to 7.5 pounds.
3. Dried albumen ---------- 2 to 2.5 pounds.

Weight obtained from 100 pounds liquid (eggs broken out).

1. Dried whole eggs ---------- 30 pounds
2. Dried yolk ---------- 45 pounds.
3. Dried albumen ---------- 13 pounds.

(Epstein)

Egg equivalents:

1-lb. egg yolk requires 2.0 dozen eggs.
1-lb. egg white requires 1.5 dozen eggs.
1-lb. dried whole egg requires 3 dozen eggs.
1-lb. dried egg yolk requires 4 dozen eggs.
1-lb. dried egg white requires 12 dozen eggs.

(Somner)

Preparation of Poultry for Market

Poultry meat should be placed on the market in an attractive manner and possessing distinct quality. This entails correct fattening or finishing in order to produce a quality carcass. For information on fattening refer to Chapter 16. After birds are finished and ready for market a study of local conditions will indicate whether they will market more advantageously live or dressed.

Methods of Killing:

Beheading and dislocating the neck are satisfactory methods of killing poultry, provided the birds are to be dressed and drawn and used for family consumption. Both are well adapted to scald or slack-scald picking, but other methods of killing should be relied upon for dry picking.

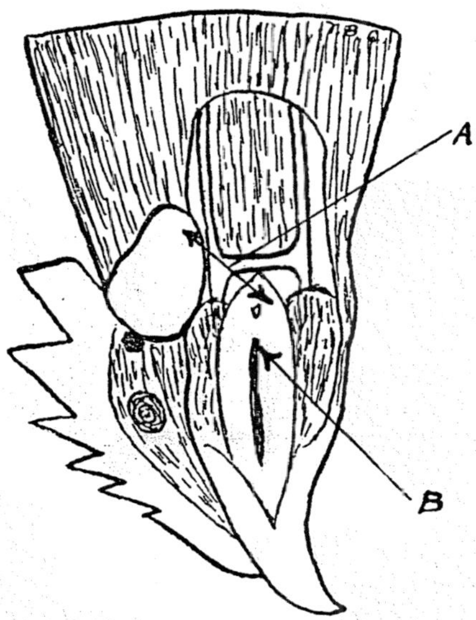

Fig. 132—Method of Bleeding (A) and Debraining (B) Poultry.

Beheading, as the term implies, is the cutting off of the head; dislocation, the separation of the vertebrae of the neck from the base of the skull. The birds' legs are held by the left hand and its head by the right hand, with the back of the skull between the thumb and index fingers. With the bird outstretched, the head is given a short backward turn. Dislocation kills the bird and the stretching of the neck permits blood to drain into the separation, effecting a blood clot. The neck is disfigured and the blood clot soon causes discoloration, but this is not a serious objection where the birds are immediately prepared for cooking.

The most satisfactory commercial method of killing is sticking and debraining. This lends itself readily to dry picking, due to the fact that the muscular control of the feather follicles is released. In preparation for the operation the bird is suspended by the feet with a weighted can attached to the head to hold the body taut and to collect the blood flow. The jugular veins are then severed and the brain properly pierced.

In debraining, the knife is placed in the cleft in the roof of the mouth and forced upward and backward on a line parallel with the beak so that the point pierces the brain about midway between the ears. The knife is then given a quick twist to mangle the brain. A satisfactory stick is indicated by a convulsive movement of the bird and oftentimes a distinct squawking.

Poor bleeding is usually indicated by reddish appearance of the feather tracts on the neck, breast, thighs and wings. This

Fig. 133—Picking machine quickly removes feathers. Cones in background for sticking and killing.

[Courtesy Poultry Dept., N. H. Univ.]

reddish discoloration deepens, finally turning greenish when the bird is held for a period of time.

Dressing:

There are many methods of dressing birds for market, but the three principal ones adapted to practical use are dry picking, scalding and semi-scalding.

Birds which have been bled and debrained are ready for immediate picking by the dry method. This is effective because it leads to few abrasions of the skin and leaves the carcass with a good market appearance. The sooner the feathers are removed after debraining the more efficient the operation. Special care should be exercised not to tear the skin, particularly with younger birds which have tender skin. Dry picking machines are in the experimental stage and when perfected will facilitate dressing operations.

In scalding birds preparatory to picking, the water temperature should approach boiling and the period of emersion should be as brief as possible for proper release of feathers. Overscalding partially cooks the skin, resulting in easy abrasing and tearing. Injured areas become dark and discolored as the bird cools, detracting from the market value. Scald picking does not require so much skill as dry picking and is most applicable where the birds are to be used for local consumption.

Semi-scalding gives a carcass an appearance approaching that of dry-picked birds, and at the same time provides greater speed than dry picking. By this method birds are emersed in water at a temperature of 125 to 130 degrees Fahrenheit for a period varying from one-fourth to one minute, depending on their age. Experience will indicate the most desirable period of emersion and temperature. It is desirable to change the water frequently because of pollution. Mechanical devices are on the market for both dry picking and semi-scalding. Picking machines, after experience has been gained, produce good-looking carcasses and save time.

In any dressing operation a complete job is desirable and includes the removal of any pin feathers on the carcass. A special "pinning knife" can be used in grasping the short slippy pin feathers. Singeing of hairy carcasses is advisable and can be done with alcohol or gas flames. Any material which tends to carbonize should not be used, however, or the carcass will appear smoky.

After dressing it is recommended that the birds be properly chilled. Dry-picked birds should be dry chilled; semi-scald birds either dry or wet chilled. By wet chilling is meant the immediate emersion of the carcass in cold water to remove animal heat. This has the additional advantage of shrinking the skin and giving the bird a more plump appearance. Dry chilling is accomplished by placing the bird in a cold compartment or atmosphere. Chill the birds immediately but slowly, so that all of the animal heat is removed from the interior of the carcass before the exterior freezes.

Wax Picking of Poultry

Wax picking is in general better adapted to commercial killing and dressing plants, but can readily be adapted for use on the poultry farm where 100 or more birds are dressed off at one killing period.

Preparation of Bird for Waxing:[1]

The bird must be killed in the usual manner by bleeding through the mouth and piercing the brain. The next operation is that of rough plucking or roughing. In rough plucking the greater part of the readily removable feathers are stripped from the carcass. Roughing may be done in one of two ways. The preferred method is dry roughing (dry plucking). The chief advantage in this method is the good condition in which the skin is left. Most birds can be successfully wax plucked following dry roughing. However, there are two exceptions: (1) Very pinny young birds and (2) moulting birds, where the pin is small and poorly developed. These two classes do not give uniformly satisfactory results unless the bird is semi-scalded prior to roughing. Semi-scalding is the second method of roughing birds.

(a) Dry Roughing for Waxing

The bird is rough plucked immediately on killing. Following this the carcass is hung in a moderately cool room until the body temperature is very considerably lowered. The time of hanging to cool will vary for different sets of conditions. In a room of average temperature about two hours will be required. In some cases a period of one hour is sufficient, but the body

[1]Excerpts from "The Use of Wax in the Plucking of Poultry." National Research Council of Canada, and Dominion Department of Agriculture, Ottawa.

Fig. 134—Dressing methods, left to right: Semi-scald and rough picked; wax dipped; wax picked; complete semi-scald.
[Courtesy Roy Jones, Extension Service, Conn. State College.]

temperature should be reduced to about 70 degrees Fahrenheit. The temperature may be readily determined by holding the thermometer between the wing and the body for a brief interval.

The necessity of proper cooling. If the bird is not cooled sufficiently, there is a tendency for the wax to leave smears. Therefore, if smears of wax are noticed, simply allow the birds to hang a little longer prior to wax treatment. As noted above, about two hours' hanging should usually be adequate to ensure cooling. In some cases, it may be advisable to extend the cooling a little, as the wax peels more readily from cool birds. The operator would do well to try various cooling periods with, say, three birds; the best results will set the conditions for the large group to be processed.

(b) Semi-Scalding for Waxing

Great care must be exercised in the use of the semi-scald as the skin of the bird is apt to suffer abrasions through carelessness. The bird is killed, allowed to bleed for two or three minutes, the large tail and wing feathers are pulled, and the carcass is then immersed in warm water for a period of about one minute. The water temperature should be 128 degrees Fahrenheit for broilers, around 129 degrees for chickens and 130 degrees for fowl. These temperatures must be maintained with care as small variations either way will seriously affect the results. Too high a temperature will scald the skin and too low a temperature will render the semi-scald ineffective. Temperature control of the water in semi-scalding is the most exacting feature of the whole process of wax plucking. During the minute in water the bird should be gently agitated to ensure penetration of the warm water to the base of the feathers.

Roughing. After the one minute in the water the rough feathers are removed with care, as barking of the skin will occur very easily.

After roughing, the bird is hung and the body heat will evaporate residual moisture in from one to two hours. This hanging period both dries and cools the bird, and thus prevents the smearing which occurs when very warm birds are waxed. Reference was made to this point under "Dry roughing for waxing." *The operator is warned against wax coating the bird before it is thoroughly dried.* If moisture is present the wax cannot adhere to the feathers and pins.

The structure of pin feathers and the need for semi-scalding. It has already been stated that dry roughing is preferred. This is true for all except the very pinny birds. Careful examination of pin feathers of a dry-roughed bird under a magnifying glass shows that they have a scaly surface, particularly near the top. If the pin is short this scale prevents good contact between the wax and the quill. When the wax is removed the feather tuft pulls out, the scale lifts and the hollow quill is left in the bird. If the pin feathers are well developed there is sufficient contact between the wax and quill to effect satisfactory pulling. The semi-scald treatment, however, tends to loosen the fine scale on the pin and enables the wax to harden directly on the smooth quill. This results in excellent pulling. In addition to this surface action on the quill, there is a tendency for the warm water to loosen the pin. These factors are important aids in pulling and are necessary in the case of the very pinny bird. However, wherever possible, dry-roughing is advisable, as the dry plucking job is well done in most cases and the appearance of the bird is unquestionably superior. If semi-scalding is absolutely necessary the very greatest care is required in the operation.

Wax Coating the Bird:

A suitable wax coat is formed on the bird by immersion in molten wax. There is no difficulty in building up a coat which removes pin feathers and hairs and at the same time comes away readily. The wax coat is responsible for the pulling action as the liquid wax penetrates to the base of feathers and pins, hardens there and takes them with it on removal.

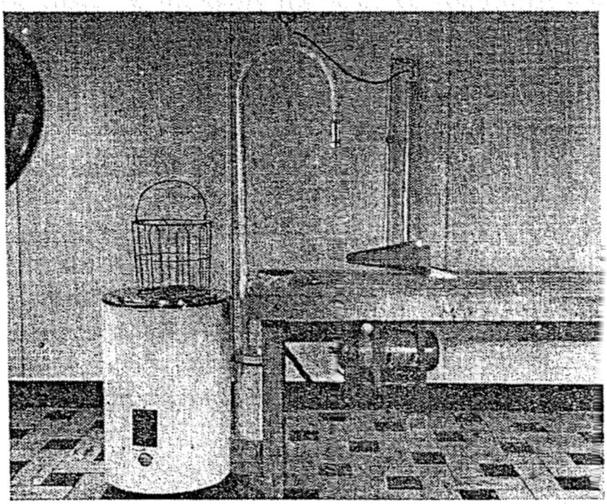

Fig. 135—Equipment such as this Cry-O-Vac packaging unit enhances appearance and customer appeal.
[Courtesy of Dewey & Almy Chemical Co.]

Wax Temperatures for Dipping

The wax can be used over a fairly wide range of temperatures. Dry roughed birds should be dipped in wax at a temperature of from 127-137 degrees Fahrenheit. About 130 degrees should be a fair average, though tough specimens may give better results with a slightly higher temperature. Semi-scalded birds can be processed at wax temperatures of from 120-128 degrees. The higher temperatures are not required as the pulling operation is not as difficult in this case. An average figure for semi-scalded birds would be around 125 degrees.

It should be mentioned that one or two extra dips will be required to build up the coat if the wax is at 135 degrees instead of 125 degrees. After a little experience the operator will be able to modify the number of dips according to the temperature of the wax. All other things being equal, the average temperature values are preferred, for it is easier to build up the coat at a temperature of 130 degrees than at 137 degrees.

Hardening the Wax Coat

The wax coat must be allowed to harden partially before its removal is attempted. The waxed bird may be allowed to hang about fifteen minutes. It is then dipped in cold water for from 30-60 seconds and the wax removed; or if the bird is hung for from twenty minutes to half an hour in a room at about 70 degrees Fahrenheit, the coat should be sufficiently hard to be removed without the use of water. If great speed is desirable, the coated bird may be hung a minute or two in the air and then dipped for about two minutes in cold water, after which the wax may be removed. In this way the wax comes off the bird four minutes after application. It is suggested that the operator try the cooling process with a few test birds under different conditions; he can then follow the best set of conditions noted. It may be pointed out that there is quite a range over which the wax may be pulled. While the hardening does not require very precise attention, the wax must not become too hard, for brittleness will make its removal more difficult.

Removing the Wax Coat

This process should be carried out over a large box or other container into which the used wax may be put for storage until it is re-melted and strained for future use. The first movement in removing the wax layer, the wings are pulled down and the wax breaks, some of it coming away on the first pull. The second movement is forcing up the body which cracks the wax and loosens it along the sides. After the coat is broken, the wax should come away in large pliable sheets.

If the coat is of suitable thickness and properly hardened, the wax layer should come away quickly. It is easier to remove the wax from a semi-scalded bird. In the case of the rough dry-plucked bird quite a pull may be required but there is no serious difficulty.

Steps Involved in Wax Plucking Poultry

1. Melt wax. This should be started several hours before it is to be used.
2. Kill the birds. Do this while wax is being melted.

3. Rough plucking
   (a) Dry roughing may be practised in most cases. Cool the bird by hanging for one hour or more after plucking.
   (b) Semi-scalding for very pinny young birds or moulting birds. Hang the bird for an hour or two to dry.
4. Adjust temperature of wax.
5. Dip the birds.
6. Harden wax coat by hanging bird or by dipping in cold water.
7. Remove wax coat. Keep container handy for collecting used wax.
8. Melt down used wax and strain off feathers and other waste.

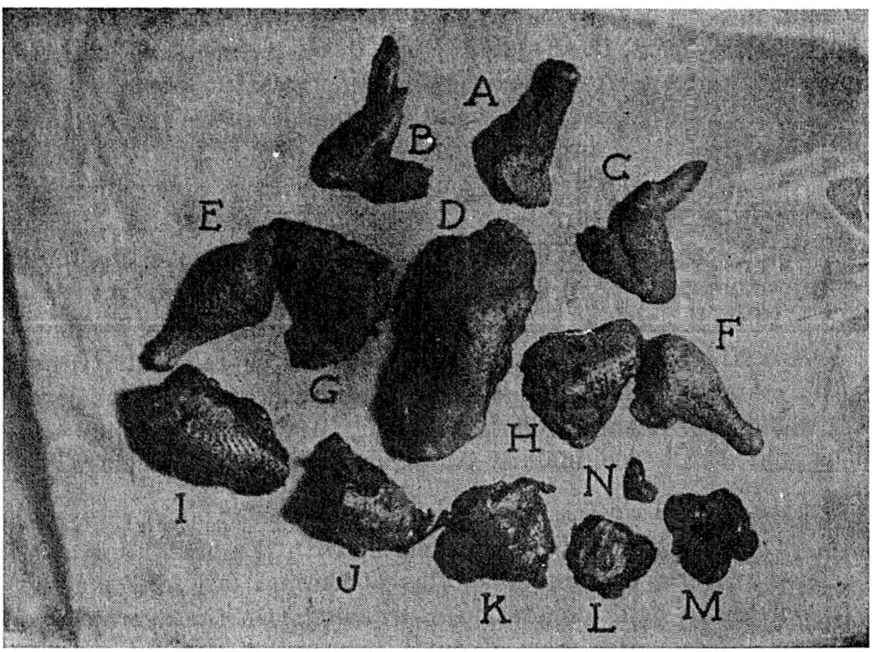

Fig. 136—Chicken cut up showing the parts as they appear ready for cooking and canning. A, neck; B and C, wings; D, breast, not yet cut up; E and F, drumsticks; G and H, thighs; I, rear portion of back; J, front portion of back; K, surplus fat; L, gizzard; M, liver, and N, heart.

[Courtesy N. E. Agr. Exp. Sta.]

Specially prepared wax carefully compounded is essential for satisfactory results and is available on the market for those wishing to utilize this process.

Drawing:

The bird should be drawn immediately before cooking. The shanks should also be removed and, in the case of old birds and roosters, the tendons too.

This is done by breaking the shanks in two, about midway between the hock joint and foot and then suspending the bird by the feet from a convenient cast-iron hook and pulling its body downward, tearing out the tendons.

The removal of the head is the next step and is followed by the slitting of the skin at the base and back of the neck. Through this the crop is loosened and removed and the neck bones severed. Next an incision is made in the abdomen to permit placing the hand inside the body cavity. All the internal organs, including the lungs, heart, liver, gizzard, and the intestines, are loosened and withdrawn. A cut around the vent releases the intestines. Then the kidneys, ovary or testes, and the oil sac are removed and the body cavity thoroughly washed out.

The gizzard, heart, liver and neck should be replaced within the carcass if the bird is to be retailed. First split the gizzard just through the muscular wall, however, and remove the lining with contents.

With broilers it is common practice to split the backbone lengthwise to reveal the internal organs for removal. Other than this, the bird is prepared as discussed in the previous paragraphs. In drawing broilers, however, the breast bone may also be split lengthwise, making two equal portions for serving.

Dressing and Drawing Loss:

Dressing and drawing loss varies with the breed, age, condition of flesh, amount of surplus abdominal fat, sex, and the completeness of the operation of drawing. On the average the dressing loss varies from 10 to 15 per cent of the live weight, and dressing and complete drawing, from 25 to 35 per cent.

The accompanying tables illustrate the loss in dressing and drawing for several classes of poultry:

A Type of No. 1 Quality

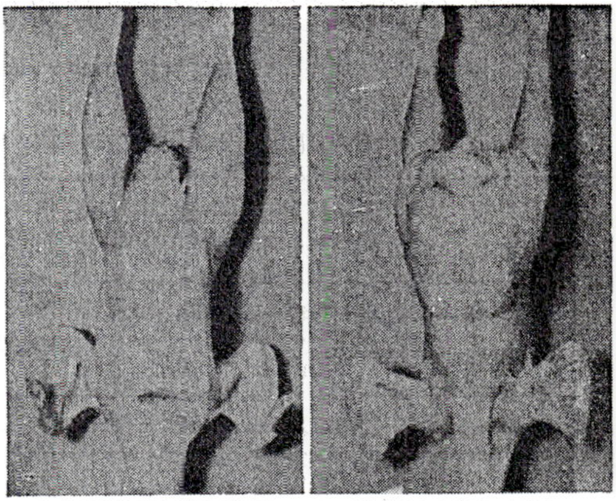

Breast and Back of No. 1 Quality

A Type of No. 2 Quality

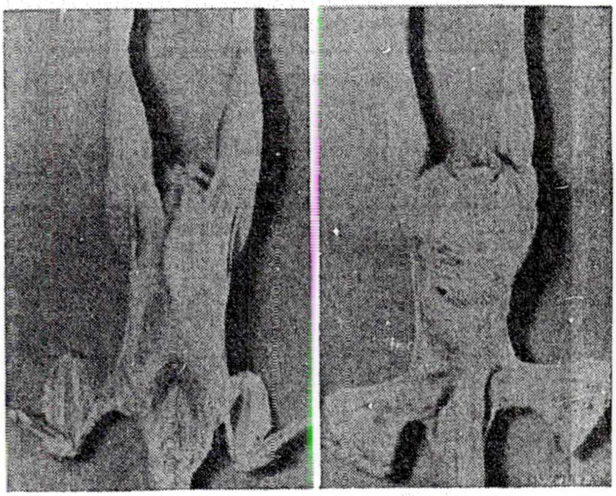

Breast and Back of No. 2 Quality

A Type of Reject

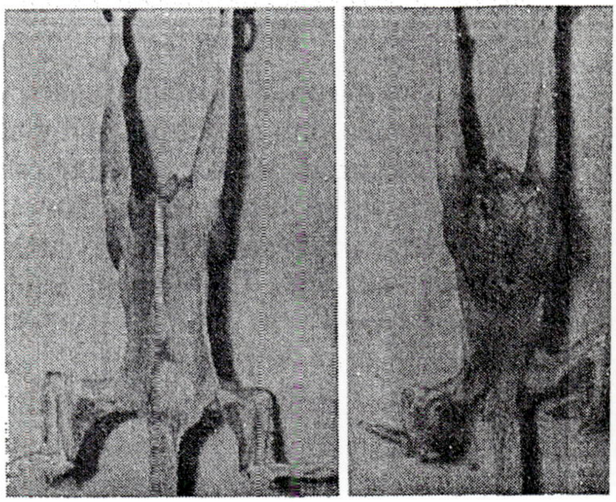

Breast and Back of Reject

Fig. 137

[Courtesy of Paul Mandeville, Institute of American Poultry Industries.]

TABLE 42—DRESSING AND DRAWING LOSSES

Birds	No. Birds	Live Weight Pounds	Dressed Weight		Drawn Weight	
			Pounds	% Loss	Pounds	% Loss*
Leghorn Broilers	21	52.96	45.43	14.21		
Leghorn Broilers	40	130			84.87	34.71
Leghorn Hens	63	214.49	184.99	13.75	149.12	30.47
Leghorn Cockerels	14	48.13	42.69	11.30	34.69	27.92
Leghorn Cockerels	29	91.18	76.28	16.34		
B. Rock Capons	38	277.56	247.68	10.76		

*Per cent loss is based on decrease from live weight.

TABLE 43—DRESSED AND DRAWN LOSS FOR DIFFERENT BREEDS†

	W. Wyan. C.	W. Wyan. P.	W. Legh. C.	W. Legh. P.	L. Brahma C.	L. Brahma P.	B. Rock C.	B. Rock P.
Average Live Wt.	5.46	4.53	3.36	2.54	4.50	3.84	3.23	3.05
Per Cent Loss in Dressing and Drawing	31.59	28.26	36.39	34.83	40.21	38.81	37.46	34.75

†Figures adapted from Penna. Agr. Exp. Sta. Bul. No. 87.

Selling Live Versus Dressed Poultry

The manner in which poultry meat should be disposed of will depend on many factors. Obviously it is necessary to consider the market in which one desires to sell. If it demands dressed and drawn poultry ready for the oven, this demand must be catered to for satisfactory outlets. On the other hand, if the market supplies the Jewish trade for Kosher killed poultry, the type desired will need to be furnished in a live condition. The distance from market outlets, as this affects shrinking in shipment, the available labor for dressing, and the volume of product—all these must receive attention in arriving at a decision to sell birds alive or dressed. In addition the price differential between live and dressed poultry, the type of business conducted, and the facilities for dressing birds are also governing factors.

TENTATIVE U. S. STANDARDS FOR CLASSES AND GRADES FOR LIVE POULTRY

Tentative U. S. Classes for Live Poultry

CHICKENS:

Broilers—Young, soft-meated chickens of either sex with tender skin, smooth bright shanks, and flexible breastbones.

Fryers—Young, soft-meated chickens of either sex with tender skin, smooth bright shanks, and flexible breastbones. Maximum weight 4 pounds.

Roasters—Young, soft-meated chickens of either sex with tender skin and smooth bright shanks. Breastbones somewhat more rigid than in broilers and fryers but with some remaining flexible cartilage over the end.

Capons—Young, soft-meated, unsexed male chickens, showing practically no comb development.

Stags—Young male chickens with somewhat toughened flesh and considerable hardening of the breastbones. Combs may be well developed and spurs may show considerable, but not full, development. Stags show, in general, an intermediate condition of flesh and sexual development between roasters and cocks.

Fowl—Mature female chickens with hardened breastbones. This includes all female chickens that have lost the soft flesh condition characteristic of young chickens. The fact that some fowl may be roasted by special or prolonged cooking is not sufficient grounds for classifying them as roasting chickens.

Cocks (Old Roosters)—Mature male chickens with toughened flesh and hardened breastbones.

TURKEYS:

Young Hen Turkeys—Young soft-meated female turkeys with tender skin (usually less than 1 year old).

Young Tom Turkeys—Young soft-meated male turkeys, with tender skin. May show slight pouchy condition of breast (usually less than 1 year old).

Old Hen Turkeys—Mature female turkeys, with toughened flesh. May have coarse or dry skin and patchy areas of surface fat (usually more than 1 year old).

Old Tom Turkeys—Mature male turkeys, with toughened flesh (usually more than 1 year old).

DUCKS:

Young Ducks—Young, soft-meated ducks of either sex, bills not completely hardened, easily dented windpipes. Their feathers may not be fully developed so they may have a somewhat downy appearance.

Old Ducks—Mature ducks of either sex, with toughened flesh, hardened bills, and hardened windpipes.

GEESE: Young or old, either sex.

GUINEAS:

Young Guineas—Young, soft-meated guineas of either sex.

Old Guineas—Mature guineas of either sex with toughened flesh.

SQUABS:

Extra soft-meated, immature pigeons of either sex.

PIGEONS:

Mature pigeons of either sex with toughened flesh.

Specifications for Tentative U. S. Standards for Individual Birds and Buying Grades for Live Poultry

U. S. Grade No. 1—A bird of this grade must be vigorous and free from external evidence of disease. Must be well-fleshed or fairly well-fleshed and fairly well-feathered. Must be soft-meated, if of a class in which soft meat is a requirement. It must be fairly well covered with fat and in this connection, proper consideration should be given to age and sex. May have slight but not serious defects or serious deformities. Considerable abdominal fat allowed in fowl.

U. S. Grade No. 2—Any edible bird below the quality of U. S. Grade No. 1 qualifies for this grade. It must be free from external evidence of disease or other conditions that might render the bird unwholesome for human food. A bird of thin flesh, or lacking in fat covering, or only partially feathered may be included if healthy. May have serious defects, serious deformities, or excessive abdominal fat.

Rejects—Any bird below U. S. Grade No. 2 that is extremely emaciated or badly crippled or with other external evidence of disease or unhealthy conditions that might make it unwholesome for human food.

Specifications for Tentative U. S. Wholesale Grades for Live Poultry

U. S. Grade No. 1—Each lot of live poultry must contain not less than 90 per cent of birds of the quality of U. S. Grade No. 1 the balance to be U. S. Grade No. 2, provided no individual containers in the lot shall have more than 15 per cent U. S. Grade No. 2 birds and shall contain no Rejects.

U. S. Grade No. 2—Each lot of live poultry shall consist of U. S. Grade No. 2 birds or better and shall contain no Rejects.

No Grade—Lots of poultry containing Rejects shall be classed as "No Grade."

Note—All poultry having excessive feed in the crops shall be considered "overcropped" and may be subject to dockage. "Excessive" feed shall be interpreted to mean not more than an average of (a) 1 ounce of feed in crops of birds weighing not more than 2½ pounds, (b) 2 ounces of feed in crops of birds (except turkeys) weighing more than 2½ pounds, (c) 3 ounces of feed in crops of turkeys.

EXPLANATION OF TERMS USED IN THE TENTATIVE U. S. STANDARDS FOR GRADES FOR LIVE POULTRY

VIGOROUS:
Bird having bright eyes and healthy appearance.

WELL-FLESHED:
Bird, which for its class, shows normally developed breast, with some fat covering over thighs and back and with a good covering of flesh over all bones.

FAIRLY WELL-FLESHED:
Bird with breast that may be slightly lacking in development with some evidence of fat over thighs and back and with all bones fairly well covered with flesh.

FAIRLY WELL COVERED WITH FAT:
Bird, which for its class, shows reasonable amount of fat in feather tracts, and some fat on back.

FAIRLY WELL-FEATHERED:
Feathers covering all parts of the body quite thoroughly except that bare backs are permitted if they are not badly sunburned or scabby. Bird may show a moderate number of pin feathers that may be scattered over all parts of the body.

DEFORMITIES:
Dented, notched, or slightly crooked breastbone, slightly crooked back, or slightly misshapen wings or legs.

SERIOUS DEFORMITIES:
Hunchback, crooked breastbone, or other definite deformities.

SLIGHT DEFECTS:
Scratches, slight bruises, slight breast blisters, slight calluses, or slightly scaly legs.

SERIOUS DEFECTS:
Large skin bruises, flesh bruises, severe breast blisters, heavy calluses, seriously scaly legs, one broken wing bone without fever, or the bird may be slightly crippled.

UNWHOLESOME:
Evidence of disease or other conditions that would render the bird unsuitable for human food.

TABLE 44

SUMMARY OF SPECIFICATIONS FOR TENTATIVE U. S. STANDARDS FOR CLASSES AND GRADES FOR LIVE POULTRY

Individual Birds

Grade Factor	U. S. Grade No. 1	U. S. Grade No. 2
VIGOR	Must be vigorous	May show lack of vigor but must be healthy
FREEDOM FROM DISEASE	No external evidence	No External evidence
FLESHING	Well fleshed or fairly well fleshed	May be poorly fleshed
CHARACTER OF FLESH	Soft meated (if of a class where soft meat is a requirement)	May be poor quality flesh but must be edible
FAT	Fairly well covered considering age and sex	May be lacking in fat covering
FEATHERING	Fairly well feathered	May be only partially feathered
DEFECTS	Slight defects permitted	Serious defects permitted
DEFORMITIES	Slight deformities permitted	Serious deformities permitted
SCALY LEG	Slightly scaly leg permitted	Seriously scaly leg permitted
ABDOMINAL FAT IN FOWL	Considerable allowed	Excessive permitted

Marketing Eggs

It is desirable and necessary to understand the market classification of eggs in order to produce, grade and dispose of them according to the market demands. Since various markets have somewhat different grade requirements, due to lack of standardization, it is advisable to thoroughly study them and cater to the best one available.

In the study of egg quality it is necessary to take into consideration the size and condition of the air cell and the condition of the yolk, albumen and shell.

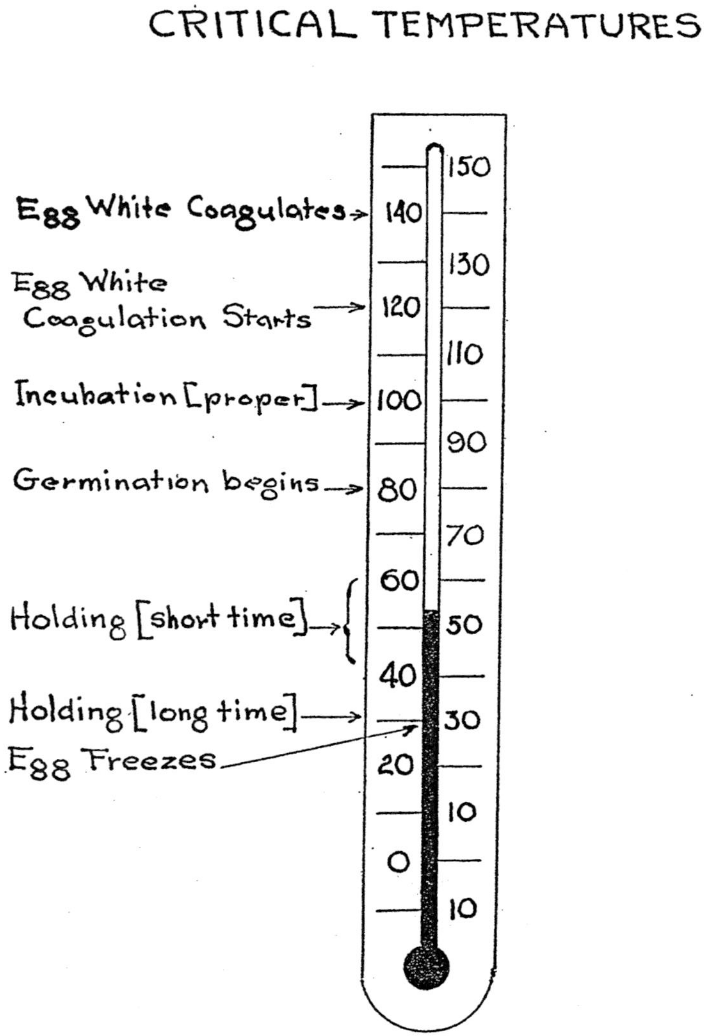

Fig. 138—Humidity levels for egg quality at low temperature (long time) 90-95% relative humidity (with ozone and CO[2]); at low temperature (long time) 85-88% relative humidity (without ozone and CO_2); at moderate temperature (short time) 80-85% relative humidity; at higher temperature 75-80% relative humidity.

[Higgins: NEPPCO Egg Grading & Marketing School, 1948]

SPECIFICATIONS FOR OFFICIAL UNITED STATES STANDARDS FOR QUALITY OF INDIVIDUAL SHELL EGGS

U. S. Standards for Quality of Individual Eggs with Clean Unbroken Shells:[1]

AA Quality—The shell must be clean, unbroken, and practically normal. The air cell must not exceed 1/8 inch in depth and be practically regular. The white must be clear and firm so that the yolk appears well centered and its outline only slightly defined when the egg is twirled before the candling light. The yolk must be free from apparent defects.

A Quality—The shell must be clean, unbroken and practically normal. The air cell must not exceed 2/8 inch in depth and must be practically regular. The white must be clear and at least reasonably firm so that the yolk appears at least fairly well centered and its outline only fairly well defined when the egg is twirled before the candling light. The yolk must be practically free from apparent defects.

B Quality—The shell must be clean, unbroken and may be slightly abnormal. The air cell must not exceed 3/8 inch in depth and may show total movement not in excess of 3/8 inch. However, an air cell not over 2/8 inch in depth may be free. The white must be clear but may be slightly weak so that the yolk may appear off center with its outline well defined when the egg is twirled before the candling light. The yolk may appear slightly enlarged and slightly flattened and may show other definite but not serious defects.

C Quality—The shell must be clean, unbroken, and may be abnormal. The air cell may be over 3/8 inch in depth and may be bubbly or free. The white may be weak and watery so that the yolk may appear off center and its outline plainly visible when

---

[1]Interior egg quality specifications for the Official Standards for Quality of Individual Shell Eggs are based on the use of a candling light delivering approximately 350 to 450 foot candles of light at the candling opening. The usual box type of candling light, without reflector, using a clean 40-watt frosted bulb about 1½ inches from and in direct line (direct light) behind the opening which should be approximately 1⅛ inches in diameter or a clean 60-watt frosted bulb immediately above and 1½ inches behind the opening (indirect light), provides approximately 310 foot candles of light at the opening. A 60-watt bulb in a direct light candler or a 75-watt bulb in an indirect light candler provides approximately 380 foot candles of light at the opening.

Reference to "usual box type of candling light" should not be construed as restricting use to that type only. Any type or make of candler may be used so long as the resulting light is the same.

the egg is twirled before the candling light. The yolk may appear dark, enlarged, and flattened and may show clearly visible germ development but no blood due to such development. It may show other serious defects that do not render the egg inedible. Small blood clots or spots may be present.

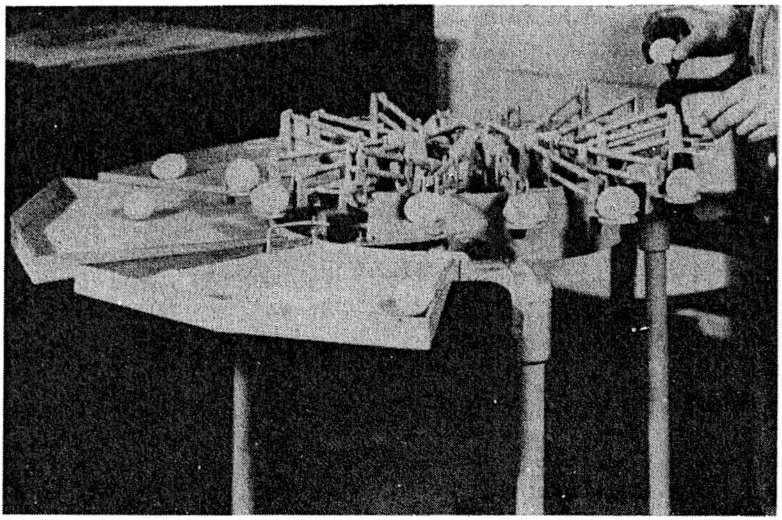

Fig. 139—Rotating type egg grader.
[Courtesy Elmore and Company]

U. S. Standards for Quality of Individual Eggs with Dirty Unbroken Shells

Stained—Individual egg that has no adhering dirt and no more than a combined total of 1/8 of the shell surface stained or soiled.

Dirty—Individual egg that has adhering dirt or more than a combined total of 1/8 of the shell surface stained or soiled.

U. S. Standards for Quality of Individual Eggs with Checked or Cracked Shells:

Check—Individual egg that has a broken shell or crack in the shell but with no leakage of the contents.

Leaker—Individual egg that has a broken shell or crack in the shell and shell membranes with the contents exuding or free to exude through the shell.

Explanation of Terms

The Official United States Standards for Quality of Individual Shell Eggs are applicable to eggs that are the product of the domesticated chicken hen and are in the shell.

Terms Descriptive of Shell

Clean—A shell that is free from foreign matter and from stains or discolorations that are readily visible. Eggs with only very small specks or stains may be considered clean, if such eggs are not of sufficient number in a package to detract appreciably from its appearance. Eggs that show traces of processing oil on the shell are considered clean unless the shell is otherwise soiled.

Stained—A shell with stained or soiled spots that together cover not more than 1/8 inch of shell surface but without adhering dirt.

Dirty—A shell with adhering dirt or with stained or soiled spots that together cover more than 1/8 of the shell surface.

Unbroken—A shell that is free from checks or breaks.

Checked or Cracked—A shell that has an actual break but its membranes are unbroken and its contents do not leak.

Leaker—A leaker is an egg in which the shell and shell membranes are broken to the extent that the egg contents are exuding or free to exude through the shell.

Practically Normal (AA), (A)—A shell that approximates the usual shape and that is of good even texture and strength and free from rough areas or thin spots. Slight ridges and rough areas that do not materially affect the shape, texture and strength of the shell, are permitted.

Slightly Abnormal (B)—A shell that may be somewhat unusual in shape or that may be slightly faulty in texture or strength. It may show definite ridges but no pronounced thin spots or rough areas.

Abnormal (C)—A shell that may be decidedly misshapen or faulty in texture or strength or that may show pronounced ridges, thin spots or rough areas.

Terms Descriptive of the Air Cell

Depth of Air Cells—(air space between shell membranes, normally in the large end of the egg)—The depth of the air cell is the distance from its top to its bottom when the egg is held air cell upward.

Practically Regular (AA) (A)—An air cell that maintains a practically fixed position in the egg and shows a fairly even outline, with no more than 1/8 inch movement in any direction as the egg is rotated.

Movement Not in Excess of 3/8 Inch (B)—An air cell that shows a total movement not in excess of 3/8 inch in any direction as the egg is rotated.

Free Air Cell (B) (C)—An air cell that moves freely toward the uppermost point in the egg as the egg is rotated slowly.

Bubbly Air Cell (C)—A ruptured air cell resulting in one or more small separate air bubbles usually floating beneath the main air cell.

Terms Descriptive of the White

Clear (AA) (A) (B)—A white that is free from discoloration or from any foreign bodies floating in it. (Prominent chalazas should not be confused with foreign bodies such as spots or blood clots).

Firm (AA)— A white that is sufficiently thick or viscous to permit but limited movement of the yolk from the center of the egg, thus preventing the yolk outline from being more than slightly defined or indistinctly indicated when the egg is twirled.

Reasonably Firm (A)—A white that is somewhat less thick or viscous than a firm white. A reasonably firm white permits the yolk to move somewhat more freely from its normal position in the center of the egg and approach the shell more closely. This would result in a fairly well defined yolk outline when the egg is twirled.

Slightly Weak (B)—A white that is lacking in thickness or viscosity to an extent that permits the yolk to move quite freely from its normal position in the center of the egg. A slightly weak white will cause the yolk outline to appear well defined when the egg is twirled.

Weak and Watery (C)—A white that is thin and generally lacking in viscosity. A weak and watery white permits the yolk to move freely from the center of the egg and to approach the shell closely, thus causing the yolk outline to appear plainly visible and dark when the egg is twirled.

Blood Clots and Spots (Not due to germ development)—Blood clots or spots on the surface of the yolk or floating in the white. These blood clots may have lost their characteristic red color and appear as small spots or foreign material commonly referred to as meat spots. Such blood clots or spots are incorporated in the egg during its formation or after the yolk leaves the ovary. If they are small (not over 1/8 inch in diameter) the egg may be classed as C quality. If larger and/or showing diffusion of blood in the white surrounding them, the egg shall be classified as loss.

Bloody White (Loss)—An egg, the white of which has blood diffused through it. Such a condition may be present in new-laid eggs. Eggs with bloody whites are classed as loss.

Terms Descriptive of Yolk

Well Centered (AA)—A yolk that occupies the center of the egg and moves only slightly from that position as the egg is twirled.

Fairly Well Centered (A)—A yolk that is not more than one-fourth of the distance from its normal central position toward the ends of the egg and swings not more than one-half of the distance from its normal position toward the sides of the egg as it is twirled.

Off Center (B) (C)—A yolk which is distinctly above or below center and swings close to the sides of the egg as it is twirled.

Outline Slightly Defined (AA)—A yolk outline that is indistinctly indicated and appears to blend into the surrounding white as the egg is twirled.

Outline Fairly Well Defined (A)—A yolk outline that is discernible but not clearly outlined as the egg is twirled.

Outline Well Defined (B)—A yolk outline that is quite definite and distinct as the egg is twirled.

Outline Plainly Visible (C)—A yolk outline that is clearly visible as a dark shadow when the egg is twirled.

Slightly Enlarged and Slightly Flattened (B)—A yolk in which the yolk membranes and tissues have weakened somewhat causing it to appear slightly enlarged and slightly flattened.

Enlarged and Flattened (C)—A yolk in which the yolk membranes and tissues have weakened and moisture has been absorbed from the white to such an extent that it appears definitely enlarged and flat.

Free From Defects (AA)—A yolk that shows no spots or areas on its surface indicating the presence of germ development or other defects.

Practically Free From Defects (A)—A yolk that shows no germ development but may show other very slightly defects on its surface.

Definite But Not Serious Defects (B)— A yolk that may show definite spots or areas on its surface but with only slight indication of germ development or other pronounced or serious defects.

Other Serious Defects (C)—A yolk that shows well developed spots or areas and other serious defects, such as olive yolks, which do not render the egg inedible.

Clearly Visible Germ Development (C)—A development of the germ spot on the yolk of a fertile egg that has progressed to a point where it is plainly visible as a definite circular area or spot with no blood in evidence.

Blood Due to Germ Development (Inedible)—Blood caused by development of the germ in a fertile egg to the point where it is visible as definite lines or blood ring. Such eggs are classified as inedible.

General Terms

Loss—Eggs that are inedible, smashed, broken so that contents are leaking, contaminated, or containing bloody whites, large blood spots, large unsightly meat spots, or other foreign material are classed as "Loss."

Inedible Eggs—Under the Food, Drug, and Cosmetic Act, eggs that are filthy, putrid, or decomposed, or otherwise unfit for food in whole or in part, are adulterated. Eggs of the following descriptions are classed as inedible: black rots, white rots, mixed rots (addled eggs), sour eggs, eggs with green whites, eggs with stuck yolks, moldy eggs, musty eggs, eggs showing blood rings, eggs containing embryo chicks (at or beyond the blood ring stage), and any other eggs that are filthy, decomposed, or putrid.

Tentative U. S. Specifications and Weight Classes For Wholesale Grades For Shell Eggs
(Based on U. S. Standards for Quality of Individual Shell Eggs)

U. S. Specials ....% AA Quality—Shall consist of eggs of which at least 20 per cent are AA Quality. The balance may be A Quality except for a permitted tolerance of 27 edible eggs (7.5 per cent) per 30-dozen case, which may be B Quality, C Quality, Stained, Dirties, or Checks in any combination and 6

TABLE 45

SUMMARY OF UNITED STATES STANDARDS FOR QUALITY OF INDIVIDUAL SHELL EGGS

Quality Factor	Specifications for Each Quality Factor							
	AA Quality	A Quality	B Quality	C Quality	Stained	Dirty	Check	Leaker
Shell	Clean, unbroken, practically normal	Clean, unbroken, practically normal	Clean, unbroken, may be slightly abnormal	Clean, unbroken, may be abnormal	Unbroken may be stained or soiled	Unbroken may be dirty	Checked or cracked but not leaking	Broken so contents are leaking
Air Cell	1/8 inch or less in depth, practically regular	1/4 inch or less in depth, practically regular	3/8 inch or less in depth may show movement not over 3/8 inch, If not over 1/4 inch may be free	May be over 3/8 inch in depth, may be free or bubbly				
White	Clear, Firm	Clear, May be reasonably firm	Clear, may be slightly weak	Clear, may be weak and watery, small blood clot or spots may be present				
Yolk	Well centered, outline slightly defined, free from defects	May be fairly well centered, outline may be fairly well defined, practically free from defects	May be off center, outline may be well defined, may be slightly enlarged and flattened, definite but not serious defects	May be off center, outline may be plainly visible, may be enlarged and flattened, may show clearly visible germ development but no blood, may show other serious defects				

eggs (1.7 per cent) loss. The actual percentage of AA Quality shall be stated in the grade name.

U. S. Extras ....% *A Quality*—Shall consist of eggs of which at least 20 per cent are A Quality or better. The balance may be B Quality except for a permitted tolerance of 42 edible eggs (11.7 per cent) per 30-dozen case, which may be C Quality, Stained, Dirties, or Checks in any combination and 8 eggs (2.2 per cent) loss.[1] The actual percentage of A Quality shall be stated in the grade name.

U. S. Standards ....% *B Quality*—Shall consist of eggs of which at least 20 per cent are B Quality or better. The balance may be C Quality or Standard,[2] except for a permitted tolerance of 42 edible eggs (11.7 per cent) per 30-dozen case, which may be Dirties or Checks in any combination and 10 eggs (2.8 per cent) loss.[1] The actual percentage of B Quality shall be stated in the grade name.

TABLE 46

WEIGHT CLASSES FOR U. S. WHOLESALE GRADES FOR SHELL EGG

Size or weight class	Average net weight per 30 dozen	Minimum net weight per 30 dozen	Basic minimum weight for individual eggs	Permitted tolerance 10 percent of individual eggs (minimum weight)
	Pounds	Pounds	Ounces per dozen	Ounces per dozen
Extra large	50½	50	26	Under 26 but not under 24
Large	45	44	23	Under 23 but not under 21
Medium	39½	39	20	Under 20 but not under 18
Small	34	No limit	No limit	No limit

These weight classes apply to all wholesale grades except U.S. Dirties and U.S. Checks for which there are no weight requirements.

U. S. Trades ....% *C Quality*—Shall consist of eggs of which at least 83.3 per cent are C Quality or Stained or better. The permitted tolerance may consist of no more than 42 eggs (11.7 per cent) per 30-dozen case, which may be Dirties or Checks in any combination and 18 eggs (5 per cent) loss. The actual percentage of C Quality or Stained shall be stated in the grade name.

U. S. Stained—Eggs with stained shells but otherwise conforming to the grade specifications for U. S. Extras or U. S.

[1]For the period from August 15 through January 31 the tolerance for loss may be increased to 3.3 per cent or 12 eggs per 30-dozen for U. S. Extras, and 4.2 per cent or 15 eggs per 30-dozen for U. S. Standards on refrigerator eggs only.

[2]Not more than 40 per cent stained eggs are permitted in U. S. Standards. If the quantity is in excess of 40 per cent, the eggs shall be classified as U. S. Stained Standards.

Standards, may be classified as "Stained" in their respective U. S. Grade. (Examples of grade identification: U. S. Stained Extras ....% A Quality and U. S. Stained Standards ....% B Quality.)

U. S. Dirties—Shall consist of edible eggs which may have dirty shells except for a permitted tolerance of 42 U. S Checks (11.7 per cent) per 30-dozen case and 18 eggs (5 per cent) loss.

U. S. Checks—Shall consist of edible eggs which may have checked shells except for a permitted tolerance of 18 eggs per 30-dozen case (5 per cent) loss.

No Grade—Eggs of possible edible quality that fail to meet the requirements of an Official or Tentative U. S. Grade or that have been contaminated by smoke, chemicals, or other foreign

TABLE 47

SUMMARY OF SPECIFICATIONS FOR TENTATIVE U. S. WHOLESALE GRADES FOR SHELL EGGS

Grade	AA Quality	A Quality or better	B Quality or better	C Quality or Stained or better	Tolerance Permitted[1]				
					Edible Eggs				Loss
					B Quality, C Quality, Stained, Dirties, or Checks	C Quality, Stained, Dirties, or Checks	Dirties or Checks	Checks	
	Minimum Percent	Minimum Percent	Minimum Percent	Minimum Percent					
U.S Specials ____% AA Quality[2]	20	Balance			7.5%				1.7%
U. S. Extras ____% A Quality[2]		20	Balance			11.7%			2.2%[3]
U.S. Standards ____% B Quality[2]			20	Balance[4]			11.7%		2.8%[3]
U.S. Trades ____% C Quality[2]				83.3			11.7%		5%
U.S. Stained	Eggs with stained shells but otherwise conforming to quality specifications of grades for U.S. Extras and U.S. Standards may be classified as U.S. Stained Extras____% A Quality or U S. Stained Standards____% B Quality, respectively.								
U.S. Dirties								11.7%	5%
U.S. Checks									5%

[1]Substitution of higher qualities for the lower qualities specified is permitted.
[2]The actual percentage must be stated in the grade name.
[3]For the period from August 15 through January 31 the tolerance for loss may be increased to 3.3 percent for U.S. Extras and 4.2 percent or U S. Standards on refrigerator eggs only.
[4]No more than 40 percent stained eggs permitted in U.S. Standards. If in excess of 40 percent they shall be classified as U.S. Stained Standards.

material that has seriously affected the character, appearance, or flavor of the eggs are classed as "No Grade."

TABLE 48

WEIGHT CLASSES FOR CONSUMER GRADES FOR SHELL EGGS

Size or weight class	Minimum net weight per dozen	Minimum net weight per 30 dozen	Minimum weight for individual eggs at rate per dozen[1]
	Ounces	Pounds	Ounces
Jumbo	30	56	29
Extra Large	27	50½	26
Large	24	45	23
Medium	21	39½	20
Small	18	34	17
Peewee	15	28	—

[1]Minimum weight listed for individual eggs at the rate per dozen are permitted in various size classes only to the extent that they will not reduce the net weight per dozen below the requred minimum, consideration being given to variable weight of individual eggs and variable efficiency of graders and scales which should be maintained on a uniform and accurate basis.

Tentative U. S. Specifications and Weight Classes For Consumer Grades For Shell Eggs

(Based on U. S. Standards for Quality of Individual Shell Eggs)

U. S. Consumer Grade AA—Shall consist of edible eggs of which at least 80 per cent[1] are AA Quality, 15 per cent[1] may be A quality, and not over 5 per cent[1] may be of the qualities below A, in any combination, but not including Dirties.[2]

U. S. Consumer Grade A—Shall consist of edible eggs of which at least 80 per cent[1] are A quality or better, 15 per cent[1] may be B quality, and not over 5 per cent[1] may be of the qualities below B, in any combination, but not including Dirties. [2] [3]

U. S. Consumer Grade B—Shall consist of edible eggs of which at least 80 per cent[1] are B quality or better, 10 per cent[1] may be C Quality or Stained, in any combination, and not over 10 per cent[1] may be Dirties or Checks in any combination.[2] [3]

U. S. Consumer Grade C—Shall consist of edible eggs of which at least 80 per cent[1] are C Quality or Stained, in any combination, or better, and the balance may be Dirties or Checks in any combination.[2]

[1]Officially graded eggs shall conform as nearly as possible to the specifications of the respective standards of quality. Tolerances (a total of 20 per cent) are permitted within each grade only as an allowance for variable efficiency and interpretation of conscientious graders, normal changes under favorable conditions during reasonable period between grading and inspection, and reasonable variation from inspector's interpretation. Substitution of higher qualities for those specified is permitted.

[2]Within the total tolerance permitted an allowance will be made at receiving points or shipping destination for ½ per cent leakers in Grades AA, A, and B, and 1 per cent in Grade C.

[3]Eggs with stained shells but otherwise conforming to the specifications of Grade A or Grade B, may be classified as Grade A, Stained, or Grade B, Stained, respectively

No Grade—Eggs of possible edible quality that fail to meet the requirements of an Official or Tentative U. S. Grade or that have been contaminated by smoke, chemicals, or other foreign material that has seriously affected the character, appearance, or flavor of the eggs are classed as "No Grade."

TABLE 49

SUMMARY OF SPECIFICATIONS FOR U. S. CONSUMER GRADES FOR SHELL EGGS

U. S. Consumer Grade	At least 80 percent (lot average)[1] must be—	Tolerance permitted[2]	
		Percent	Quality
U. S. Grade AA	AA Quality	15 to 20, not over 5[3]	A B C. Stained, or Check
U. S. Grade A	A Quality or better	15 to 20, not over 5[3]	B C, Stained, or Check
U. S. Grade B	B Quality or better	10 to 20, not over 10[3]	C, or Stained, Dirty or Check
U. S. Grade C	C Quality or better	Not over 20	Dirty or Check

[1]In lots of more than 30 cases no individual case may fall below 70 percent of the specified quality and in lots of 30 cases or less the 80 percent minimum requirement shall apply to each individual case.
[2]Within tolerance permitted, an allowance will be made at receiving points or shipping destination or ½ percent leakers in Grades AA, A, and B, and 1 percent in Grade C.
[3]Substitution of higher qualities for the lower qualities specified is permitted.

Cleaning Dirty Eggs

Work conducted by Funk[1] indicates that dirty eggs may be cleaned in a weak solution of sodium hydroxide, without affecting their keeping quality in storage. He states, "these results, however, do show that by cleaning dirty eggs with water containing .7 per cent and .35 per cent sodium hydroxide their edible properties may be preserved, and that the loss in these eggs may not exceed that found in clean unwashed eggs."

The water containing sodium hydroxide used in washing eggs should be changed often so that eggs are thoroughly cleaned of all contamination. Rubber gloves must be used by those washing eggs with lye water.

Dry cleaning of eggs utilizing abrasive surfaces is frequently used by the average poultryman.

The use of detergents in the cleaning of eggs has been a recent development in the central section of New Jersey. Leslie Black, Extension Specialist reports on this system as follows:

"The eggs are collected in wire egg baskets and brought to the cellar or egg room. A clean galvanized ash barrel or other

[1]Missouri Agri. Exp. Sta. Bull. 394, 1938.

UNITED STATES STANDARDS FOR (

Illustrations of candled appearance of white-shelled eggs showing n

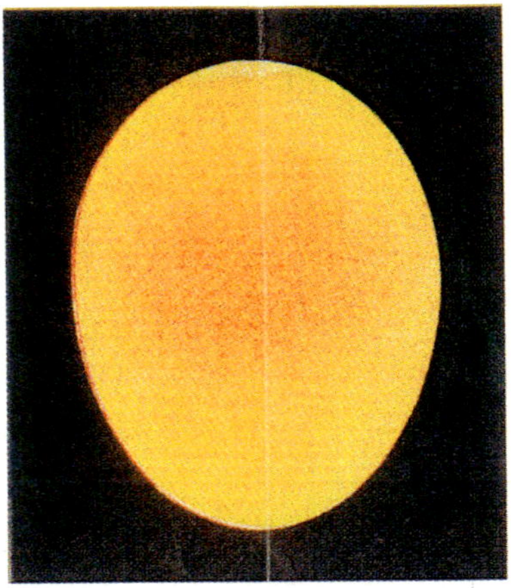

AA Quality

1. Shell—clean; unbroken; practically normal.
2. Air cell—1/8 inch or less in depth; practically regular.
3. White—clear; firm.
4. Yolk—well centered; outline slightly defined, free from defects.

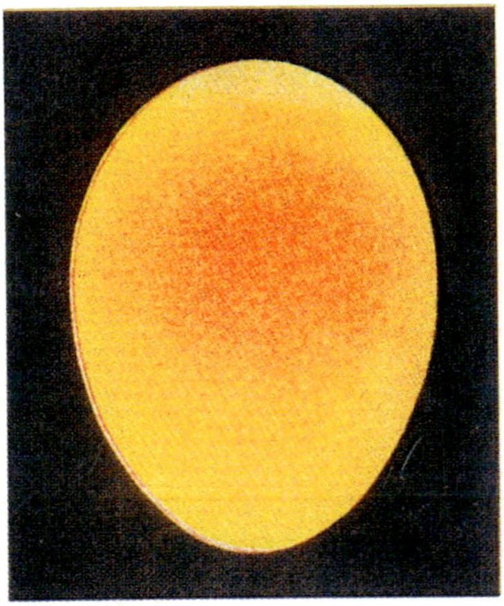

A Quality

1. Shell—clean; unbroken; practically norm
2. Air cell—2/8 inch or less in depth; pr tically regular.
3. White—clear; may be reasonably firm.
4. Yolk—may by fairly well centered; outl fairly well defined; practically free from defe

Illustrations of broken-out appearance (to

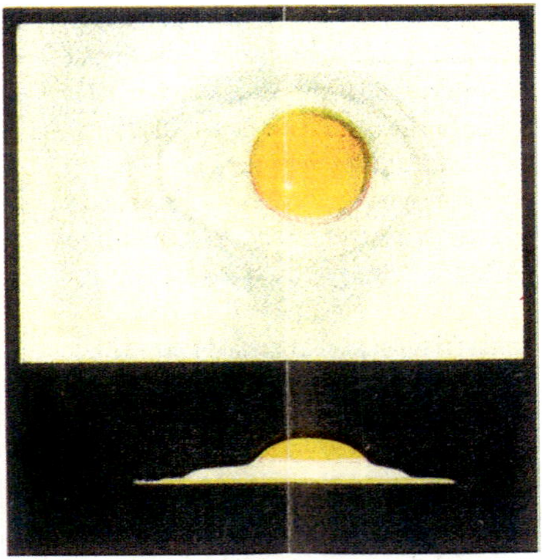

AA Egg covers small area; much thick white surrounds yolk; has small amount of thin white; yolk round and upstanding.

A Egg covers moderate area; has considera thick white; medium amount of thin wh yolk round and upstanding.

Graders should check their work by breaking out an egg occasionally and comparin

...LITY OF INDIVIDUAL SHELL EGGS

...m depth of air cell and outline and position of yolk in each quality

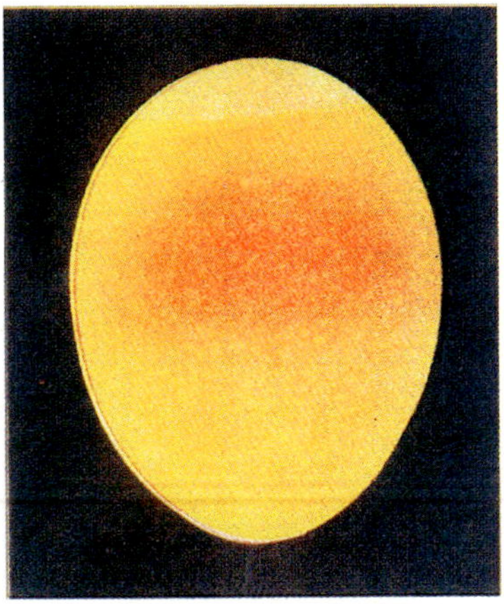

B Quality

1. Shell—clean; unbroken; may be slightly ...bnormal.

2. Air cell—3/8 inch or less in depth; may ...how movement not over 3/8 inch; if not over 2/8 ...nch, may be free.

3. White—clear; may be slightly weak.

4. Yolk—may be off center; outline well defined; ...nay be slightly enlarged and flattened; may show definite but not serious defects.

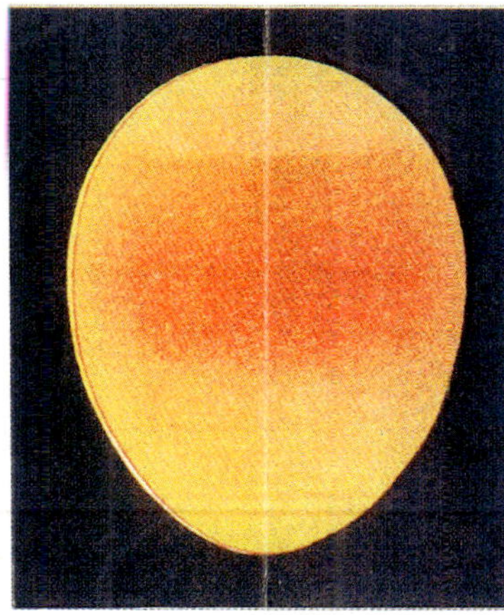

C Quality

1. Shell—clean; unbroken; may be abnormal.

2. Air cell—may be over 3/8 inch in depth; may be free or bubbly.

3. White—clear; may be weak and watery; small blood clots or spots may be present.

4. Yolk—may be off center, enlarged, and flattened; may show clearly visible germ development but no blood; may show other serious defects; outline plainly visible.

...side views) of each quality—1/3 actual size

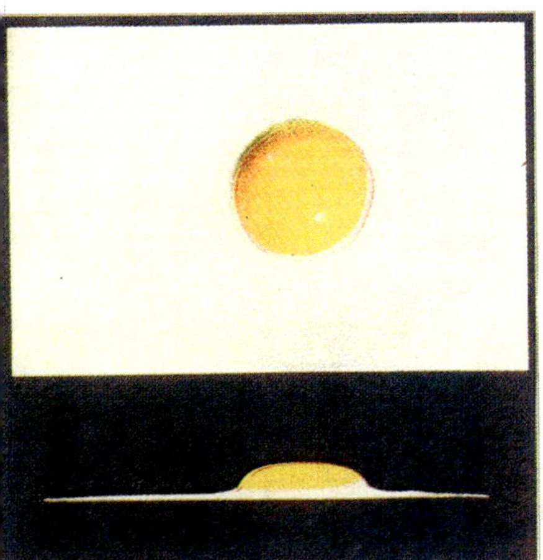

B Egg covers wide area; has small amount of thick white; much thin white; yolk somewhat ...flattened and enlarged.

C Egg covers very wide area; has no thick white; large amount of thin white thinly spread; yolk very flat and enlarged.

...his chart. U. S. Department of Agriculture, Production and Marketing Administration.

TABLE 50
SUPPLEMENT GRADE SPECIFICATIONS FOR GOVERNMENT PURCHASE ONLY

Percentage of A Quality	A Quality or better	Maximum number and percentage of eggs allowed as tolerance per 30 dozen		
		B Quality or better	Edible eggs of C Quality, Stained or Checks, or better, in any combination	Loss and inedible
	Minimum Percent			
80% A Quality (Identical with U. S. Consumer Grade A)	80	54 to 72 15 to 20%	18 (5%)	None
60% A Quality	60	108 to 144 30 to 40%	36 (10%)	None
40% A Quality	40	174 to 216 48.3 to 60%	42 (11.7%)	None
20% A Quality	20	246 to 288 68.3 to 80%	42 (11.7%)	None

Tolerance for leakers will be applied only at receiving points or shipping destination.

suitable container is filled with enough warm water (140 to 160 degrees Fahrenheit as it comes from the broiler) to cover a filled basket of eggs. A teaspoonful of a detergent (Dreft, Vel or similar product) is added for each gallon of water. A basket of eggs is then dipped so that all eggs are covered and the basket is then removed and set on the floor to drain. The process is continued until all the baskets have been dipped. As many as twenty or thirty baskets can be immersed in the treated water. When all baskets have been dipped, the first basket is dipped again and set upon an elevated stand. A hose connected to the hot water boiler and equipped with a hand controlled spray valve is then used to wash the loosened dirt from the eggs by applying the

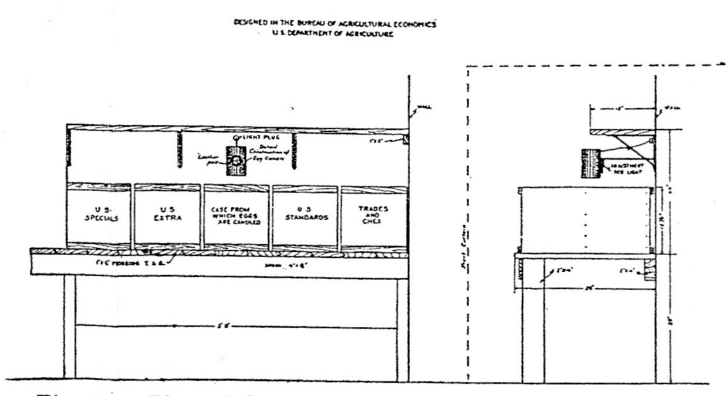

Fig. 140—Plan of Construction of an Egg Candling Booth.

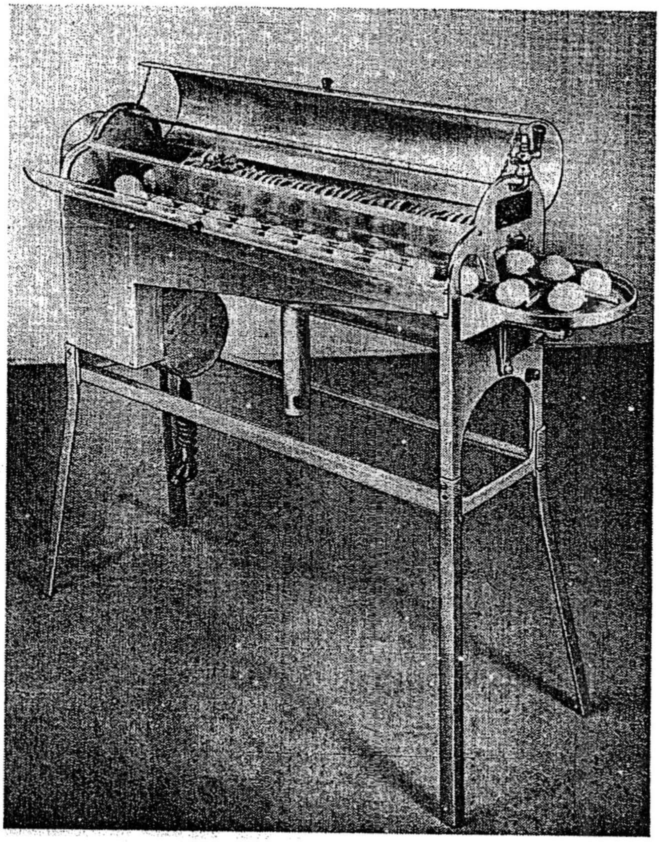

Fig. 141—Wright 101 Egg Washer.
[Courtesy of Coop. Grange League Federation Exchange, Inc.]

spray of warm water to all sides of the basket. Each basket is dipped and washed in turn. The washed baskets of eggs are then placed in the cool room to dry until the next morning when the eggs are graded and packed in cases. Some soiled eggs may be found in the center of the basket where the warm water spray failed to penetrate. The dirt is readily removed with a hand buffer due to the loosening action of the detergent."

Grading Eggs and Candling:

The classification of eggs by grade and the defects observable by candling form a basis for the grading and candling practice.

In grading and candling eggs it is necessary that both interior and exterior qualities be considered. Before one can recognize abnormal conditions he must have thorough knowledge of the normal.

An egg is made up of several components; namely, yolk, germinal disk, chalazae, albumen, shell membranes, air cell, and shell. When a normal fresh egg is whirled before a candling apparatus it will present a clear appearance, the yolk will be dimly visible and the air cell will be small and localized.

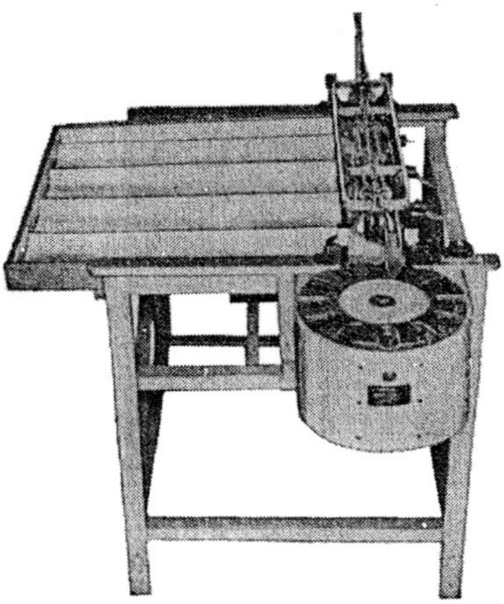

Fig. 142—Model "B," Philadelphia Egg Grader. Output 5 cases per hour, sorting into 5 grades by weight.
[Courtesy Philadelphia Metal Works.]

Fig. 143—Automatic Grading and Packing Machine.
[Courtesy S-W-S Manufacturing Co.]

Candling is the process of determining interior egg quality by visibility obtained through illumination. Various types of candling devices can be made or purchased. There are two general types, one adapted to use in a lighted or semi-lighted room and the other to a relatively dark room. The latter is better suited for average use because it will enable the operator to judge interior quality more easily.

In candling, the egg is placed directly in the shaft of light and given a quick twirl to rotate its contents. This rotation makes possible careful observation of all its interior characteristics and identification of abnormal conditions. In whirling the egg the appearance of the chalazae should not be mistaken for a defect. It can readily be distinguished through its reflection of light.

The aperture in the candler should be located at a convenient height to facilitate manipulation of eggs. Most commercial operators make a practice of handling two eggs simultaneously in each hand, rotating them in turn before the candler, thus increasing materially the speed of the operation.

Fig. 144—Five grades of eggs with automatic grading unit.
[Courtesy Guy A. Leader, Leaders Leghorn Farms.]

Candling makes it possible to determine interior quality of eggs. In the marketing of eggs it will also be found advantageous to consider the exterior factors of size, shape, cleanliness and color. Eggs may be graded for these various factors while they are being handled for candling.

It is well to bear in mind that much research work has been done, and will continue, until better candling devices will be developed. This detailed study will undoubtedly result in a decided change in both methods of candling and grading. This should be of benefit to the producer, the middleman, as well as the ultimate consumer.

Cartoning and Packing:

After eggs are candled and graded they are placed in cartons or packed in 30-dozen shipping cases. The use of cartons is quite largely restricted to eggs going to retailers or direct to the consumer. When used they should also serve as an advertising medium for the product. They should be attractive, substantially constructed to prevent breakage, and sealed to avoid substitution, particularly when the eggs are not placed directly in the hands of the consumer. It is desirable that the eggs be graded so that the contents of the cartons are uniform and standardized.

If cartoned eggs are sold through retail stores, they should be furnished the retailer only in sufficient quantity to take care of his immediate needs. This precaution will prevent deterioration of eggs when sales are slow.

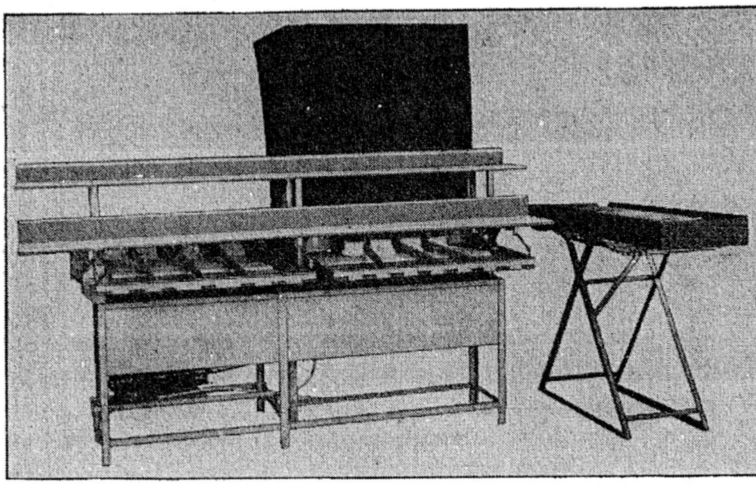

Fig. 145—Automatic Grading and Candling Machine.
[Courtesy National Industries, Inc.]

Packing the eggs in 30-dozen cases is usually done when eggs are wholesaled. The use of new or undamaged fillers is recommended to prevent side shifting of the contents of the case, thus eliminating a source of breakage. The cup-type flats are in general use and express records show that they have decidedly reduced breakage in shipment. Cushioning should be provided for both the top and bottom of the case, through the use of cup flats or excelsior pads. Take care that all packing material is dry and free from objectionable odors.

In packing the eggs, place the pointed end downward so that the weight of the egg contents will not bear on the shell membrane at the air cell. This method will give the packed eggs a better appearance. The eggs in all layers should be representative of the case contents so that the grade can readily be determined by inspection of the top layers. Extremely long eggs which would protrude above the top of the fillers and be likely to break should not be packed but instead used for local requirements.

After the case is completely filled and the top cushioning has been placed, apply the top of the case, nailing only at the ends. Cardboard cases are also in general usage. In some sections, returnable wooden cases with hinged tops are in use. Consult your local railroad or express agent regarding express classifications and any special packing regulations.

Egg Marketing Programs

Many states have adopted state grades for the classification, standardization, and identification of market eggs. These grades have been promulgated by the state departments of agriculture and their use may or may not be optional with producers and organizations.

For eggs that are specially packaged, some states have developed a recognized and distinctive carton. In addition, gummed labels identifying the producer are used to seal the product in their cartons. This seal serves a double purpose: (1) to identify the product and the producer, and (2) to give the separate state departments of agriculture an opportunity to check on the adherence to grade. The seal in addition prevents substitution of other eggs in the carton and assures the consumer of a quality product.

Salesmanship

The satisfactory disposal of poultry produce, as in other lines of endeavor, is largely dependent on the sales ability of the producer in finding satisfactory market outlets. The producer who does not exert energy to secure these favorable outlets must be content to take a lower price for his commodities. Progressive effort must be exerted not only to make proper contacts but to maintain uniform quality of produce so that demand may be held constant. To make proper market contacts it is essential that the producer become acquainted with the market to which he is catering and study the specific requirement as to type, quality and grade of product demanded and the effect, if any, of seasonable demands as affecting these factors.

The development of the local market with direct-to-the-consumer sale will be found to the advantage of the producer where such contacts can be made. The producer is mainly interested in the net returns and these are likely to be higher than where it is necessary to ship to distant wholesale markets. Any surplus over local demands can be disposed of through regular wholesale market channels.

Questions on Chapter XV

1. List the market channels available for the disposal of poultry products.
2. What are some of the requirements for successful cooperative marketing?
3. In killing of poultry, describe the sticking and debraining method.
4. What are the procedures in dry picking, semi-scalding, and waxing employed in dressing birds for market?
5. What is the percentage loss in weight sustained in dressing?
6. What is the loss in weight where birds are not only dressed but completely drawn?
7. What is meant by the following market terms: fowls, stags, broilers, chickens, capons, fryers and roasters?
8. Define a U. S. Grade No. 1 bird.
9. Give the minimum weight for individual eggs at the rate per dozen in consumer grades.
10. What is meant by the following market terms: U. S. Grade AA, U. S. Grade A, U. S. Grade B, and U. S. Grade C.
11. List the component parts of an egg.
12. How is the interior quality of an egg determined?
13. Describe precautions to be taken in packing eggs in cases.

CHAPTER XVI

FATTENING*

Subject Matter

Systems of Fattening
Feed Ingredients Commonly Used
Size of Birds
Rations
Marketing Fattened Birds

Little fattening is practiced on the farm; the bulk of intense fattening is done by commercial packing interests and market receivers. Commercial packing plants collect birds at concentration points in areas where large numbers are available, particularly in the mid-west. The birds are then properly graded, fattened, killed and packed. In this manner no live-bird shrinkage is entailed in shipment to consuming areas.

The birds are packed according to size and quality so that uniformity is outstanding. The volume handled on this basis makes it impossible for the individual poultryman to compete successfully. In spite of this, birds going to market certainly need to be in suitable physical condition and carry a reasonable amount of fat, preferably well distributed over the carcass, if a satisfactory price is to be secured. Birds fattened on the farm can best be used for local consumption, as live shipment of poultry results in excessive shrinkage.

Systems of Fattening

There are four systems of fattening: (1) cramming, (2) "noodling," (3) pen fattening, and (4) crate fattening. Cramming, whereby a mechanical device is used to force feed into the crop, unnaturally increases the food intake of birds. It is productive of a splendid carcass of exceptional quality and is practiced in the British Isles and elsewhere. It requires individual handling of the birds and a special cramming machine, but due to the labor involved it is not looked upon favorably in the United States.

*The term fattening, as here used in reference to poultry, designates the feeding and management of birds prior to marketing to increase the weight and improve the condition of the flesh for human consumption.

"Noodling," as a fattening practice, refers to a special method of fattening geese. The geese are confined to a small pen and fed a special cooked ration by hand several times daily.

Pen fattening refers to the confinement of a considerable number of birds in small pens to restrict exercise. Quite often

Fig. 146—In the Feeding Batteries at a Commercial Fattening Plant. Note the Eagerness for Feed.
[Courtesy E. B. Heaton, Institute of American Poultry Industries.]

a pen-fattening ration composed of half the regular laying mash and half yellow corn meal is used, although many other rations are applicable.

The practice of crate fattening is most frequently used in commercial poultry-fattening houses, and gives maximum results for the labor involved. The equipment may consist of a home-made crate, but if the enterprise is on a large scale purchased fattening batteries are advisable. Generally broilers and fryers are fed 12-14 days, roasters 10-12 days, and fowl 6-7 days. Birds should be fed sparingly for the first few days, because when put into cramped quarters and supplied unlimited food from the beginning they will develop digestive disturbances and will often come out lighter in weight than when they went into the fattening batteries. After the birds become used to confined conditions they may be fed all they will eat. A gain of one pound may be obtained with four pounds of a good fattening ration. Such a ration may be composed of two parts of corn meal, one part of rolled oats, and one part of middlings, with enough skim milk to make the mixture run freely when poured from the pail into the feeding trough. This is the recommendation of the United States Department of Agriculture.

Feed Ingredients Commonly Used

Milk and ground cereal grains are the principal foods used in fattening poultry. Besides its food value, milk adds palatibility to the ration and enables birds to stand heavy feeding for a longer period of time. Where white flesh is not particularly discriminated against on the market, oats are especially valuable, due to the fact that the fat globules are distributed throughout the flesh and less in the skin. The market preference for yellow-skin birds is met by including yellow corn meal in the ration. This gives the desirable yellow color. Wheat tends to produce more abdominal fat. Thus the cereal mixture can be adjusted to accomplish the desired results.

Size of Birds

Size of birds in addition to the quality of the stock and the management procedure in crate fattening will determine the efficiency of feed consumption. The smaller sizes of birds will show the greatest gain, both in the percentage gain over the original weight and also in the efficiency as measured by the pounds of feed necessary to produce one pound of gain. However, the larger sizes, although showing higher feed costs per pound gain,

TABLE 51—COMPARISON OF DIFFERENT SIZES OF CHICKEN IN THE ECONOMY OF MEAT PRODUCTION BY FATTENING

Initial Weight Pounds	Fattened Weight Pounds	% Gain	Lbs. Feed to Produce 1 Lb. Gain	% Flesh to Chilled Dressed Weight
1.8	2.5	41.65	3.81	43.2
2.9	3.6	23.51	5.15	50.1
4.0	4.7	15.15	5.98	52.2
5.1	5.7	11.91	6.80	54.3
5.9	6.5	11.19	6.40	56.3

will show a higher percentage of flesh on the carcass. Nevertheless, the poultryman is primarily interested in the efficiency of the feeding practice in the addition of weight. The following table from work by W. A. Maw of MacDonald College is therefore of particular interest as a demonstration of the factor of size as influencing results.

Fig. 147—Typical Fattening Battery.
[Courtesy Union Steel and Wire Co.]

Simple Versus Complex Rations for Poultry

Work conducted by E. M. Funk at the University of Missouri has demonstrated that a simple fattening ration may give as good results as a more complex one. The simple mixture (No. 1) was composed of 70 pounds yellow corn meal, 20 pounds wheat shorts, and 1 pound salt. In comparison to this a more complex ration (No. 2) was used as follows: 60 pounds yellow corn meal, 60 pounds rolled oats, 40 pounds oat meal, 40 pounds ground wheat, 10 pounds wheat bran, 2 pounds meat scrap, 2 pounds alfalfa meal, and one pound salt.

Each basal ration was supplemented with condensed buttermilk, liquid buttermilk, and 5%, 15% and 20% dried skim milk.

The results of the two rations may be briefly summarized as follows:

	Gain Live Weight	Loss in Blood and Feather Dressing	Pound of Feed Per Pound of Gain
Simple Ration No. 1	14.6%	10.6%	3.58
Complex Ration No. 2	14.4%	10.6%	3.31

Fattening Rations

Ration No. 1
- 2 parts laying mash
- 1 part yellow corn meal

Ration No. 2
- 5 parts corn meal
- 3 parts middlings
- 1 part ground oats

Ration No. 3
- 4 parts corn meal
- 2 parts ground oats
- 2 parts middlings
- 2 parts meat scraps

Ration No. 4
- 5 parts corn meal
- 3 parts oatmeal
- 2 parts ground barley
- 1 part meat scraps

The above fattening rations are all mixed with liquid buttermilk or skimmilk to the consistency of a thick batter. Dried buttermilk at the rate of 10 pounds per 100 pounds of mash may also be used and enough water added to attain the desired consistency. There may be some advantage in using a mash containing meat scraps to provide additional protein for young chickens.

Marketing Fattened Birds

When birds are fattened by any of the above mentioned methods they should be killed as soon as they have reached their maximum development or they will have a tendency to "go off their feed" and lose much of the gain already made. Birds which have gone through a special fattening period shrink rapidly in shipment and consequently should be dressed and sold locally or properly boxed or barrelled and shipped.

A special market should be developed before branching out into this phase of the poultry business. It will be necessary to test out the market prices procurable and the market demand for various types of poultry to determine whether the venture would be profitable. It has been the experience of some that where intensive fattening systems are used, an insufficient price was secured to make the operation worthwhile.

Questions on Chapter XVI

1. Describe the procedure in crate fattening birds.
2. What feed ingredients are the principal foods used in the fattening of poultry?
3. Does size of bird have any influence on the efficiency of the fattening procedure?
4. What should be the procedure in marketing birds which have been fattened?
5. What are the relative merits of simple vs. complicated fattening mixtures?

CHAPTER XVII

HOUSING AND APPLIANCES

Subject Matter

History of the domestication of animal life shows that economic demands have led to the use of artificial stimulative measures for increasing the productive value of poultry. Environment is by no means the least of these, yet, because the hen is extremely sensitive to favorable and unfavorable conditions, she reacts to it very decidedly.

Housing furnishes a means of controlling at least one phase of environment, in that it tempers adverse weather and climatic conditions that are not conducive to continual seasonal productivity. Birds in the wild state produce only during favorable seasons. Through housing man has been able to maintain production the year 'round.

In domesticating poultry we attempt to duplicate and improve on the optimum conditions under natural environment.

Under ideal conditions in nature the wild fowl finds satisfactory (1) shelter where (2) ventilation, (3) humidity and (4) temperature are right. The bird is furnished an abundance of (5) light and (6) room to move about to secure (7) food and water consistent with its requirements. These may therefore be considered as underlying principles governing the housing of poultry.

Requirements of Housing

Shelter:

The first requirement of shelter is to protect birds not only from adverse weather conditions but from their natural enemies

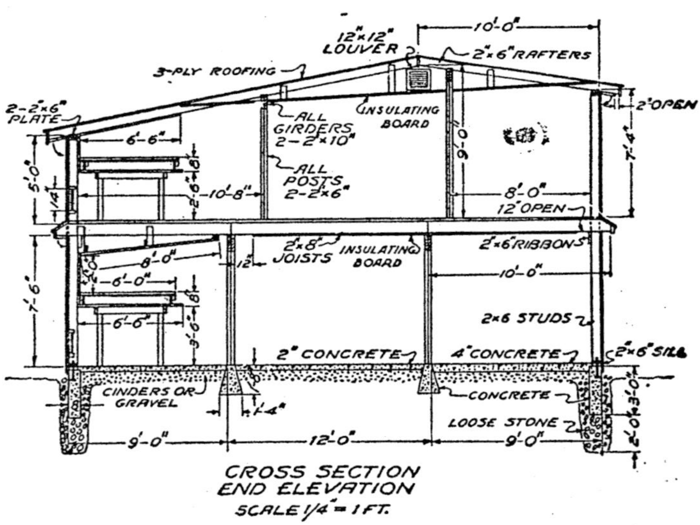

Fig. 148—Rafter Ventilation in Multiple-Story Houses.

Air rising from any point on the floor passes to the ceiling and out over the plate in both floors. The front outlet should be protected, but allowing a one or two-inch opening. The rear ventilator board fits tight and is dropped only during the summer months.

[Courtesy of Cornell University and New England Poultryman.]

and from thievery. It should furnish comfortable quarters so that the birds will respond to the management practices with increased egg production.

Substantial losses are sometimes sustained during very hot summer weather from heat prostration. Insulation of the roof helps to prevent extensive losses. Wetting down the roof, wetting down the litter, dousing the birds with water, more attention to adequate supply of drinking water, and the use of auxiliary fans in the pens as emergency measures will assist in reducing losses.

Design the house so that it is convenient for such operations as cleaning, disinfecting, feeding, watering and the general routine. These will be greatly facilitated by simplicity of design and arrangement.

The building should be substantially constructed with durable material to reduce maintenance cost. An excessive investment per bird should be avoided. Economy consistent with the purposes in mind ought to be the goal in poultry-house construction.

Ventilation and Humidity:

The moisture carrying capacity of a cubic foot of air depends on its temperature. If a given quantity of air is saturated and thus carrying all the moisture it can at a given temperature, cooling such air will reduce its capacity to hold moisture. The excess moisture at the lower temperature will be "squeezed out" and deposited on the material or surface with which the air is in contact.

TABLE 52—WATER CARRYING CAPACITY OF AIR*

One Cubic Foot of 100% Saturated Air At	
—20° F=0.166 grains of moisture	
—10° F=0.285 " " "	72% more than at—20° F.
0° =0.481 " " "	68% " " " —10° F.
+10° F=0.776 " " "	61% " " " 0° F.
+20° F=1.235 " " "	59% " " " +10° F.
	156% " " " 0° F.
+30° =1.935 " " "	56% " " " +20° F.
	302% " " " 0° F.
+40° =2.849 " " "	47% " " " +30° F.
	492% " " " 0° F.
+50° =4.076 " " "	43% " " " +40° F.
	751% " " " 0° F.
+60° =5.745 " " "	41% " " " +50° F.
	1094% " " " 0° F.

*Mass. State College. Extension Leaflet 145.

Reference to Table 52 indicates that in general a change of ten degrees in temperature changes the moisture carrying capacity of air approximately 50 per cent (40 to 72). Thus the importance of maintaining a differential of at least ten degrees between the inside and outside temperature of a laying house can be seen. This is especially true during periods when wet litter is a major problem.

If the air after it has been brought into a house can be warmed up, its moisture carrying capacity will be increased as indicated, in proportion to the change in temperature. This

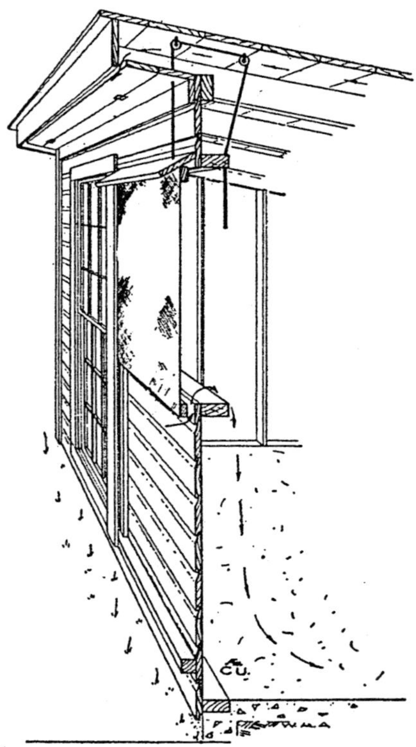

Fig. 149—Detail of the Front Ventilator Opening.

Note the one-inch opening extending entirely across the front of the building which permits the out-going air to escape, but is so constructed that the space is protected. Note the set-out curtain which provides a one-inch space along the bottom of each curtain when pulled up. If the curtain material is thin enough to allow air to pass through, the one-inch set-out is of less importance.

[Courtesy of Cornell University and New England Poultryman.]

warmed air can then serve as a "vehicle" for removing excess moisture in a hen house.

If, on the other hand, moisture laden air is brought into a pen and cooled to a lower temperature, it will deposit its excess moisture either in the litter or on sidewalls or ceiling of the pens. Under such a condition, the more air that is circulated through a pen, the damper will become the litter and sidewalls or ceiling.

It is apparent from the above that to have dry conditions in a hen pen, it is necessary not only to have the proper differential between inside and outside temperatures, but the proper rate of air movement.

Air replacement by the movement of air through the poultry house is necessary to supply oxygen for the respiratory process and for the removal of carbon dioxide and moisture. This air exchange must not cause drafts. It must be regulated according to varying climatic conditions. If too rapid or too slow it is not conducive to satisfactory conditions in the house, as correct ventilation is an attempt to minimize inside temperature variations while at the same time providing sufficient fresh air for the requirements of the birds without the accumulation of moisture.

The necessity of ventilation to remove moisture is indicated by the fact that several gallons of water are given off by 100 birds daily through respiration and excretion. Fowls require approximately 10 to 12 cubic feet of air per hour per pound of live weight.

Thus in a pen housing 300 six-pound birds, 18,000 to 21,600 cubic feet of air per hour would be required to provide proper and adequate air exchange during average conditions. Reference to tables will indicate that in order to move this amount of air per hour, an 18-inch or 10-inch out-take flue is necessary, depending on the type used. Intake flues should provide about 10 per cent greater square area than the out-take. Provision should be made for a ventilating system with capacity considerably beyond minimum requirements so that during adverse conditions a reasonable exchange of air may still occur. Heating the pens will assist in air movement, particularly during periods when low wind velocity or temperature differences makes it difficult to provide sufficient air movement to carry the moisture from the laying pens.

Reference to the tables will give a basis of computing size of flues to use for different pens, based on (1) number and weight of birds, and (2) cubic air content of the house.

TABLE 53—KONICAL-TYPE VENTILATORS*

Diameter of Neck Inches	Area in Square Inches	Exhaust Cu. Ft. Per Minute	Exhaust Cu. Ft. Per Hour
8	50	77	4,620
10	78½	121	7,260
12	113	174	10,440
14	154	237	14,220
16	201	310	18,600
18	254½	393	23,580
20	314	484	29,040
24	452	696	41,760
30	707	1089	65,340
36	1018	1568	99,080
40	1256	1934	116,040
48	1808	2784	167,040
60	2828	4355	261,300

The ratings above given are conservative and based on the same temperature within and outside the building at a normal wind. Of course a difference between the inside and outside temperatures and a higher wind rating will greatly accelerate the air movement and increase cubic feet of air exhausted.

*Courtesy of The Eller Manufacturing Company, Canton, Ohio.

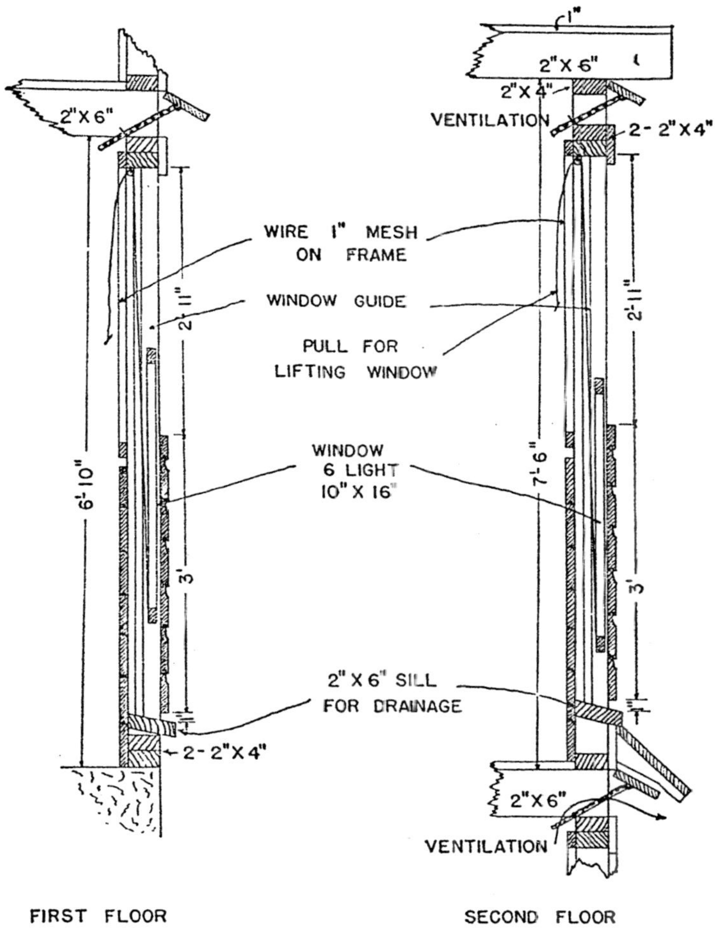

FRONT WINDOW PLAN FOR 24'X24' HOUSE

Fig. 150—Front of Connecticut 24′ x 24′ laying house, showing "slot" and "rafter" ventilators.

[Conn. Bull. 266. Courtesy of Roy E. Jones.]

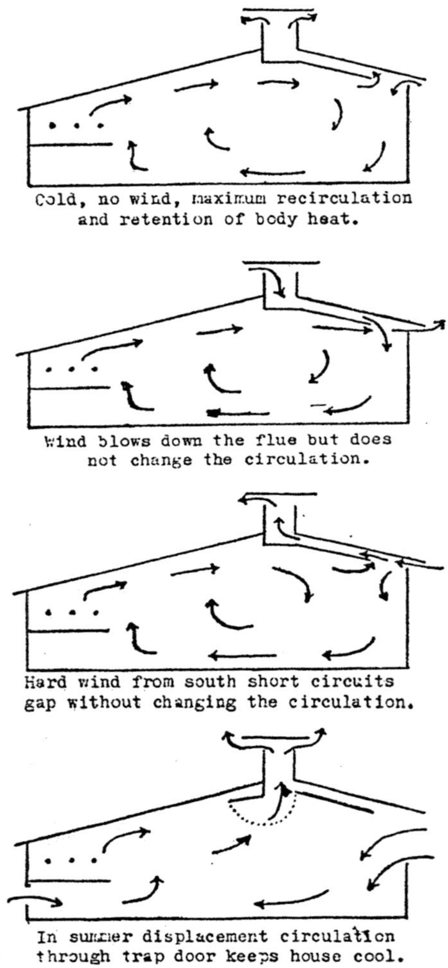

Fig. 151—Extended Flue Ventilator.
[Courtesy Mass. State College and New England Poultryman.]

TABLE 54—TURBINE-TYPE VENTILATORS*

Diameter of Turbines	Air Displacement—Cu. Ft. Per Hour	
	Wind Velocity 4 Miles Per Hour	Wind Velocity 8 Miles Per Hour
8 inches	17,300	21,600
10 "	26,500	32,500
12 "	38,600	46,200
15 "	54,000	69,000
18 "	85,200	102,100
20 "	105,100	125,600
24 "	149,000	185,000
30 "	225,000	272,000
36 "	281,000	330,000
42 "	324,000	414,000
48 "	360,000	473,000

*Courtesy of Allen Air-Turbine Ventilator Company, Detroit, Michigan.

TABLE 55—VENTICONE VENTILATORS*

Size of Flue	Area in Cu. In.	Capacities Cu. Ft. Per Minute	Capacities Cu. Ft. of Air Per Hour
12 inches	113	185	11,100
14 "	154	270	16,200
16 "	201	365	21,900
18 "	254	420	25,200
20 "	314	600	36,000
24 "	452	750	45,000
26 "	531	880	52,800
30 "	707	1,200	72,000
36 "	1,107	1,800	108,000

*Courtesy of Shenandoah Manufacturing Company, Harrisburg, Va.

Temperature:

Housing is partly an attempt to prevent rapid variations in the environmental temperature and to conserve the body heat of the birds during cold weather. The more uniformly temperature can be maintained at a suitable level and extremes avoided, the more satisfactory will be the results.

Many poultrymen have resorted to artificial heat in conjunction with ventilation and insulation of their poultry houses. In such cases it is felt that the temperature should be held in the neighborhood of 45 to 50 degrees Fahrenheit. The more constant the temperature the better are the results from artificial heating. Fluctuations of temperature are likely to react as unfavorably as fluctuations of cold weather in an unheated house.

Light:

A poultry house ought to be constructed so that the birds are furnished with sufficient light. The light should be well distributed to encourage feed and water consumption. Windows in both the front and back of the house as well as the ends, where possible and practical, aid in the proper diffusion and distribution of light, illuminating the entire floor space and avoiding dark corners and areas. The height of the front windows or openings determines how far back in the house rays of sun will penetrate. The following table, as illustrated in Farmer's Bulletin 1554 (U.S.D.A.), may be used as a guide in deciding height of windows. (Multiply height in feet at top of window by 2.3 to get distance back in pen that direct light penetrates.)

TABLE 56

The depth which the sun's rays shine on the floor of the house in the vicinity of Washington, D. C., (latitude 40° N.) on January 1 is given in the accompanying table.

Height of Top of Windows		Depth of Sunshine	Height of Top of Windows		Depth of Sunshine
Ft.	In.	Ft.	Ft.	In.	Ft.
3	6	8	6	2	14
4	5	10	7	1	16
5	4	12	7	11	18

It is advisable to have some of the openings so arranged that direct sunlight enters the house without having its effective ultra-violet rays filtered out. If the inside walls of the pen are white or light in color they will further assist in the distribution of light within the pen. In some sections of the country, due to the short winter days, artificial illumination is necessary to provide a satisfactory length of working day for the birds.

Space Allotment:

For Hens—It is generally recommended that three to four square feet of floor space be allotted per bird. The growing tendency in some of the commercialized districts is to allow even less room. The dangers of to heavy crowding are well known and should be avoided.

The amount of floor area per bird will be somewhat dependent on the yard space and the effectiveness of the poultry-house ventilation. It is foolhardy to consider but one standard of space allotment without due consideration to all requirements of the birds. For instance, 7 inches of perching space for light breeds, 8 to 10 inches for American or similar sized birds, and 10 to 12 inches for larger kinds are considered adequate. Feeding, watering and nesting space must also be suitable to optimum results.

For Chickens—The same principles of allotting space also applies to brooding chickens, except that younger birds require proportionately less, depending on their age. Allow at least one-half square foot of floor space at one day of age. This ought to be increased to at least one square foot at six weeks. Further increases of floor and yard space will be necessary as the chicks grow older.

Nutrition:

Naturally it is necessary to supply food material under domestication, due to the fact that the birds are almost entirely dependent on the operator. All of the varied requirements of the birds must be adequately met, if they are to do their best. For further detailed discussions refer to the chapters on feeding and nutrition.

Types of Houses

Poultry houses may be classified as to type of roof or as to the use to which they are to be put. Any of the types listed may be portable or stationary.

Types According to Roofs:

1. *Shed or "Flat" roof*—This is a single-slope roof, widely used and generally adapted. The pitch may vary from ½ inch to 3 inches per foot. With "flat" roof a built-up roof covering or 18 inch lap roofing should be used.

2. *Combination*—Such a roof consists of two unequal spans of rafters, the rear one representing two-thirds of the width of the house and the front, one-third.

3. *Equal Span or Gable*—Where great depth to the building or uniform ceiling height is desired, this type of roof is suitable.

4. *Monitor and Semi-monitor*—These types are adapted to wide houses which are difficult to light adequately. Such construction is likely to be drafty, however, unless the carpentry is carefully done.

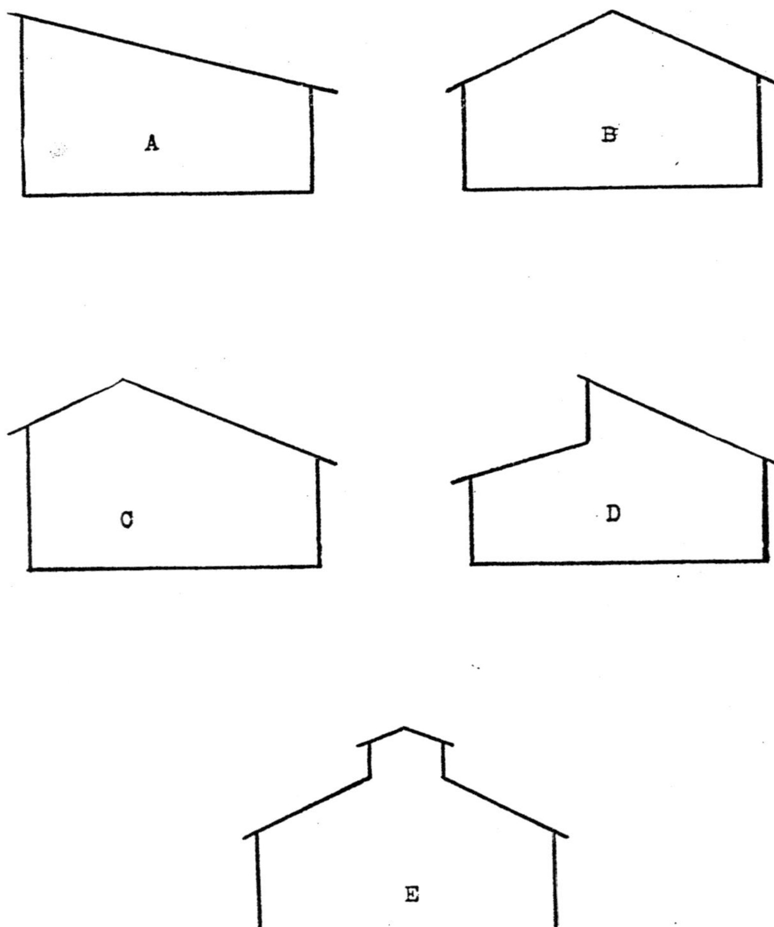

Fig. 152—Common Roof Types: A—Shed or "Flat"; B—Gable; C—Combination or ⅔ Span; D—Semi-Monitor; E—Monitor.

5. *Gambel*—This type is similarly adapted to deep houses, but is likely to be more expensive in construction. It is used mainly on single-story buildings.

6. *Igloo*—This type is semi-circular and forms both the roof and side walls of the building.

7. *Roofless*—This house provides only roosting quarters without overhead shelter and is found in sections of scant precipitation, such as in Arizona and other arid or semi-arid regions.

Types According to Use:

1. *Laying House*—For the housing of the laying flock.

2. *Special Breed House*—For the housing of special groups of birds that are to be used for breeders.

3. *Colony Brooder House*—For housing small independent units of chicks.

4. *Continuous Brooder House*—For brooding chicks in adjacent units.

5. *Shelters*—Can be used as range houses for growing stock, or for summer housing and laying quarters for adult stock.

Interior Arrangement and Equipment

In interior arrangement conservation of floor space is advisable wherever possible. This means raising interior equipment to a level where it is still accessible to the birds yet gives them maximum use of the floor.

Dropping platforms, where used, are usually placed along the rear wall so as not to interfere with distribution of light or with the attendant in his routine operations. This wall is often the tightest portion of the house and thus less liable to be drafty. Wiring the dropping boards with poultry netting is a popular scheme to keep the birds out of contact with the droppings. In long continuous houses, the manure-carrier track should be located in front of the roosts in such a manner that the manure may be scraped from the dropping boards, directly into the litter carrier. Doors are located in partition directly in front of roosts. Such labor-saving devices are recommended wherever installation is possible to reduce costs to a minimum.

Dropping pits have come into rather general usage. This arrangement eliminates dropping boards thereby reducing the cost of construction. They do not require such frequent cleaning as dropping boards but should be cleaned several times during the year at the operators convenience. Recently some houses have been constructed without providing for roosting provisions at all, some operators are claiming an advantage of dried litter.

In any long house, partitions are advisable at regular distances to divide the entire place into suitable sized units. Solid partitions are preferable to prevent drafts on the floor, the roosting quarters, and in the house in general. In partitioning the house into small units more possibilities are offered for the control of diseases. Better control of management conditions are afforded at the same time.

In general, alley ways are not advocated for poultry houses because of floor space wasted, draftiness, and difficulty of natural illumination. They are adapted, nevertheless, to some particular types of houses, such as special breed and long continuous brooder houses.

Feed hoppers and water fountains belong in the front of the pen where they will be well lighted and readily accessible. They should be large enough and correctly designed to meet their specific purpose. Hoppers from which birds can feed from both sides are preferable to wall hoppers because of the greater linear feet of feeding space. Hoppers of relatively small capacity are preferable as they keep the mash fresher and more palatable and thus secure greater feed consumption. In addition they are likely to lead to more observation and careful attention for the birds because they entail regular replenishments by the attendant.

Feed hoppers may be placed end to end lengthwise of the house. Such an arrangement will reduce labor particularly when a feed carrier is used. In several sections of the country long, continuous mechanical feeders have been installed in poultry houses to further reduce the labor requirements.

Fountains ought to be of such a capacity that water is always available for the birds. The installation of automatic waterers wherever possible will guarantee a continuous supply and reduce labor of watering. Piping water to the pens under any system will be found a great labor saver.

Auxiliary feeders, such as green-food racks, are valuable additions to tne equipment and may be suspended from the wall or in the central portion of the pen.

Nests are usually placed on the side walls or may be arranged in batteries at convenient locations in the pen, although some poultrymen prefer to install them directly under the outer edge of the dropping platform or under or between the windows on the front wall.

"Community" nests with one opening in the center of the front have been gaining in popularity in recent years.

Broody coops are sometimes placed above or below the tier of nests or may be suspended from the ceiling in the center of the pen. A more effective means of handling the broody-hen problem is to have a battery of coops or a fattening battery located in the feed room or other available space and concentrate all such hens in it. The birds are thus more readily cared for and an eye-sore is removed from the pen. Broody coops in the individual pens are usually hard to keep clean and the birds usually will not get the care and attention they require.

In the installation of a lighting system it is recommended that the light and deflectors be located sufficiently high in the pen so that not only the floor and feeders are illuminated but

Fig. 153—Houseless Method of Keeping Poultry. Trapnests, Waterer, and Feed Hoppers are Protected by a Palm Leaf Shade.
[Courtesy of Burt Heywang, U. S. D. A.]

also the roosts. One lighting unit of 40 watts is right for each 200 square feet of floor space. Space these units in such a manner that the light will be fairly uniformly distributed.

Suitable cleaning tools and equipment are also necessary in the proper handling of poultry.

In multiple decked buildings, electric hoists used as elevators are a necessity for the transportation of birds, feed, and other materials.

Auxiliary Buildings

Feed Room:

Feed storage may be constructed either as a separate unit or as part of a long continuous house. In either case let convenience decide the location with respect to the laying and brooding houses as well as the other divisions of the poultry farm which it serves. Location is particularly important because the feed room is the farm center of many routine operations. In the long-type house consisting of several pens the feed room may advantageously form the central section. This will reduce the labor of transporting feed to the birds, as compared with a feed room located at one end of the building.

TABLE 57

REFERENCE TABLE ON POULTRY MANURE*

Source of Manure	Plant Food Analysis			Tons of Manure required to equal 1 ton of 5-10-5	Total lbs. of supplement needed to make fertilizer equal to 1 ton of 5-10-5	
	Nitrogen %	Phosphorous %	Potash %	Tons	20% Super-Phosphate	60% Muriate of Potash
Hen (Fresh)	1.5	1.0	.5	4 plus	600	100
Hen (Old	1.2	2.0	1.0	5	0	0
Broiler (Dried)	4.5	3.5	2.0	1 plus	650	100
Duck	1.0	1.5	.5	5 plus	250	84
Turkey	1.2	1.5	1.0	5 plus	250	0

*(Farm Bureau News, Suffolk County, N. Y. June 1948)

Feed rooms are generally constructed inadequate in size to meet the requirements for storage. With sufficient storage space the poultryman can take advantage of the market in purchasing certain storable feed stuffs. Without it he must buy on the basis of only his immediate requirements. A storage area of 1,000 square feet is sufficient to take care of the requirements of 1,000 to 1,200 laying birds. Additional storage space will be necessary for the young stock.

If the intention is to purchase ingredients then it will be necessary to plan for mixing. Hand mixing is laborious and time-consuming and the uniformity of the mixture is often questionable. Many suitable feed-mixing machines are on the market in varying sizes to suit the size of flock and quantity of feed to be mixed.

Construction of rat-proof storage bins will facilitate the storage and handling of feed and labor-saving devices in the feed room will greatly reduce the actual cost of handling feed materials.

Egg Room:

A room or building for handling eggs previous to marketing is a necessary auxiliary unit. A room that can be maintained below 65 degrees in summer and that can be kept at 55 degrees in winter is desirable. Sufficient space and appliances are necessary for storage, candling, grading, and packing of eggs. It is desirable to equip the room with running water. A reasonable degree of humidity should be maintained to prevent loss of moisture from the eggs. The room should be located and ventilated so as to be free from objectionable odors.

Manure Shed:

A manure shed offers a means of conveniently storing poultry dung and litter. Correct capacity will prevent too long delay in the removal of the manure. Correct construction will allow the contents to be removed readily. Pits are not generally to be recommended due to the difficulty of emptying them. In locating the shed consider accessibility to the poultry pens and convenience in hauling out the manure.

Screening the shed to restrict flies from this breeding place is particularly advisable where trouble has been experienced with tapeworms. Sprays are available which greatly assist control of flies.

Killing Room:

Suitable quarters for killing, dressing, and drawing are necessary if the plan is to sell birds dressed or drawn. Most poultry farms have a certain amount of local trade requiring fowl ready for consumption.

The killing room ought to fit neatly into the layout of the farm. It should be attractive and sanitary, and equipped with retention quarters such as crates or batteries for fattening or holding the birds preparatory to killing. Hangers for suspending birds preparatory to sticking and a waste retainer to take care of the accumulation of feathers, blood and other offal, represent additional valuable equipment. All this will facilitate the killing operations. Refrigeration, with holding and deep freeze facilities, is essential where volume justifies construction of a killing room. It facilitates more orderly marketing.

Storage Building:

Provision for adequate storage facilities has been overlooked in the planning of many poultry farms. These are necessary for litter, seasonal equipment, routine equipment and supplies. Sufficient floor space will make possible the segregation of various types of equipment and supplies so that they will be readily accessible. Unless a separate garage is provided for cars or trucks, space should be set aside for them in this building.

Construction Room:

This is almost a necessity on a commercial poultry farm to reduce construction and repair costs. A poultryman needs to be a mechanic, capable of carpentry, plumbing, or in other words, skillful at other trades.

The shop should be well equipped with necessary tools. If these are kept in place and in good shape they are an incentive to do the necessary work. An advantage is to have the working space well lighted, not only with daylight but with artificial light

Fig. 154—New Hampshire Combination Type Laying House. Sixteen Pens, each 24'x24'.

[Courtesy Hubbard Farms.]

for nights and dark days. Means of heating it for work in the cold weather is desirable. Much of the shop work can be done at odd times and seasons of the year when routine duties are not so restricting.

Incubator Room:

The incubator room may be completely above ground, or partly or entirely below. But in view of the fact that ventilation is an important incubation problem, an incubator room entirely or partially above the ground level is more satisfactory. With modern insulation material and ventilating systems it is much easier to secure a satisfactory room temperature that can be maintained and adequate ventilation with the building entirely above ground. In buildings thus located transportation of eggs, chicks, waste and supplies is greatly facilitated.

A room for incubation purposes needs running water and a drainage system. A large vat into which trays may be placed for cleaning and disinfecting is essential to ease of operation. Additional requisites are artificial light for candling, provision for heating, and sufficient table room for traying eggs and packing chicks.

Fig. 155—"Pointed Arch" Type Laying House.
[Courtesy James Mfg. Co.]

Trends in Laying House Construction:

Recent years have witnessed a decided trend towards multiple story laying houses. Lower construction costs (and frequently maintenance costs) and centralization of labor activities in caring for the birds are largely responsible.

Another development has been the use of much larger pens than has been the custom in the past. Many pens 40′ by 40′ are in use. In existing buildings poultrymen have been removing every other partition where the construction of the building made this feasible. The purpose is to effectuate larger sized pens and reduce labor requirements.

Interest has also been generated in the construction of low cost summer laying quarters. Many types of inexpensive units are used for housing the very early hatched pullets and in some cases the yearling stock.

More attention to rat proofing poultry houses is also being observed. Authorities claim that poisoning is not considered satisfactory control. Their population is in direct proportion to available feed supplies. Consequently, poultrymen are including shelves in designing foundations as well as other features in house construction that assist materially in rat control. Poultrymen should consult their county agents and others for specific suggestions in this respect.

Battery-Brooding Room:

This also should be considered as a possible auxiliary building on the average poultry farm. It should be designed to fit the particular system of battery brooding selected. For further discussion on the requirements of such a room refer to Chapter IV, Battery Brooding.

Building Materials and Cost of Construction

Plan:

Before construction is started the definite plans of the building should be drawn up. They should show not only the design of the building but also complete interior fixtures and all specifications. This plan may then be submitted to a local contractor or lumber company for estimate of cost of material. Naturally in planning a building, the proprietor should consider and take advantage of standard or stock lengths of lumber to avoid extra charge for unusual lengths, or excessive waste. The plan should provide for proper spacing of studding, rafters, and joists so that insulation boards can be conveniently applied, either at the time of construction or in the future.

State colleges and universities can be of assistance in planning suitable poultry houses for your needs. They have developed plans based on observations in the area. Poultrymen planning on the construction of new buildings or the renovation of existing buildings should consult them.

Bill of Material:

After the plan has been drawn up it is necessary to work out a list of all material to be used, with the exact size, length and amount of each item. This list is then used in writing a bill of material of the size and length commercially available.

Delivered dressed lumber does not contain full width of material as listed under standard grading. Further waste comes with sawing and fitting it into the construction of the building. For these two reasons an allowance is necessary in order to have sufficient material to complete the job. This varies from as high as one-third in three-inch matched lumber to as low as one-sixth in the six-inch. In ten-inch shiplap the allowance is one-tenth; in drop siding and framing lumber, one-fifth.

In figuring the concrete requirements the cubic volume must be determined. Next decide on the ratio of cement to sand and gravel or coarse aggregate. If the volume of concrete is 1,000 cubic feet, for example, and a 1:2:4 mixture is wanted, it will require one-fourth of this total amount or 250 cubic feet of cement, two-fourths or one-half of the total volume or 500 cubic feet of sand, and four-fourths or 1,000 cubic feet of gravel. This is the approximate method of figuring the requirements for concrete construction. If more accurate figures are required, the water-cement ratio should be used.

On the average about 15 pounds of nails will be needed per 1,000 board feet. For flooring, sheathing or bridging 20 to 25 pounds of nails per 1,000 board feet will be right.

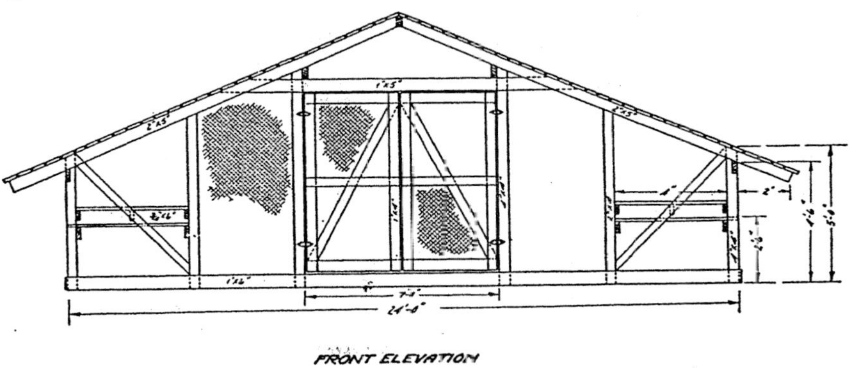

Fig. 156—Front Framing Plan for 24′ x 24′ Laying Shelter.
[Courtesy N. E. Poultryman.]

Fig. 157—Framing Plan for 24′ x 24′ Permanent Summer Laying Shelter.
[Courtesy N. E. Poultryman.]

Requirements on roofing are obviously based on the surface area of the roof. A square of roofing material is usually sufficient to cover 100 square feet of roof surface. Shingles are variable, depending on the amount of shingle exposed. There are approximately 1,000 shingles to every four bundles.

In figuring paint requirements it is generally considered that one gallon of ready-mixed paint will cover 500 square feet of surface with one coat.

In the bill of material should be listed in addition all hardware, poultry netting, window sash complete with glass and frames, electrical wiring and appliances.

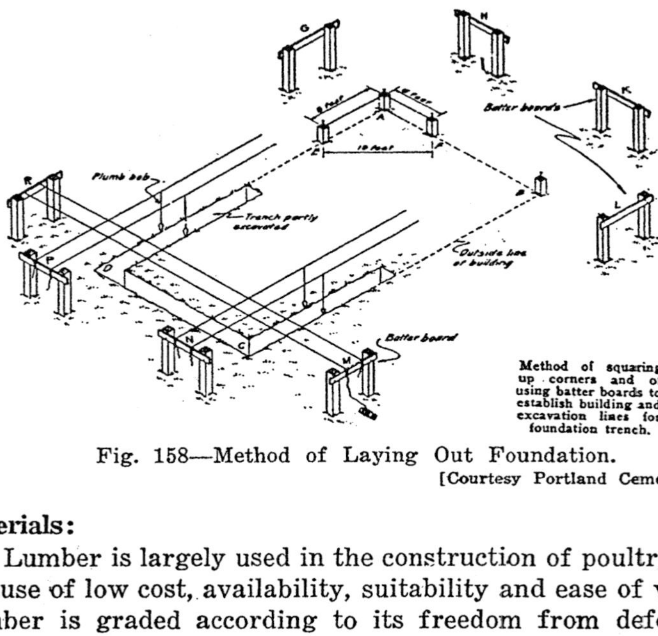

Fig. 158—Method of Laying Out Foundation.
[Courtesy Portland Cement Co.]

Materials:

Lumber is largely used in the construction of poultry houses because of low cost, availability, suitability and ease of working. Lumber is graded according to its freedom from defects and priced accordingly. The grade used will depend on its desirability, appearance, and price, and the permanency of the structure. The lower grade of lumber is usually not desirable for permanent construction because of weakness and waste. Planed material is best for a poultry house, except in the case of subflooring or roofing, and studding where it is to be covered with insulation board.

Concrete can be used to advantage for foundations and floors. Where ground is extremely sloping, thus materially increasing the cost of concrete floors, wooden floors are preferable.

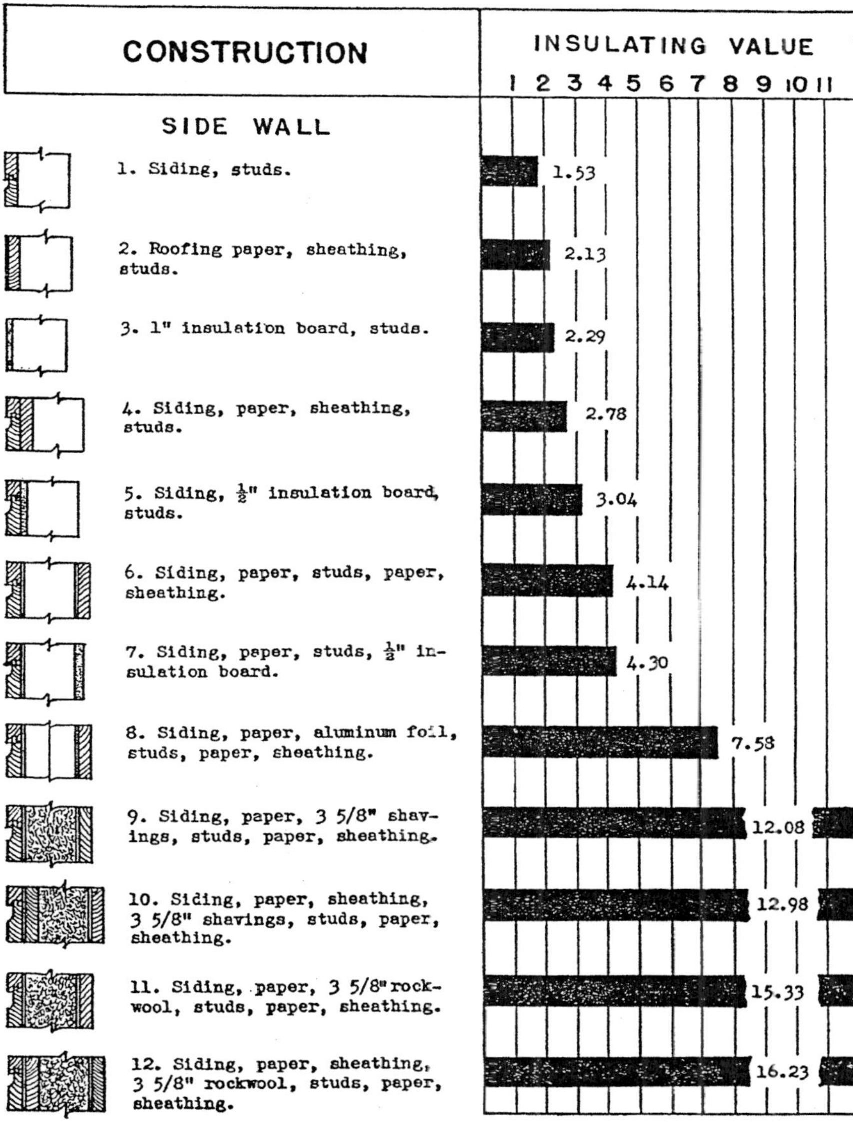

Fig. 159—Side Wall Insulation Values.

[Conn. Bull., 266. Courtesy of Roy E. Jones.]

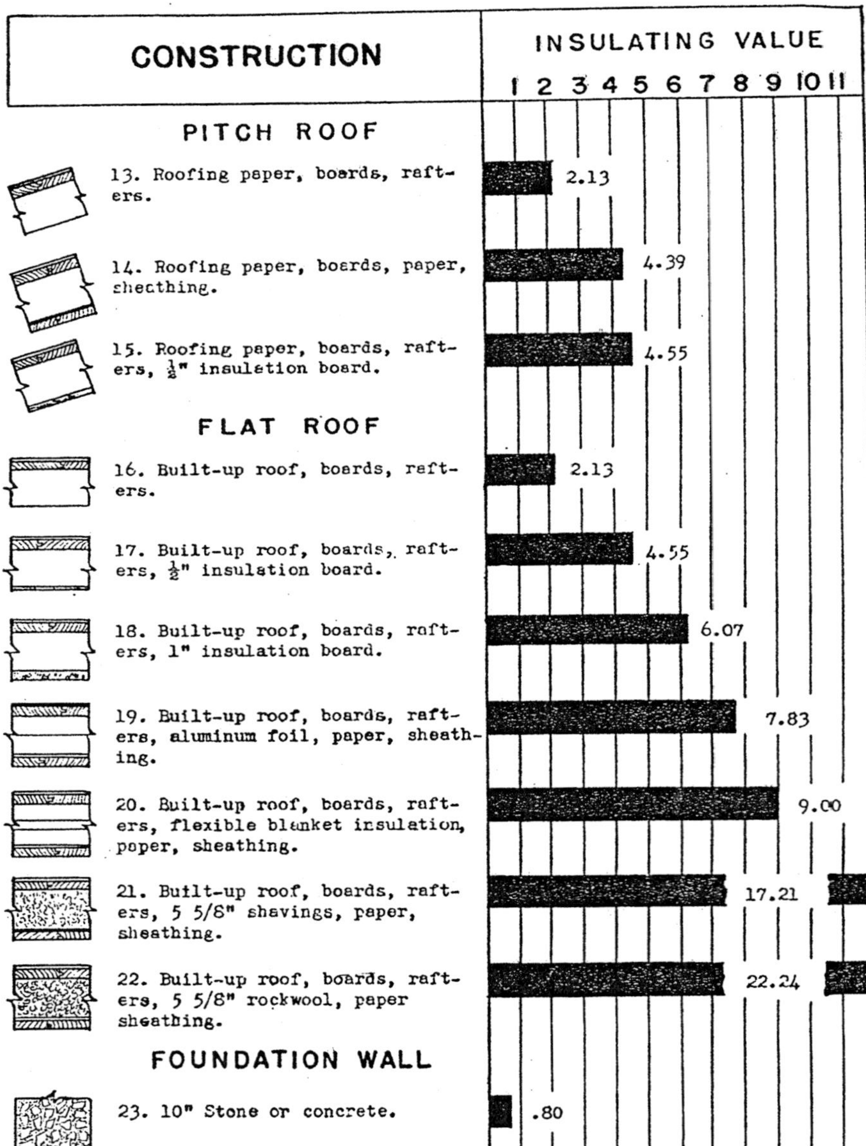

Fig. 160—Roof Insulation.

[Conn. Bull., 266. Courtesy of Roy E. Jones.]

Hollow tile and cinder blocks have been used to some extent for side-wall construction. They have the advantage of being non-inflammable and of providing insulation, structural strength, and permanency at low-maintenance cost.

Insulation boards on the market and suitable for such house construction should be purchased on the basis of tensile strength, insulation value, ease of application, durability and cost. More attention is being given to their use than ever before, due to their minimizing effects on climatic fluctuations. Most of these boards must be protected from picking by the birds within areas the birds can reach. This may be done by plastering, or covering with metal or wood sheathing.

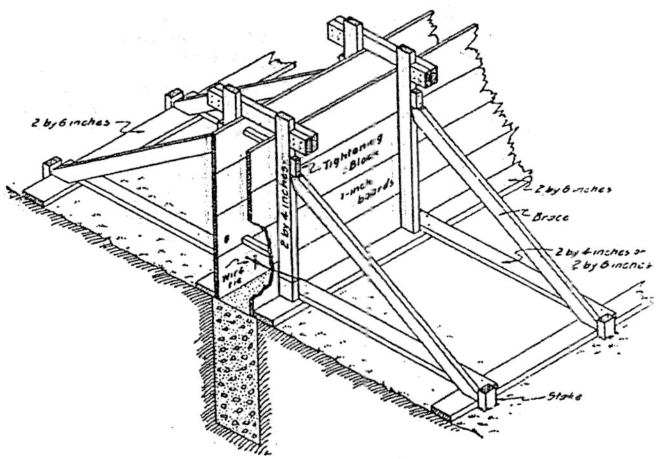

Fig. 161—Method of Constructing and Reinforcing Forms for Concrete. [Courtesy Portland Cement Co.]

Cost and Permanency:

The factors of cost and permanency go hand in hand in deciding the ultimate cost of construction per bird-capacity. Careful planning will result in economics in both labor and materials. If a man will put in his own labor on the structure the cost may be proportionately lowered. Labor may constitute as much as thirty to fifty per cent of the total construction bill, indicating that a saving in this item will materially reduce the ultimate cost. A saving in material may be effected by the construction of multi-story buildings. Many times it is advisable to remodel existing barns into suitable multi-story houses. In such cases, the cost of housing is greatly lowered.

TABLE 58—TABLE OF RECOMMENDED MIXTURES AND MAXIMUM AGGREGATE SIZES*

Mixture	Use	Max. Size Agg.
1:1:1½—Mixture for:	Wearing course of two-course pavements	¾ in.
1:2:3 —Mixture for:	Reinforced concrete roof slabs	1 in.
	One-course walks, floors and pavements	1½ in.
	Fence posts	¾ in.
	Sills and lintels without mortar surface	¾ in.
	Watering troughs and tanks and pits subject to water pressure	1 in.
	Reinforced concrete columns	1 in.
1:2:4 —Mixture for:	Reinforced concrete floors, walls and beams	1 in.
	Foundations for engines	2 in.
	Concrete work in general subject to vibration	1½ in.
1:2½:4—Mixture for:		
	Silo walls, grain bins, coal bins and similar structures	1½ in.
	Building walls above ground	1½ in.
	Walls of pits or basements exposed to moisture	1½ in.
	Base of two-course floor or pavement	3 in.
1:1½—Mixture for:	Inside finish of walls and tanks where additional protection against moisture is necessary	No. 8 screen
1:2—Mixture for:	Facing blocks and concrete products	½ in.
	Wearing course of two-course walks, floors and pavements where loads are light	¼ in.

TABLE 59

MATERIALS REQUIRED FOR 100 SQ. FT. OF SURFACE FOR VARYING THICKNESSES OF CONCRETE OR MORTAR*

C. = Cement in Sacks.
F.A. = Fine Aggregate (Sand) in Cu. Ft.
C.A. = Coarse Aggregate (Pebbles or Broken Stone) in Cu. Ft.

Quantities may vary 10 percent either way according to aggregate used. No allowance made for waste.

Proportion	1:1		1:1½		1:2		1:2½		1:3		1:1:1			1:1:1½		
Thickness in Inches	C.	F.A.	C.	F.A.	C.	F.A.	C.	F.A.	C.	F.A.	C.	F.A.	C.A.	C.	F.A.	C.A.
⅜	2.7	2.2	1.8	2.7	1.5	3.0	1.3	3.2	1.1	3.5						
½	3.0	3.0	2.4	3.5	2.0	4.1	1.7	4.3	1.5	4.6	2.1	2.2	2.2	1.8	1.9	2.7
¾	4.5	4.3	3.6	5.1	2.9	5.9	2.5	6.2	2.2	6.8	3.0	3.0	3.0	2.7	2.7	4.1
1	6.0	6.0	4.8	7.0	3.9	7.8	3.3	8.4	3.0	8.9	4.2	4.1	4.1	3.7	3.8	5.4
1¼	7.3	7.3	6.0	8.9	4.9	9.7	4.2	10.5	3.7	11.1	5.2	5.1	5.1	4.6	4.6	6.8
1½	8.9	8.9	7.3	10.8	5.9	11.0	5.1	12.7	4.5	13.5	6.3	6.2	6.2	5.5	5.4	8.1
1¾	10.5	10.5	8.4	12.4	6.9	13.5	6.0	15.1	5.4	16.2	7.3	7.3	7.3	6.4	6.5	9.7
2	12.2	12.2	9.6	14.3	7.9	15.7	6.9	17.2	6.2	18.6	8.3	8.4	8.4	7.3	7.3	11.3

Proportion	1:1½:3			1:2:3			1:2:4			1:2½:4			1:2½:5		
Thickness in Inches	C.	F.A.	C.A.	C.	F.A.	C.A.	C.	F.A.	C.A.	C.	F.A.	C.A.	C.	F.A.	C.A.
2½	5.9	8.9	17.6	5.4	10.8	16.2	4.6	9.2	18.4	4.3	10.8	17.0	3.9	9.7	19.4
3	7.0	10.5	21.1	6.5	13.0	19.4	5.6	11.1	22.1	5.2	13.0	20.8	4.6	11.6	23.2
3½	8.2	12.4	24.6	7.5	15.1	22.7	6.5	13.0	25.9	6.0	15.1	24.0	5.4	13.5	27.0
4	9.4	14.0	28.1	8.6	17.3	25.7	7.4	14.9	29.7	6.9	17.3	27.5	6.2	15.4	30.8
4½	10.5	15.9	31.6	9.7	19.4	28.9	8.3	16.7	33.2	7.7	19.4	30.8	6.9	17.3	34.6
5	11.7	17.6	35.1	10.8	21.6	32.1	9.3	18.6	37.0	8.6	21.6	34.3	7.7	19.2	38.6
5½	12.9	19.4	38.6	11.8	23.8	35.4	10.2	20.5	40.5	9.5	23.5	37.8	8.5	21.1	42.5
6	14.0	21.1	47.1	12.9	25.9	38.6	11.1	22.1	44.2	10.3	25.7	41.3	9.2	23.2	46.4

Bill of Materials

(Summarized for Lumber Dealer)

The framing may be of yellow pine, hemlock, spruce, or fir, depending on availability and price.

Ventilator boards, casing, window sills, cornice, and the like, may be of white pine.

Siding and roofing may be of a cheap grade of white pine, spruce, hemlock, fir, or yellow pine.

---

*How to make concrete. Portland Cement Association.

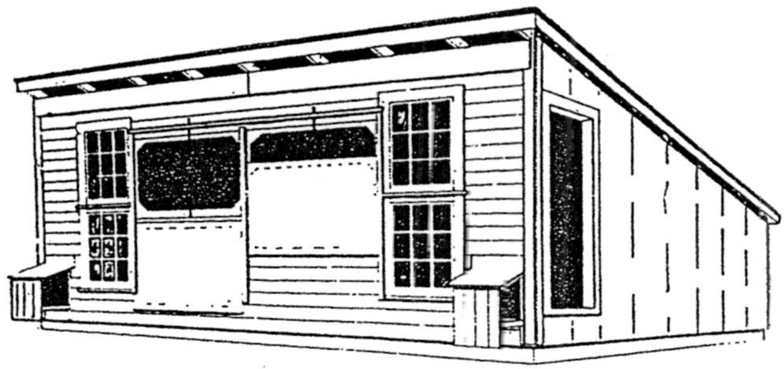

Fig. 162—Cornell Shed Roof Laying House, Pens 20′ x20′, Siding Curtains, Rafter Ventilators, and Protected Hen Exits.

[Cornell Ext. Bulletin.]

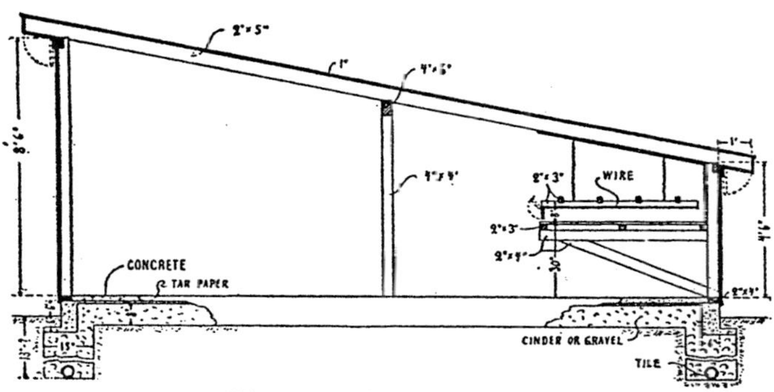

Fig. 163—Side Framing.

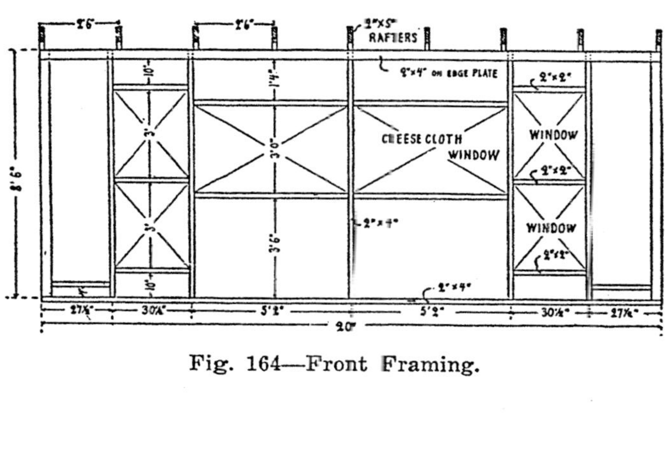

Fig. 164—Front Framing.

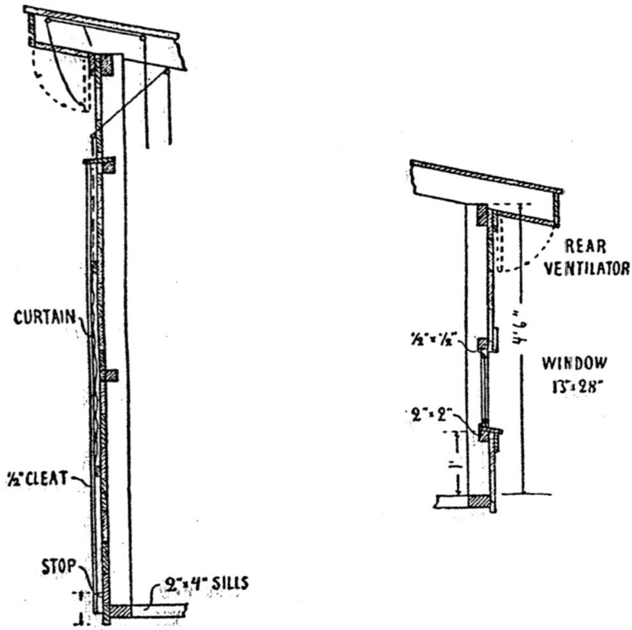

Fig. 165—Window and Ventilator Arrangement.

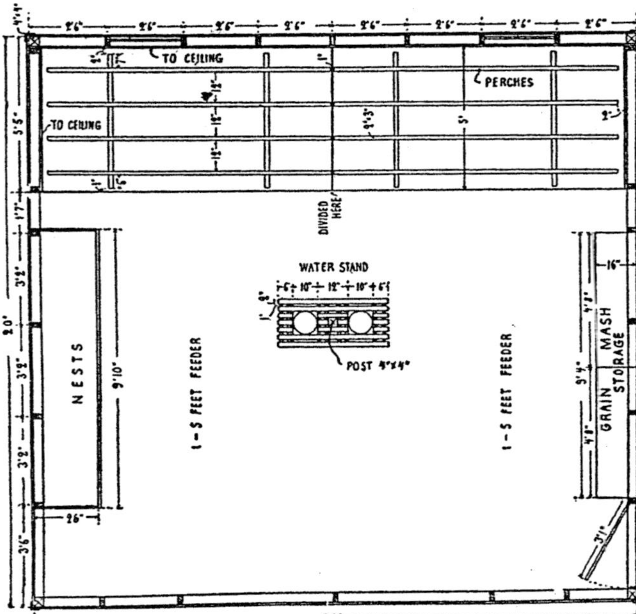

Fig. 166—Floor Plan.

LIST OF MATERIAL FOR CORNELL LAYING HOUSE*

Framing Material

2"x6"
- 3—10'
- 1—14'

2"x5" [3]
- 18—12'

2"x4"
- 13—14'
- 3—12'
- 17—10'

2"x3"
- 10—10' [1]
- 4—14' [2]

2"x2"
- 2—16'
- 4—14'
- 2—12'
- 7—10' [4]

$\frac{1}{2}$"x$\frac{1}{2}$" material
- 3—10'

$\frac{7}{8}$" Material (Planed, Unmatched)

12" wide
- 1—10'

10" wide
- 4—10'

8" wide
- 1—14'
- 4—10'

6" wide
- 3—14'
- 6—12'
- 36—10'

5" wide
- 1—12'

4" wide
- 3—16'
- 4—14'
- 7—10'
- 5—12'

3" wide
- 2—14'
- 1—12"
- 16—10'

2" wide
- 3—16'
- 6—14'

Foundation and Floor

5 yards gravel
2$\frac{1}{2}$ yards sand
26 sacks cement
4 rolls 1-ply tar paper[5]
8—8"x$\frac{1}{2}$" bolts

Matched Lumber

$\frac{7}{8}$"x6"-12'—633 board feet
$\frac{7}{8}$"x6"-10'—568 board feet
$\frac{7}{8}$"x4"-10'—329 board feet

Hardware

4 pair 4" butts
10 pair 3" strap-hinges
5 pair 3" T-hinges
14 feet $\frac{1}{2}$-inch chain
8—1" single pulleys
12 yards, 2" mesh, 5' poultry wire
4 yards, 1" mesh, 30" poultry wire
7 yards, 1" mesh, 36" poultry wire
8 heavy iron hooks
8—2" iron buttons
25 pounds 8-penny nails
4 pounds 6-penny nails
8 pounds 10-penny nails
5 pounds 16-penny nails

Miscellaneous

1—40' clothes line
4 yards 36" cheesecloth
5 rolls 3-ply roofing paper
4 rolls 2-ply roofing paper
4—9-light 8"x10" sash
2—3-light 8"x10" sash

---

*Cornell Ext. Circ.

[1]If 2"x4" are cheaper, four of the ten pieces may be 2"x4".

[2]Use 2"x4" if cheaper.

[3]Use 2"x6" rafters if cheaper.

[4]If 2"x4" are cheaper, five of the seven pieces may be 2"x4".

[5]If the house is built on a well-drained location, the paper need not be used.

(Summarized for Carpenter)

Lumber

- 9—2"x4"–14', 2—2"x4"–12' studs
- 18—2"x5"–12' rafters[1]
- 8—2"x4"–10' sills
- 4—2"x4"–10' plates
- 4—2"x4"–14' corner posts
- 1—2"x4"–12' center posts
- 1—2"x4"–10' door framing and roosting-closet cleat
- 4—2"x6"–10' stringer
- 5—2"x2"–10' window framing[2]

Casing

- 2—⅞"x3"–14', 2—⅞"x4"–14' corners
- 6—⅞"x4"–10', 6—⅞"x4"–12' frieze
- 1—⅞"x4"–16', 2—⅞"x4"–14', 1—⅞"x4"–12', 1—⅞"x4"– 6' door and windows

Matched Lumber

- 568 board feet—⅞"x6"–10' siding
- 633 board feet—⅞"x6"–12' roofing
- 209 board feet—⅞"x4"–10' roosting closet

Foundation and Floor

- 5 yards gravel
- 2½ yards sand
- 26 sacks cement
- 4 rolls 1-ply tar paper[3]
- 8—8"x½" bolts

Curtain Frame and Cleats

- 1—⅞"x2"–14', 2—⅞"x2"–12' frame
- 2—⅞"x2"–10', 2—⅞"x4"–12' cleats
- 4 yds. 36" cheesecloth
- 7 yds., 1" mesh, 36" poultry wire
- 4—1" single pulleys

Doors and Window Sills

- 1—2"x6"– 4'
- 2—⅞"x6"–10'

Cornice

- 6—⅞"x6"–10'
- 6—⅞"x6"–12'

Drip Boards

- 1—⅞"x5"–12'
- 3—½"x½"–10' window cleats

Miscellaneous

- 1—40' clothes line for regulating curtains and ventilators
- 5 rolls 3-ply roofing paper
- 4 rolls 2-ply roofing paper
- 1 pair 6" T-hinges (door)

Nails

- 25 pounds 8-penny nails
- 4 pounds 6-penny nails
- 8 pounds 10-penny nails
- 5 pounds 16-penny nails

FIXTURES

Wind Guards

- 2—2"x2"–14'
- 2—2"x2"–12'
- 32 board feet ⅞" matched lumber

Storage Bin

- 2—2"x2"–16'
- 88 board feet ⅞" matched lumber
- 4 pair 3" strap-hinges

Nests and Broody Coops

- 1—2"x2"–10'
- 2—2"x2"–14'
- 8—⅞"x3"–10'
- 4—⅞"x8"–10'
- 3—⅞"x6"–14'
- 6—⅞"x 6"–10'
- 1—⅞"x 4"–16'
- 1—⅞"x12"–10'
- 2—⅞"x 2"–16'
- 5 pair 3" T-hinges
- 5 yds., 2" mesh, 5' poultry wir
- 4 yds., 1" mesh, 30" poultry wir
- 4 pair 4" butts

Window Sash

- 4—9-light 8"x10" sash
- 2—3-light 8"x10" sash

Ventilators

- 4—⅞"x10"–10'
- 4 pair 3" strap-hinges
- 4—1" single pulleys
- 8—2" iron buttons

[1]Use 2"x6" if cheaper.

[2]Use 2"x4" if cheaper.

[3]If the house is built on a well-drained location, the paper need not be used.

Droppings Boards

4—2"x4"–10'
6—2"x3"–10'
24—$\frac{7}{8}$"x6"–10'

Dry-Mash Feeders[1]

1—$\frac{7}{8}$"x3" –12'	legs
2—$\frac{7}{8}$"x3" – 9'	sides
1—$\frac{7}{8}$"x6" –8' 8"	ends
4—$\frac{7}{8}$"x3" –10'	platforms
2—$\frac{7}{8}$"x3" –10'	feeder tops
1—$\frac{7}{8}$"x6$\frac{7}{8}$"–10'	feeder side
1—$\frac{7}{8}$"x6" –10'	feeder side
1—2"x2" –10'	wheel
4—7$\frac{3}{4}$"x7$\frac{3}{4}$"	feeder ends

Perches

4—2"x3"–14' [2]
4—2"x3"–10' [2]
2—$\frac{7}{8}$"x4"–10'
4—$\frac{7}{8}$"x2"–14'
2 pair 3" strap-hinges
7 yards, 2" mesh, 5' poultry wire
14'—$\frac{1}{2}$" chains
8 heavy iron hooks

---

[1]Use 2"x6" if cheaper.
[2]Use 2"x4" if cheaper.

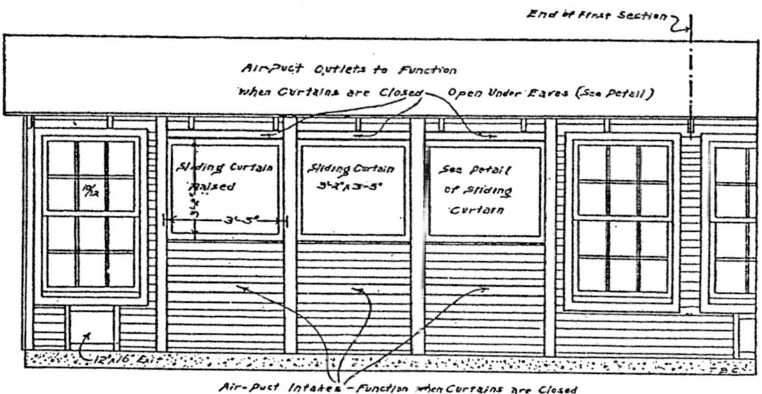

Fig. 167—Front View Penn. State 20' x 20' Laying House.
[Ext. Circ. 115, by T. B. Charles.]

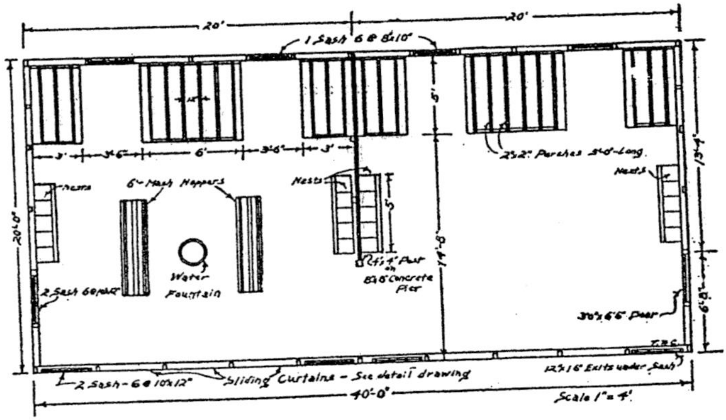

Floor Plan.

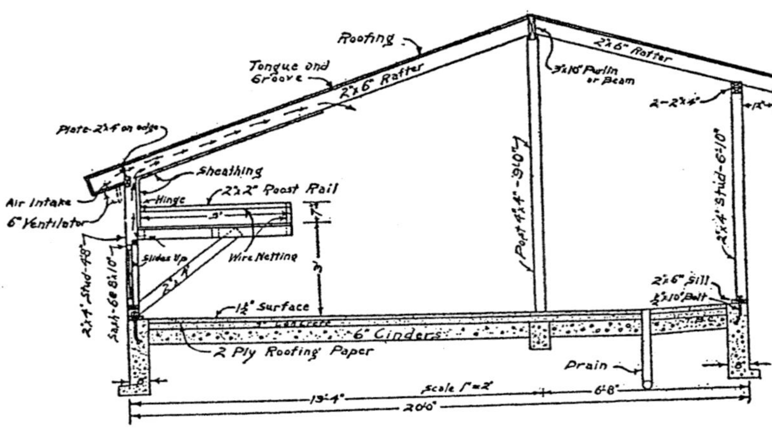

End View.

Fig. 168—Penn. State 20′ x 40′ Laying Unit.

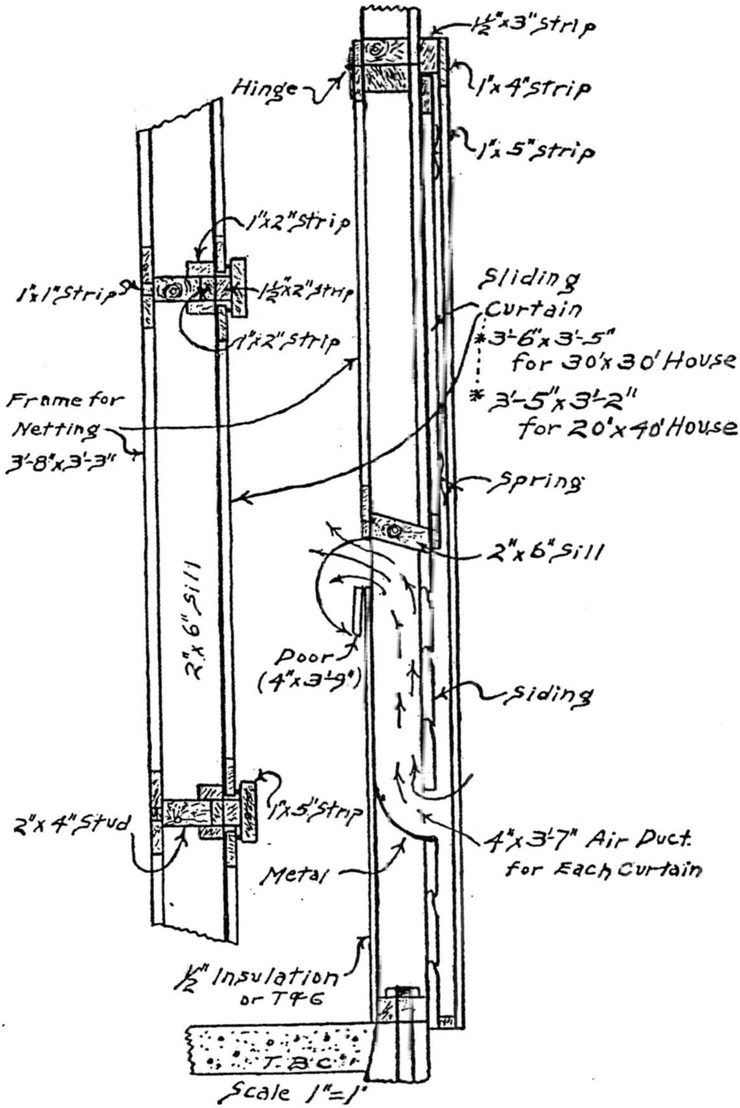

Fig. 169—Front Ventilation and Sliding Curtain Arrangement of Penn. State Laying House.

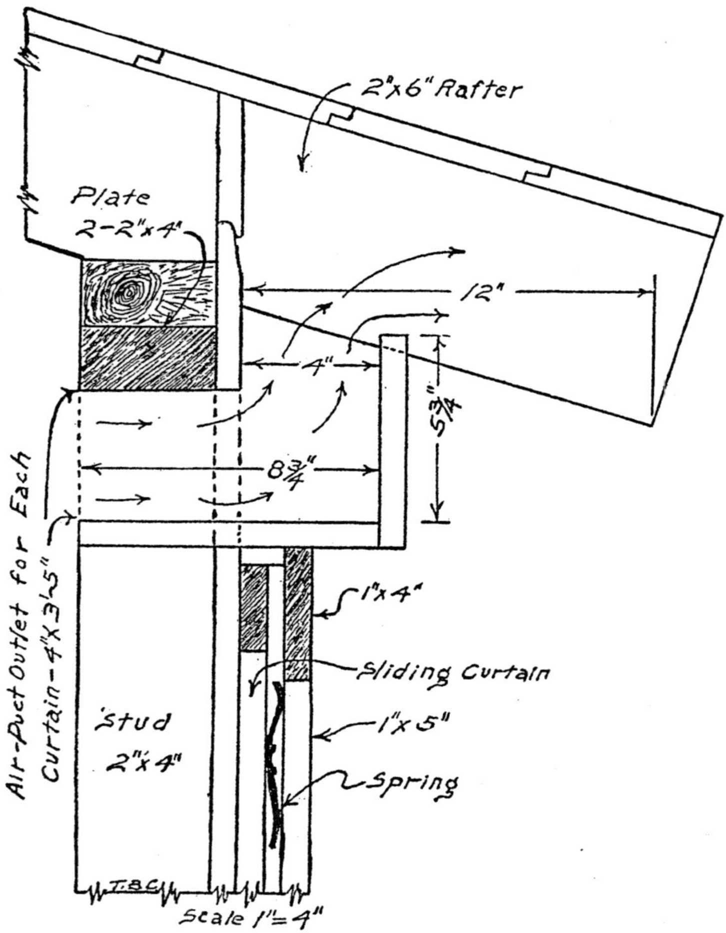

Fig. 170—Plate Ventilator of Laying House.

LIST OF MATERIALS FOR THE PENN. STATE 20′x40′ UNIT LAYING HOUSE*

Use—	No. or Amt. Required	Size	Length	
Sills	6	2″x6″	12′	
Sills	6	2″x6″	10′	
Sills	1	2″x4″	14′	
Plates	6	2″x4″	10′	
Plates	6	2″x4″	12′	
Plates	3	2″x4″	8′	
Studding	11	2″x4″	10′	
Studding	2	2″x4″	12′	
Studding	6	2″x4″	14′	
Studding	2	2″x4″	16′	
Posts	3	4″x4″	10′	
Framing	4	2″x4″	12′	
Framing	1	2″x2″	12′	
Window sills	3	2″x6″	12′	
Window sills	1	2″x6″	15′	
Rafters	17	2″x6″	16′	(rear span)
Rafters	17	2″x6″	8′	(front span)
Ridge pole or purlin	2	3″x10″	20′	
Roosting rails	12	2″x 2″	10′	Upper edge rounded
Roosting rail supports	3	2″x 2″	12′	
Roosting platform	150 bd. ft.	1″x 4″	¾″	T and G flooring
Roosting platform supports	8	2″x 4″	10′	
Roosting platform supports	7	2″x 4″	12′	
Siding	700 bd. ft.	Novelty for back, ends, and front		
Roof boards	1050 bd. ft.	1″x8″	10′, 12″, 14′, ship lap	
Two-thirds partition	100 bd. ft.	1″x6″	10′, 14′ T & G sheathing	
Ceiling for roost cabinet	400 bd. ft.	T & G sheathing		
Trim (windows, corners, etc.)	10	1″x4″	10′	No. 2 white pine or select yellow pine
Trim (windows, corners, etc.)	8	1″x4″	12′	No. 2 white pine or select yellow pine
Trim (windows, corners, etc.)	3	1″x4″	14′	No. 2 white pine or select yellow pine

Muslin Curtains and Wire Frames for Windows

Frames	290 linear ft.	1″ x 3″	
Slides and stops	60 linear ft.	1½″ x 2″	
Slides and stops	48 linear ft.	1½″ x 3″	
Slides and stops	50 linear ft.	1″ x 4″	
Slides and stops	32 linear ft.	1″ x 5″	
Slides and stops	100 linear ft.	1″ x 2″	
Slides and stops	12 linear ft.	1″ x 1″	
Auxiliary ventilators	80 sq. ft.	½″ Insulating board or T & G ceiling: Inside under curtain openings	
Auxiliary ventilators	2	1″ x 8¾″	12′ No. 2 white pine
Auxiliary ventilators	2	1″ x 5¾″	14′ No. 2 white pine

Foundation and Floor

Cement	92 bags
Sand	8 cu. yards
Stone	16 cu. yards
Roofing paper (2-ply)	8 rolls (insulating floor)

*Penn State Circ. No. 115, by T. B. Charles.

Hardware, Etc.

Roofing paper (3-ply)	11	rolls (roof)
Door	1	3′ x 6′ 6″
Windows	10	sash, 6 light 10″ x 12″
Windows	10	sash, 6 light 8″ x 10″
Bolts	12	½″ x 10″
Wire netting	45	ft. 3 ft. wide for frames inside windows and curtain openings
Muslin	7½	yards (36″ wide)

Nails, hinges, tacks, lock, paint, etc.

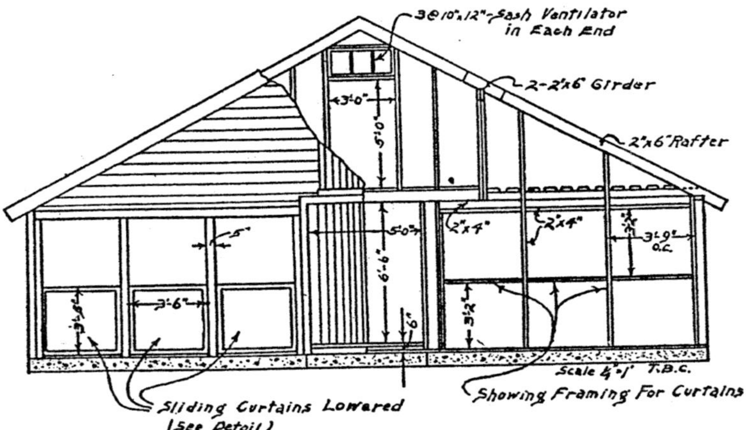

Fig. 171—Front View Penn. State 30′ x 30′ Laying House.
[Ext. Circ. 115, by T. B. Charles.]

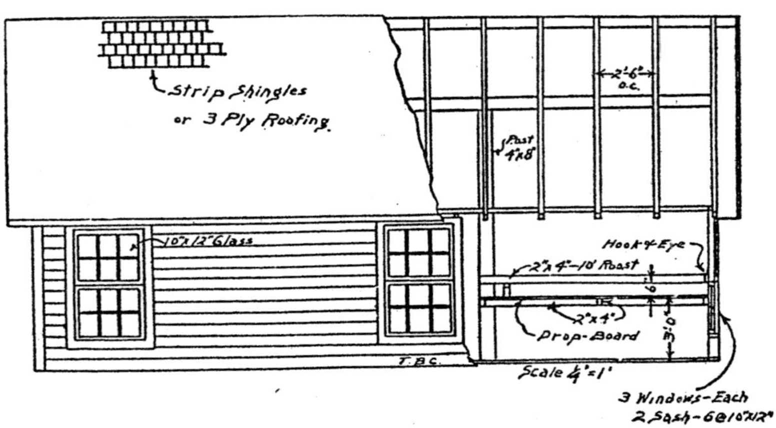

Fig. 172—Side View Penn. State 30′ x 30′ Laying House.

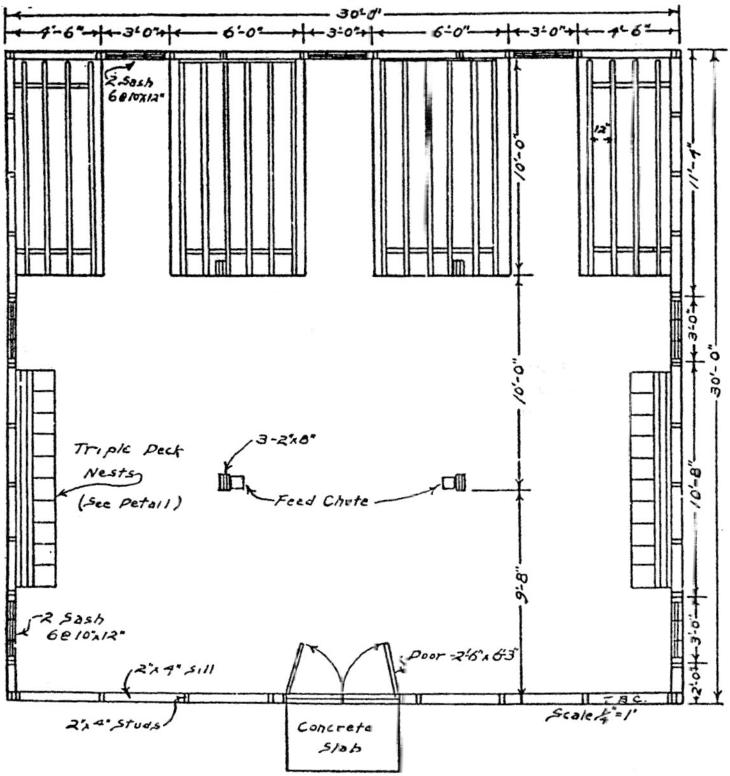

Fig. 173—Floor Plan Penn. State 30' x 30' Laying House.

LIST OF MATERIAL FOR 30 BY 30-FOOT POULTRY HOUSE*

Use—	No. or Amt. Required	Size	Length
Sills	4	2" x 4"	16'
Sills	4	2" x 4"	14'
Plates	4	2" x 4"	16'
Studs	6	2" x 4"	10'
Studs	9	2" x 4"	12'
Studs	21	2" x 4"	14'
Framing	1	2" x 4"	10'
Framing	6	2" x 4"	12'
Framing	3	2" x 4"	16'
Posts	10	2" x 8"	14'
Girders	6	2" x 6"	10'
Girders	2	2" x 6"	14'
Girders	2	2" x 6"	16'
Rafters	39	2" x 6"	14'
Joists	16	2" x 6"	10'

*Penn State Circ. No. 115, by T. B. Charles.

Use—	No. or Amt. Required	Size	Length
Joists	16	2" x 4"	10'
Window sills	4	2" x 6"	12' No. 2 white pine or select yellow pine
Feed storage floor	330 bd. ft.	1" x 6"	T&G sheathing or flooring
Straw loft strips	225 bd. ft.	1" x 6"	T&G sheathing or flooring
Roof boards	1300 bd. ft.	1" x10"	Shiplap or 6" T & G
Siding	1000 bd. ft.	1" x 6"	Novelty siding
Trim, windows and door casings	27	1" x 4"	12' No. 2 white pine or select yellow pine
Trim and slides	10	1" x 4"	14' No. 2 white pine or select yellow pine
Windows	252 linear ft.	½" x 1"	window strip
Curtain frames and slides	6 linear ft.	1" x 3"	14' No. 2 white pine
Curtain frames and slides	2	1" x 5"	14' No. 2 white pine
Curtain frames and slides	100 linear ft.	1" x 2"	
Curtain frames and slides	14 linear ft.	1" x 1"	
Curtain frames and slides	60 linear ft.	1½"x 2"	
Curtain frames and slides	48 linear ft.	1½"x 3"	
Wire netting frames	226 linear ft.	1" x 3"	12' No. 2 white pine or select yellow pine
Grain and mash bins and chutes	2	1" x 6"	6'
Grain and mash bins and chutes	2	1" x 8"	6'
Grain and mash bins and chutes	4	2" x 2"	10'
Grain and mash bins and chutes	100 bd. ft.	1" x 4"	flooring
Roosts	20	2" x 4"	10' rounded on upper edge
Roosts	1	2" x 4"	12' for center roost —rail support
Roosts	1	1" x 4"	12'
Roosts	1	2" x 2"	16' for side roost—rail support
Frame for roosts	3	2" x 4"	10'
Frame for roosts	6	2" x 4"	12'
Droppings Board	250 bd. ft.	1" x 4"	flooring or 1" x 6" sheathing
Windows	14 sash, 6 light	10" x 12"	
Windows	2 sash, 3 light	10" x 12"	
Roofing	13 squares of strip shingles or 13 rolls of 3-ply roofing		

Hardware

20 lbs.	20 d. nails
10 lbs.	16 d. nails
30 lbs.	8 d. nails
5 lbs.	6 d. nails
5 lbs.	8 d. nails (finishing)
4 sets	Hooks and eyes for roost
14	Bolts ½" x 12"
5 prs.	4-inch butt hinges
65 ft.	1" mesh poultry netting 36" wide
Hasp, lock, etc.	

Material for Concrete Foundation and Floor

92 bags cement
8.5 cu. yards sand
17 cu. yards stone
9 rolls 2-ply roofing (insulation)

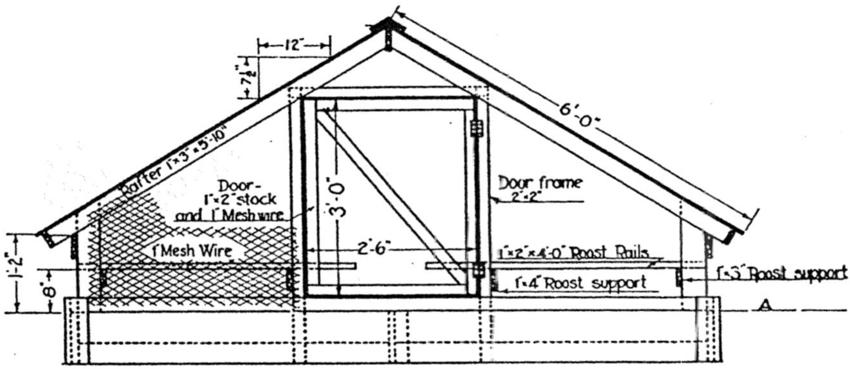

Fig. 174—Front Design of Range Shelter.

[Courtesy N. H. Agr. Exp. Sta.]

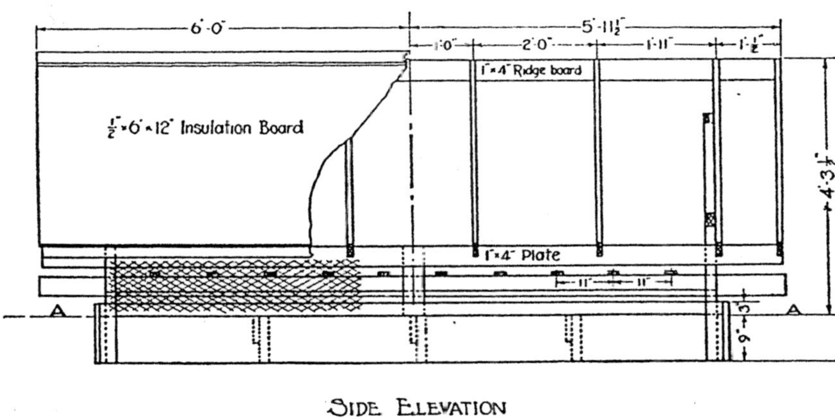

Fig. 175—Roof and Side Design of Range Shelter.

[Courtesy N. H. Agr. Exp. Sta.]

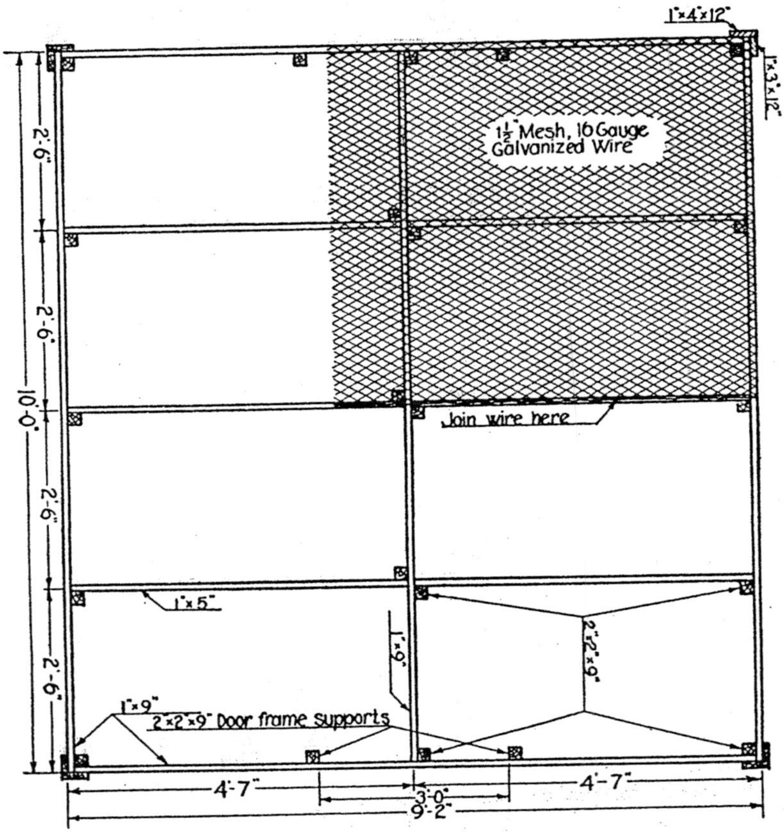

Fig. 176—Plan of Floor.

MATERIALS FOR N. H. TYPE RANGE SHELTER, 9′ x 10′

Section 1—Wire Floor—

5—1″ x 9″–10′ Sides and middle
1—1″ x 4″– 4′ Corner guides
1—1″ x 3″– 4′ Corner guides
3—1″ x 5″–10′ Floor joists
1—2″ x 2″–18′ Corner & posts
20′—5′ wide, 1½″ mesh, 16 gauge galvanized wire

Section 2—Roosting Quarters—

2—6′ x 12′ sheets ½″ insulating board (or metal roofing)
8—1″ x 3″ x 12′ Rafters
1—1″ x 4″ x 12′ } Saddle
1—1″ x 3″ x 12′ } boards
1—1″ x 4″ x 12′ Ridge board
1—2″ x 4″ x 10′ Posts
2—2″ x 2″ x 10′ Door framing
2—1″ x 4″ x 12′ Rafter plates
2—1″ x 3″ x 10′ Outer roost rail supports
2—1″ x 4″ x 12′ Inner roost rail supports and handles
4—1″ x 3″ x 10′ Sills
7—1″ x 2″ x 12′ Roost rails
2—1″ x 2″ x 16′ Door, both ends
2—1″ x 3″ x 12′ Facia

Hardware—

1 pound 8d finish nails
2 pounds 8d common nails
1 pound 6d common nails
2 pounds 10d common nails
3 pounds 1½″ galvanized roofing nails
25′—1″ mesh, 19 gauge galvanized poultry netting 3′ wide
2 pounds ⅞″ galvanized wire staples
4—3″ butt hinges with screws
2—locks and hasps

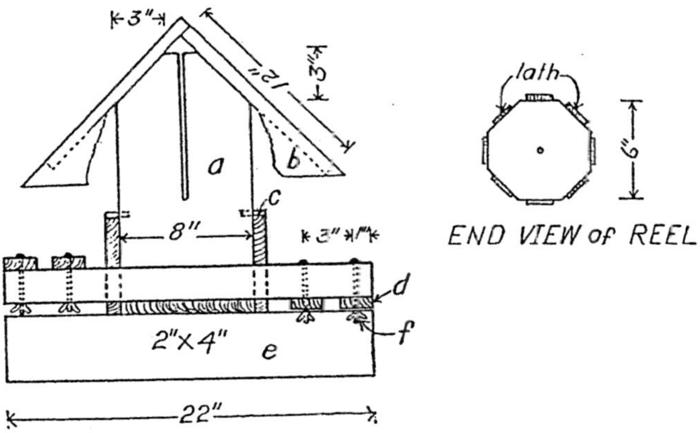

Fig. 177—End View Penn. State Hanging "Reel-Type" Hopper.

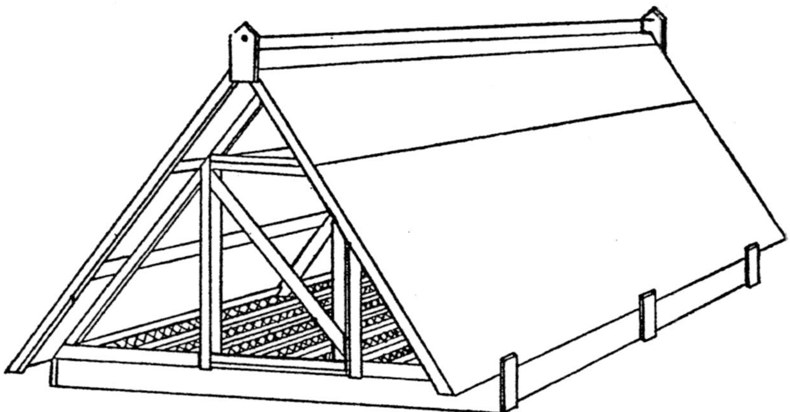

[Rhode Island]

Fig. 178—A-Type Shelter.

RANGE SHELTER BILL OF MATERIALS

FRAMING

Floor

7—2"x2"x8'	roosts
2—1"x8"x8'	end boards
3—1"x8"x7' 10"	middle and side boards
6—1"x4"x1'	side cleats

Roof

6—2"x3"x6'	rafters
1—2"x2"x10'	ridge board
2—1"x6"x9' 10"	lower purlins
4—1"x4"x9' 10"	middle purlins
2—1"x4"x7' 10"	spreaders
2—1"x4"x10'	peak boards
2—1"x4"x1'	wire supports
4—1"x2"x6'	edge boards

Door and Frame

2—1"x2"x34"	door sides
2—1"x2"x25"	door ends
1—1"x2"x39"	door cross brace
2—1"x2"x38"	door opening
1—1"x4"x30"	cross piece

ROOFING MATERIALS (Choose one)

3—3/16"x4'x10'	pressed wood, or
6—.027"x26"10'	V-crump, or corrugated gal. iron, or aluminum sheets, or

120 sq. ft. 2"x2" poultry netting plus 140 sq. ft. roofing paper
Nails and cement according to material selected

HARDWARE

2—	butt or strap hinges and screws
1—	button or hook and eye
2 lbs.	6 d. box nails
1 lb.	10 d. finishing nails
2 lbs.	small staples

12 ft. 48" wide 1" poultry netting
11 ft. No. 9 gauge wire
16 ft 48" wide, 1½" or 1"x2" mesh, 16 gauge wire

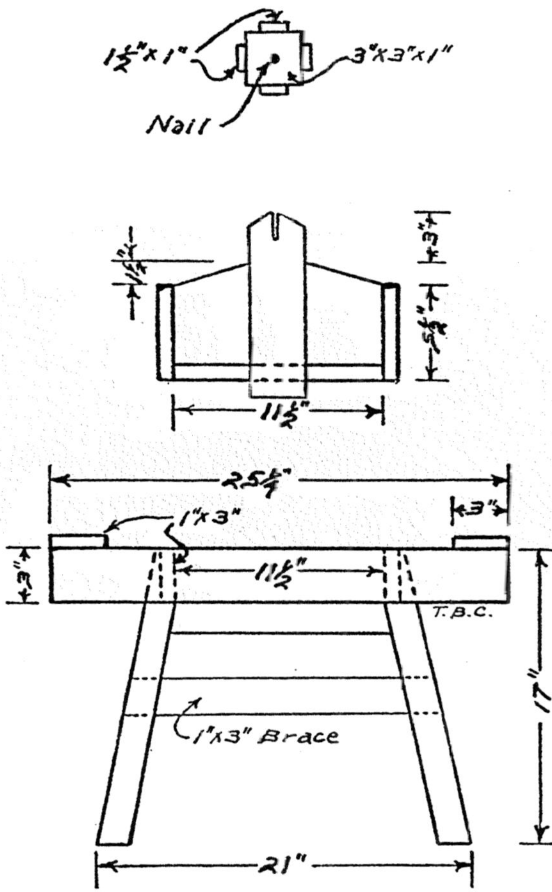

Fig. 179—Plan of Feed Hopper and Stand.

[Penn State Circ. 115.]

Fig. 180—A metal range shelter differing in design from conventional types.
[Courtesy of Hawkins Million Dollar Hen, Inc.]

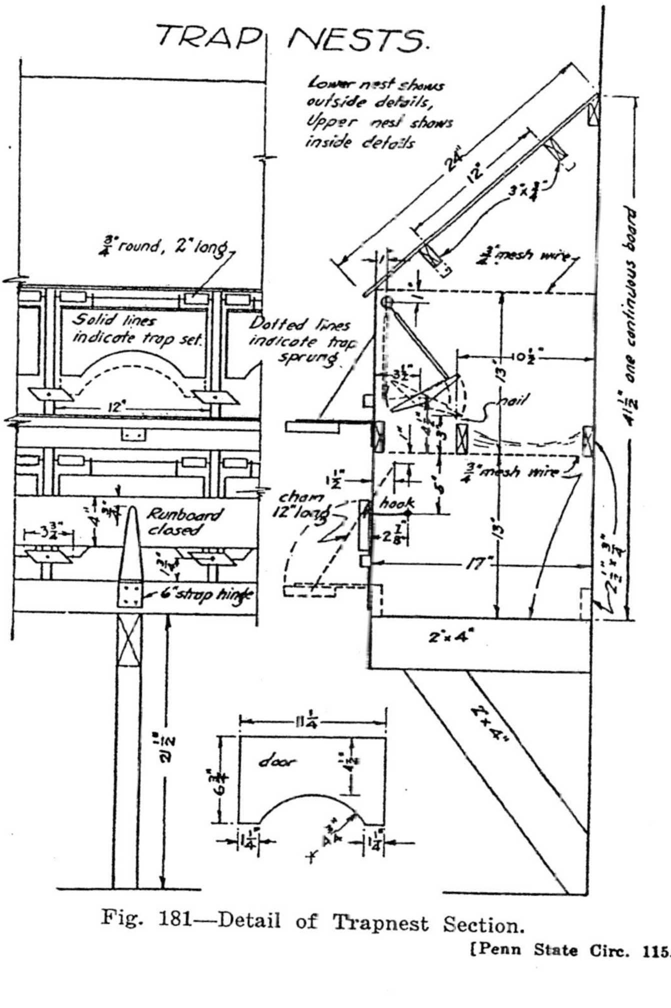

Fig. 181—Detail of Trapnest Section.

[Penn State Circ. 115.]

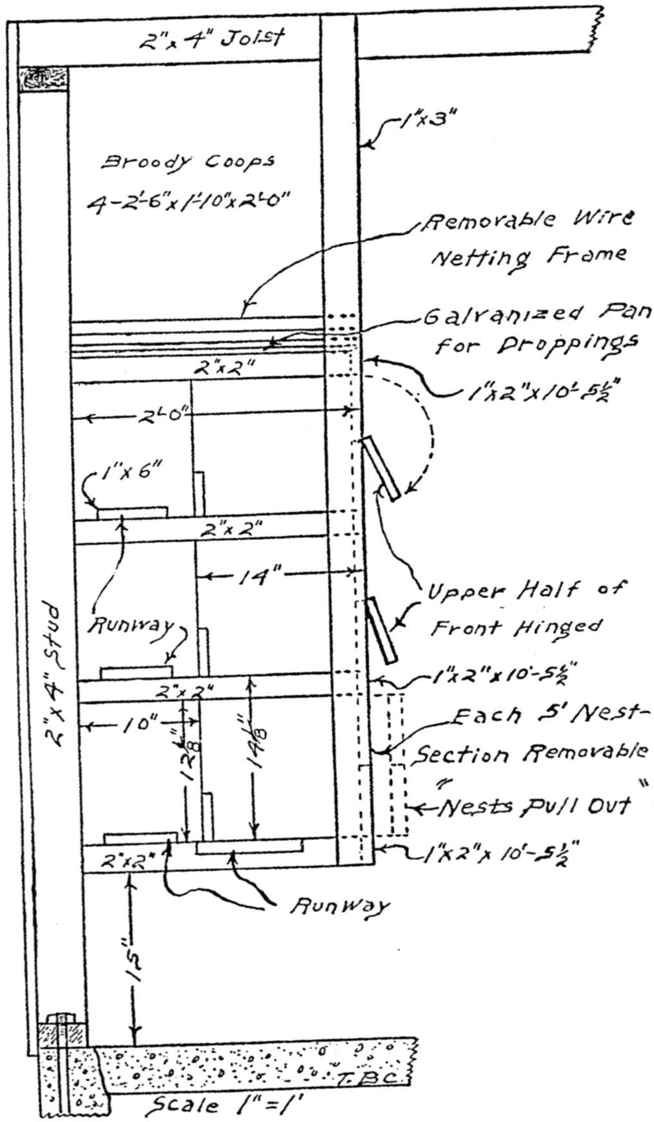

Fig. 182—Details of Nests with Rear Exit.
[Penn State Circ. 115.]

Questions on Chapter XVII

1. List the requirements of housing.
2. How much floor space should be allotted per laying bird?
3. How much roosting space will a bird require?
4. Do laying birds require an abundance of fresh air?
5. Is it desirable that windows be high or low if sunlight is to penetrate well back into the pen?
6. By drawing show what is meant by the following roof descriptions: shed, combination, equal span, monitor, and semi-monitor.
7. Describe an acceptable interior arrangement for the laying pen.
8. What auxiliary buildings are frequently needed on a commercial poultry farm?
9. By illustration show how the cement, sand, and gravel requirements are calculated for concrete work.
10. Why is lumber so largely used in the construction of poultry buildings?

CHAPTER XVIII

MANAGEMENT FACTORS

Subject Matter

Poultry management is the conduct of the enterprise with a view to securing desired results. It means a continual change in practices due to variable factors that must be effectively coped with for success. Some of the more important factors affecting management practices are here listed.

1. Market conditions and requirements.
2. Seasonal variations.
3. New developments.
4. Educational and extension activities.
5. Changing price levels of products produced and sold.
6. Purchasing power of products as compared with general index of commodities.
7. Variability of disease problems.
8. Variability of live stock.
9. Stress of competition.

It is just as true of the poultry business as of any other that the operator must keep abreast of the times by altering his

Fig. 183—New front arrangement of Penn. State laying house (20′ x 20′) showing slot and rafter ventilation.

[Courtesy H. C. Krandel]

management practices in the light of new information, if he is to meet competition and continue in the lead.

Since all of the practices involved in poultry keeping are managerial in nature, it is the purpose of this chapter to discuss some not considered elsewhere in this text.

Throughout this chapter reference is made to some outstanding farm management studies in order to draw some conclusions and to focus the attention of the reader to relationships that exist. Labor returns vary appreciatively under varying economic situations. Current price levels are not comparable with those reported but they serve the purpose of pointing out rela-

TABLE 60

CAPITAL INVESTMENT AND SIZE OF FLOCK
NEW HAMPSHIRE AGR. EXP. STATION BULLETIN NO. 265 (1930)

SIZE OF FLOCK	Number of farms	Average inventory	Inventory per bird
500–800	6	$ 7,432.66	$19.44
801–1000	3	12,563.43	13.19
1001–1200	3	11,182.90	10.75
1201–1400	5	13,432.91	10.20
1401–1600	....	....	....
1601–1800	2	13,592.48	7.99
1801 and over	4	24,642.76	9.97
Average for all farms	....	$13,424.06	$10.40

tionships between capital investment and size of flock, relation of capital investment and labor income, relation of size of flock and net returns, effect of average egg production on labor return, etc.

Fig. 184—Double-decked Houses Provide Economical Housing and efficient use of labor.
[Courtesy N. E. Poultryman.]

Capital Requirements

As in any business from which a satisfactory income is derived, the poultry enterprise requires a definite amount of capital. Cost-account records in several states indicate that this averages from $10 to $20 per hen on a poultry farm equipped to

TABLE 61

DISTRIBUTION OF CAPITAL INVESTMENT ON POULTRY FARMS*

	Connecticut 1917		New Jersey 1918		N. Y. Up-State 1927		N.Y Long Island 1927		New Hampshire 1930	
	Amount	Per cent	Amount	Per cent	Amount	Per cent	Amount	Per cent	Amount	Per cent
Real estate	$8,304	74.1	$5,572	76.9	$12,365	73.3	$25,324	86.6	$7,433	55.3
Poultry	1,790	16.0	1,146	15.8	2,109	12.5	2,986	10.2	3,854	28.6
Other livestock	....	.....	....	.....	764	4.5	59	.2	168	1.2
Machinery and equipment	649	5.8	310	4.3	1,470	8.7	829	2.8	1,684	12.5
Feed and supplies	398	3.5	109	1.5	163	1.0	44	.2	285	2.1
Cash	48	4	105	1.5	....	.....	....	.....	....	.....
Total capital	$11,189	.....	$7,242	.....	$16,862	.....	$29,242	.....	$13,424	.....

*Cents have been dropped but actual totals given.

TABLE 62

RELATION OF CAPITAL INVESTMENT TO LABOR INCOME

	Connecticut, 1917	New Jersey, 1918	New York Up-State, 1927	New York Long Island, 1927	New Hampshire, 1930
Capital investment	$11,189.00	$ 7,243.00	$16,862.00	$29,242.00	$13,424.00
Livestock	1,790.00	1,146.00	2,864.00	3,045.00	3,852.00
Receipts	3,519.00	2 818.00	7,527.00	10,092.00	10,088.00
Expenses and interest	2,207.00	2 088.00	5,873.00	7,621.00	8,689.00
Labor income	752.00	730.00	1,654.00	2,471.00	1,399.00
Average number of birds	592.00	737.00	1,020.00	1,322.00	1,020.00
Labor income per bird	1.27	0.99	1.62	1.86	1.37
Investment per bird	18.90	9.82	16.53	22.11	13.16

return a fair labor income to its owner or operator. Of course, there are many poultry enterprises where the original investment per bird may be either below or above these average figures, due to fortunate circumstances on the one hand, or possibly to high real-estate values or poor judgment on the other.

Table 62 illustrates a relationship between total capital investment and total labor income. Where capital investment is highest it will be observed that labor income is also highest and vice versa. The same is true of capital investment and average

Fig. 185—Capacity May Be Increased at Minimum Expense by the Conversion of Existing Buildings.

TABLE 63

AVERAGE CAPITAL INVESTMENT PER FARM FOR 23 FARMS

	At beginning of year		At end of year*		Average inventory		Average inventory per bird	Per cent of total inventory	Inventory gain	Inventory loss
	No.	Value.	No.	Value.	No.	Value.				
Real estate, acres of land	82.3	$1631.48	82.3	$1631.48	82.3	$1631.48	$1.26	12.15		
Dwelling house		2052.13		2052.13		2052.13	1.59	15.29		
Poultry buildings		3290.09		3278.53		3284.31	2.55	24.46		$11.56
Other buildings		460.87		469.67		465.27	.36	3.47	$8.80	
Total real estate		7434.57		7431.81		7433.19	5.76	55.37		2.76
Equipment, Poultry		1296.28		1346.94		1321.61	1.02	9.84	50.66	
Equipment, Other		372.46		352.69		362.58	.28	2.70		19.77
Sup. & Feed (Poultry)		175.40		152.00		163.70	.13	1.22		23.40
Supplies & Feed (Other)		127.42		116.73		122.08	.09	.91		10.69
Hens	472.8	668.55	624.1	721.16	548.45	694.86	.54	5.18	52.60	
Pullets	1328.2	2592.41	1555.3	2956.68	1441.75	2774.54	2.15	20.67	364.27	
Cockerels	180.2	278.48	144.7	285.56	162.75	282.02	.22	2.10	7.08	
Cocks	3.7	8.13	2.2	5.35	2.95	6.74		.05		2.78
Miscellaneous poultry	137.4	82.28	123.4	106.03	130.4	94.16	.07	.70	23.75	
Value other stock		161.30		175.87		168.58	.13	1.26	14.57	
Total capital		$13197.29		$13650.82		$13424.06	$10.40	100%	$453.53	

*Inventory figures for end of year allow for a 10% depreciation charge on the poultry buildings and equipment. Average inventory per bird is based on number of birds housed in fall.

N. H. Agr. Experiment Station, Bulletin 265

labor income per bird. In general, labor income per bird is about one-tenth the investment per bird. In cases where it is less it is possibly influenced by real-estate values. In most of the cases presented, there is evidence of a close relationship between the investment in livestock and the labor income per bird.

These figures on capital and income are usually the result of maturity of experience. One should bear in mind that the poultry business is readily adapted to small beginnings and gradual development. Many of our most successful poultrymen have started in the business in a small way and have expanded slowly, while gaining experience and additional capital.

Location and Type of Farm

Location of farm will influence to a large extent the total capital investment. Farms near dense centers of population will be adversely affected in this respect by real-estate values, but the advantage of nearness to market will help to offset this disadvantage, and if in addition to being near good markets, the farm is advantageously located on a main highway, roadside sale

Fig. 186—Triple-deck House, 106′ x 38½′. Capacity 3500 Leghorns, allowing 3.43 sq. ft. per bird. One man takes care of routine work.
[Courtesy N. E. Poultryman.]

of products may possibly be established to capitalize on the advertising value of such a location. Obviously if these advantages are not to be used, real estate of lower valuation may be found in a less desirable location.

The type of poultry enterprise is another element determining the capital investment. There are at least four types of poultry farming, in addition to many combinations of these types:

1. Market-egg production.
2. Breeding and hatching egg production.

3. Meat raising—broilers, capons, etc.
4. Hatchery operation.

Equipment requirements vary with the type of enterprise, some necessitating greater investment in permanent outfit than others. Overhead expense per unit must be based on total investment, productive season over which the enterprise is conducted, and total volume of business.

The type and size of business determine the amount of livestock required and, therefore, the capital needed. Other factors influencing capital investment include the amount of consumable supplies such as feed necessary for the particular type of business, and the amount of money the prospective proprietor has to invest.

Factors Influencing Profit

Type and Size of Business:

Not all types of poultry enterprises are profitable in all sections, nor are they equally profitable in the same locality. The margins of profit on poultry products are not constant nor identical.

As would be expected the income obtained will be related to the size of the business.

TABLE 64

RELATION OF SIZE OF FLOCK TO NET RETURNS*

	Average number pullets, hens and males	Number of farms	Average number pullets and hens per farm	Average eggs per hen	Average labor income	Cost of producing a dozen eggs	Value per dozen	Gain	
								Per hen	Per dozen
Less 700	531	40	523	135	$1,223	$.402	$.408	$.08	$.007
700–1,200	900	52	885	129	1,592	.380	.410	.32	.030
More than 1,200	1,908	29	1,872	132	2,361	.388	.434	.49	.047

*From Economic Studies in Poultry Farming in New York State, Cornell University, Ithaca, N. Y., 1927.

With the larger sized flocks, it is noted, a greater total labor income was secured on New York farms (Table 64). On the other hand a large flock does not necessarily indicate a high average labor income per bird. For instance in the flocks averaging 531 birds the labor income per bird was $2.30, while in flocks averaging 1,908 birds the figure is $1.24.

Breed and Strain:

There is a very great difference between breeds as to the profit each will return in a particular enterprise. For this reason a breed selected for any one of the types of poultry farming should have suitable characteristics and be capable of efficient production.

Similarly wide differences exist among various strains of the same breed or variety, and the special economic characteristics of one strain over another should be capitalized. Two men with identical breeds and with everything else similar except the strain of the birds, may make widely different profits. This emphasizes the importance of selecting not only the breed and variety, but the strain best adapted for the purpose.

Rate of Egg Production:

If this is large the enterprise naturally is likely to be more profitable.

TABLE 65

VARIATION IN LABOR RETURN ASSOCIATED WITH DIFFERENCE IN ANNUAL EGG PRODUCTION PER BIRD*

Annual Egg Production per Bird	Labor Return per Bird 1926	1927
80	$1.00	$0.69
100	1.58	1.29
120	2.16	1.89
140	2.74	2.49
160	3.32	3.09
180	3.90	3.69

It will be observed in Table 65 that labor income increased with production. The number of eggs from each bird is one of the most important factors bearing on the profitableness of the business.

Size of flock has some effect on the number of eggs per bird but is not of major importance. As will be noted in the following table covering seven years' records of flocks in New Hampshire, there was no significant difference in production per bird.

*Mass. Agr. Exp. Sta. Bul. No. 251.

Fig. 187—Five egg grades, with double grading unit.
[Courtesy Guy A. Leader, Leaders Leghorn Farms.]

TABLE 66

EFFECT OF SIZE OF FLOCK ON EGGS PER BIRD[1]

Classes	1934-35	1935-36	1936-37	1937-38	1938-39	1939-40	1940-41
1-100	172.34	177.29	183.83	180.51	181.75	180.22	184.34
100-500	169.88	178.86	174.18	181.58	177.95	187.29	186.98
500-1000	169.83	174.69	177.05	182.00	181.23	183.02	189.43
1000-1500	174.63	175.79	180.78	182.18	190.54	177.19	179.04
1500-2000	164.87	170.13	171.86	180.12	184.80	189.14	187.33
2000-Over	158.20	162.77	176.38	182.04	176.95	185.39	187.24
Average	166.99	171.99	177.00	181.40	180.85	184.93	183.73

The average number of eggs for the year 1932-33 was 162.54 as compared with 184.93 eggs in 1939-40 or a difference of 22.39 eggs. It is interesting to note that the average egg production per bird for the period has tended to increase the last few years, with the highest average being in 1944-45.

[1]N. H. Agr. Exp. Sta.

TABLE 67

SHOWING THE RELATION OF EGGS PRODUCED PER HEN TO LABOR INCOME ON NEW YORK COMMERCIAL POULTRY FARMS, 5 YEARS, 1929-1933. (1937)

	Production per Hen			Production per Hen		
	Low	Average	High	Low	Average	High
	Eggs per Hen			Labor Income		
1929	113	143	180	$2.042	$2,602	$2,352
1930	112	143	176	632	1,608	2,308
1931	117	147	172	619	1,259	1,200
1932	111	147	177	88	687	1,094
1933	112	144	179	—107	427	1,009
5 yr. av.	113	145	177	$ 655	$1,317	$1,593

Note: An increase of 1 dozen eggs produced per hen was accompanied by an increase in the labor income of $176 or of $14.65 per egg, as a five years' average.

Efficiency of Labor Utilization:

Many poultry farms are poorly managed from the standpoint of utilization of available labor. The poultry farm enterprise should be so balanced that each person connected with it will produce the greatest possible return as measured by man-work units.

The accompanying comparison (Table 68) indicates that with better labor efficiency, the return per hour for labor expended on the laying flock may be materially increased. The ef-

TABLE 68

COMPARISON OF LABOR EFFICIENCY*

	Average for 121 Farms Up-State	Average for 32 Long Island Farms
Number of fowls per man	537	778
Number of chicks raised per man	1,153	1,721
Dozens of eggs produced per man	5,720	9,056
Gross income per man	$3,962	$5,936
Labor returns per man	$1,353	$1,975
Hours of labor per 100 fowls in laying flock	198	212
Hours of labor per 1000 pullets raised	497	504
Returns from laying flock per hour of labor	$ 0.431	$ 0.926

* From "Economic Studies of Poultry Farming in New York State," Cornell University, 1927.

ficiency of labor is greatly affected by the use of labor-saving appliances and the convenient arrangement of buildings.

One of the most practical indicators of labor efficiency is the number of fowls per man. According to (Table 70) a survey of New York poultry farms it was necessary to have high labor efficiency in order to make a satisfactory labor income. The

Fig. 188—Triple deck house provides efficient use of labor. [Courtesy Roy E. Jones]

comparisons indicated in this table show that it was necessary to handle 1200 or more birds per man. The average for their 5-year study was 671 fowls cared for per man, which was too low for a satisfactory income.

Another useful measure of the degree of labor efficiency is the dozens of eggs produced per man. The total annual egg production should amount to more than 12,000 dozens per man, if a good labor income is to be expected. E. G. Misner states, "As a five years' average, the inefficient farms, producing only 4,314 dozens of eggs per man, returned a labor income of $606 to the operators. Moderately efficient farms, producing 7,556 dozens per man, returned a labor income of $1,154 and exceptionally efficient farms, producing 13,104 dozens per man, returned a labor income of $1,959 per operator." For each additional 1000 dozens of eggs produced per man, the labor income increased $154.

Still another measure of labor efficiency is the proportion of income required to satisfy the labor charges on a poultry farm. When more than 25 per cent of the receipts is required

TABLE 69

SEASONAL DISTRIBUTION OF THE LABOR REQUIREMENTS ON POULTRY[1]

Month	Hours for 100 Mature Hens	Hours for Replacements for 100 Hens
January	12.09	.21
February	13.91	.28
March	12.22	5.46
April	13.13	13.86
May	12.22	13.72
June	10.53	10.57
July	10.01	7.07
August	8.71	5.81
September	7.41	5.81
October	8.84	5.39
November	9.62	1.68
December	11.31	.14
Total Hours	130.00	70.00

TABLE 70

SHOWING RELATION OF LABOR EFFICIENCY AS MEASURED BY NUMBER OF BIRDS PER MAN TO LABOR INCOME ON NEW YORK COMMERCIAL POULTRY FARMS, 1929-1933.*

	Birds per Man			Labor Income		
	Low Efficiency	Average Efficiency	High Efficiency	Low Efficiency	Average Efficiency	High Efficiency
1929	349	700	1,333	$1,786	$2,486	$2,943
1930	372	722	1,242	748	1,623	2,654
1931	376	696	1,247	517	1,105	1,441
1932	365	709	1,199	85	906	1,300
1933	396	672	1,300	112	518	1,290
5 yr. av.	372	700	1,264	$ 650	$1,328	$1,926

Note: For each increase of 100 in the number of birds per man the labor income increased $143. When chicks were sold the labor income increased $182, when no chicks were sold it increased $99, for each 100 birds per man.

---

[1]Prof. B. H. Frame, University of Missouri.

*Cornell University Agr. Exp. Sta. Bull. 684, 1937.

to equal all charges for labor, including the value of the operator's time, the business is not liable to be particularly successful.

The Chore-Route Factor

The chore-route factor is one of the outstanding causes of reduced labor efficiency. Scudder, Burnier, Lunn, and Knowlton in Oregon State Bulletin No. 287, state:

"The route of travel followed in going to and from the laying house, the brooder house, the pullet range, etc., in doing the daily chores connected with the enterprise.

These routes were measured on a sufficiently large number of farms to show the situation clearly, and the average number of trips per day to each point was recorded.

TABLE 71

THE CHORE-ROUTE FACTOR*

Distance to buildings one way	Dwelling to laying house				Dwelling to brooder house				Laying house to range			
	Percentage of total farms	Average distance	Miles per year	Hours per year	Percentage of total farms	Average distance	Miles per year	Hours per year	Percentage of total farms	Average distance	Miles per year	Hours per year
	Per cent	Feet	Miles	Hours	Per cent	Feet	Miles	Hours	Per cent	Feet	Miles	Hours
100 feet and less....	20.8	89	62	37	21.1	80	14	8	28.6	70	14	8
101 to 200 feet.....	36.0	174	120	72	35.6	169	28	17	31.1	184	34	20
201 to 300 feet.....	26.4	280	194	116	21.1	287	48	29	13.0	290	54	32
301 to 500 feet.....	10.4	407	282	169	14.4	431	74	44	9.1	456	86	52
501 feet and over	6.4	650	450	270	7.8	814	138	83	18.2	996	188	113
Average	100.0	239	166	100	100.0	263	44	26	100.0	338	64	38
Number of farms ..	125	......	......	......	90	......	......	......	77	......	......	......

Average trips per day: To laying house, 5; to brooder house, 8; to range house, 4.
Number of days per year: To laying house, 365; to brooder house, 56; to range house, 124.
Rate of travel to all houses: 15 miles per day (9 hours)
*Oregon Station Bulletin, No. 287.

The results of this study are rather astonishing. Because of becoming habituated to it, few farmers realize the large amount of time and labor expended annually in traveling these chore routes and the extent this factor can lower efficiency.

Turning to Table 71, we find that the group with the least distance to travel to the laying house covers only 62 miles per year, requiring at a normal rate of 15 miles per nine-hour day the use of only 37 hours per year for this duty; while, at the other end of the scale, the group with an average of 650 feet distance from the dwelling to the laying house travels a total of 450 miles per year, requiring 270 hours of time. Even greater contrasts are noted in the chore routes to the brooder house and

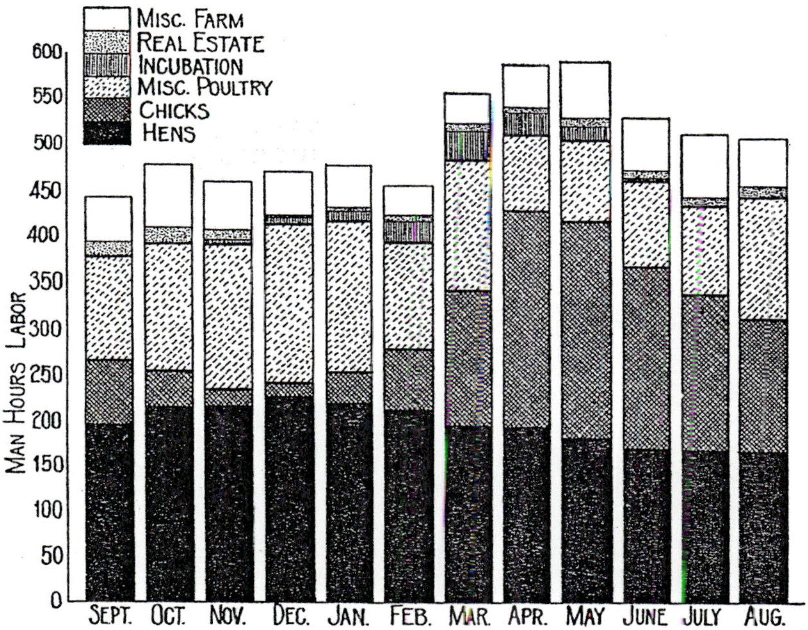

Fig. 189—A comparison of monthly labor requirements of hens, chicks, etc., on 23 farms.[1]

to the pullet range. The long-distance men on these different routes traveled from seven to fourteen times as far as the short-distance men with a correspondingly greater expenditure of time and labor.

It is obvious, of course, that the increased distance traveled on the chore route does not increase the egg production. As a matter of fact, the short-chore-route operator can take care of a good many more hens in the time saved that the long-distance man spends in just traveling.

Unnecessary labor is drudgery. When it is remembered that the poultryman does most chore-route travel early in the morning before breakfast, in the heat of the day at noon, or when tired at night, day in and day out through all kinds of weather, often with bad under-footing, and usually while carrying some burden, unnecessary chort-route travel becomes double drudgery. Naturally, it lowers the efficiency and increases the cost of production."

---

[1] N. H. Bull. 265.

Efficient Handling of Capital:

To be efficiently used capital must be ample to properly finance the business and must be wisely expended. With insufficient capital an enterprise will not provide a satisfactory living, even though it may pay a high rate of interest.

Capital should be properly allocated between real estate, buildings and equipment, with an adequate amount left for investment in the livestock requirements of the farm and for run-

NOMENCLATURE

Unit:

- A Dwelling and Yard
  - Garden
  - Home Orchard
  - Chick Starting Field
- B Breeding House and Yards
- C Pullet House and Yards
- D Range for Growing Chicks
- E Range for Growing Chicks
  - (Each Unit One Acre)

Houses

- 1 Dwelling
- 2 Garage
- 3 Service Building
- 4 Breeding House, 20x60
- 5 Breeding Pen, 20x20
- 6 Temporary Feed Room, 20x20
- 7 Pullet House, 20x100
- 8 Colony Houses, 10x12

Fields

- a Dooryard
- b Garden
- c Orchard and Small Chicks
- d Special Breeding Yard, 40x122
- ee Alternate Breeding Yards
- ff Alternate Yard for Layers
- g Range for Growing
- h Chicks

Size of Farm

20x40 rods

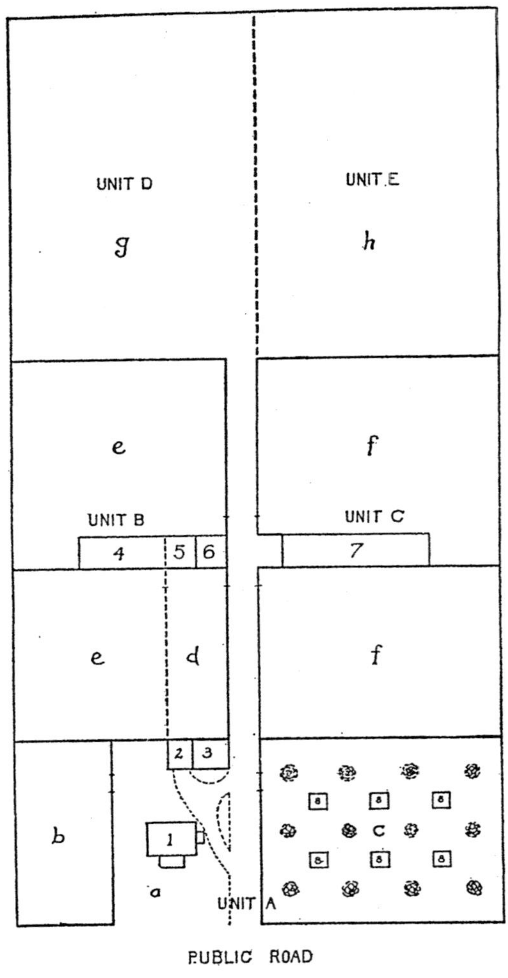

Fig. 190—Layout of 5-acre Poultry Farm.

ning expenses, so that the earning power is not restricted by too small a proportion in productive form.

TABLE 72

RELATION OF CAPITAL TO RETURN ON INVESTMENT ON 150 POULTRY FARMS IN NEW JERSEY

New Jersey Agr. Exp. Sta. Bulletin 329, 1918

Capital—	No. of Farms	Returns per $1,000 Invested	Per Cent Returns
$ 3,000 or less	3	$129	12.9
$ 3,001 to $ 5,000	29	132	13.2
$ 5,001 to $ 7,000	45	138	13.8
$ 7,001 to $ 9,000	38	143	14.3
$ 9,001 to $11,000	22	179	17.9
$11,001 and over	13	148	14.8
Average		$151	15.1

Under New Jersey conditions, according to this study (Table 72), capital efficiency reached its maximum when the investment ranged between $9,001 and $11,000. This study also indicates that a man must invest a reasonable amount of money before he can earn a satisfactory living in the poultry business.

Efficiency of Feeding:

The efficiency of feed in the production of eggs is determined largely by the quantity of eggs produced per unit of feed, by the cost of the feed, cost of labor, breeding of the birds, climatic conditions, and managerial skill displayed by the operator.

The effect of the level of production on the efficiency of feed consumption is illustrated by the following table. It will be noted that the low producers required 11.3 pounds of feed to produce one dozen eggs while the highest producing group required 5.2 pounds.

Many poultrymen are unable to obtain efficient use of feed because of their inability to induce high feed consumption. As an example, consider laying birds, each consuming approximately 1½ pounds of feed weekly, 75 to 80 per cent of which is utilized for maintenance. The owner feeds the birds so that they are consuming but little over a pound per bird per week, and in this case is merely meeting maintenance requirements. High egg production is possible only with high feed consumption.

Heavy feed consumption may be induced by (1) use of artificial lights, (2) moist mash feeding, (3) use of regular laying mash in pellet form, or (4) any other supplementary feeds that will add variety and result in getting more feed into the birds regularly.

TABLE 73

FEED NECESSARY TO PRODUCE A DOZEN EGGS ACCORDING TO THE NEW JERSEY EXPERIMENT STATION*

Season	101–125 Egg group	126–150 Egg group	151–175 Egg group	176–200 Egg group
	lbs.	lbs.	lbs.	lbs.
Winter—November, December, January, February	20.8	10.5	8.3	5.3
Spring—March, April, May	5.8	5.1	4.4	4.8
Summer—June, July, August	6.2	5.7	4.5	4.6
Autumn—September, October	13.4	12.6	7.7	6.3
Average for the year	11.3	8.4	6.3	5.2

*Hints to Poultrymen, Vol. 11, No. 4.

Efficiency Program:

The factors discussed indicate that in order to be successful a poultryman must have a definite program mapped out that will produce results. A sample of such a system is well outlined by Misner and Lee, in "Economic Studies of Poultry Farming in New York," and may well serve as a guide to success.

"Favorable size of business:

1. An average of 1500 birds or more. (House 1800 or more.)
2. Annual egg production of more than 18,000 dozens.
3. Combination of chicks and pullets to sell, with market egg production.
4. As little capital as necessary to provide a favorable size of business. Usually $10,000 or more is required.

Good rates of production:

5. Egg production of more than 150 eggs per hen, based on the average number of birds for the year.
6. More than 25 per cent production during October-December.
7. Hatchability of more than 60 per cent of total eggs incubated.
8. Successful rearing of more than 80 per cent of the original number of chicks.
9. Mortality of less than 20 per cent of the original number, or less than 25 per cent of the average number of layers.

10. Value of the eggs more than double the laying-flock feed costs.

Diversity or Balance of the poultry enterprise:

11. More than 20 per cent of the income from the poultry enterprise derived from the sale of pullets and chicks.
12. More than 60 per cent of the flock comprised of pullets.

Efficiency in the use of labor:

13. More than 1000 layers per man.
14. A gross income of more than $5000 per man.
15. A labor charge, including the value of operator's time, of less than 25 per cent of the receipts.

Efficiency in use of capital:

16. A rapid turnover, requiring less than two years for receipts to equal capital."

Culling:

Systematic culling should be practiced to weed out unprofitable individuals. The aim should be to house such high-quality pullets that few will need to be removed during the laying year.

Mortality:

Mortality affects the profitableness of the poultry enterprise, of course, and is a particularly important problem in laying flocks where the rate is in general about one per cent a month. It may range considerably higher, especially where disease epidemics occur.

Fig. 191—Dormer Windows Allow the Utilization of Roof Peak Areas of Barns.

[Courtesy New England Poultryman.]

In many cases systematic culling salvages for meat purposes birds which are slightly out of condition. This undoubtedly tends to reduce the mortality rate. For a more comprehensive view of the situation, therefore, an analysis of the reduction of the flock through both culling and mortality is advisable. The following records from the New Hampshire Home Egg-Laying Contest illustrate the importance of such a study.

TABLE 74

CULLING AND MORTALITY

Months	Culling	%Mort.	Culling	%Mort.	Culling	%Mort.	Culling	%Mort.
	1937–38	1937–38	1938–39	1938–39	1939–40	1939–40	1940–41	1940–41
October	4.14	.64	4.68	.69	6.00	.67	8.05	.70
November	2.24	.71	2.89	.76	2.21	.80	3.10	.80
December	3.44	.71	2.89	.87	4.04	.86	2.29	.82
January	3.56	.80	3.13	.84	5.04	.94	2.72	.96
February	3.74	.83	3.74	.94	3.27	1.01	3.41	1.04
March	3.64	.87	5.20	1.08	3.80	1.17	2.89	1.23
April	7.62	1.26	3.55	1.04	4.12	1.16	3.77	1.35
May	7.63	1.32	5.80	1.37	6.48	1.35	4.87	1.04
June	8.63	1.44	6.54	1.20	7.68	1.42	7.70	1.23
July	11.52	1.53	9.67	1.43	12.31	1.47	10.66	1.19
August	16.63	1.49	10.46	1.24	14.40	1.39	12.90	1.34
September	16.13	1.42	21.02	1.46	17.77	1.33	21.90	1.41
Yearly Total	93.08	12.82	79.57	12.92	87.12	12.79	81.26	13.09
Monthly Av	7.41	1.08	6.63	1.08	7.26	1.13	6.77	1.09

The fifteen year average (1926-27 to 1940-41) by months was October, 4.67; November, 2.92; December, 3.62; January, 3.32; February, 3.78; March, 4.29; April, 5.29; May, 5.95; June, 7.88; July, 10.12; August, 14.30; September, 19.02.

From the above table it may be observed that the heaviest amount of culling has taken place during June, July, August and September during the fourteen year period as these are months that most poultrymen begin making ready housing room for their new pullets. Also, it is a period of normal decrease in production. There has not been much variation in the average monthly percent culls for the period, and the monthly average for all years was 6.74% and a yearly average of 87.43%.

Personal Factor:

In the final analysis the personal factor is of major importance in the success or failure of a poultry enterprise. It often explains the degree of failure or success. For knowledge of the business, executive ability and initiative are rewarded in poultry farming as in other lines of business. The person who instinctively has good "chicken sense" is most likely to succeed.

Like other livestock industries poultry husbandry requires the ability to observe. This talent is decidedly improved by experience, as illustrated in the following (Table 75) extracted from Bulletin 329 of the New Jersey Experiment Station:

TABLE 75

EXPERIENCE AS AFFECTING LABOR INCOME

Years of Experience	Labor Income
1.0 to 2	$ 362
2.1 to 4	392
4.1 to 6	720
6.1 to 8	722
8.1 to 10	1,002
Over 10	1,344

Management Considerations:

Anticipation and Equalization of Income—In the systematic operation of a business a schematic plan is essential for guiding future development. Changes in conditions will inevitably make modification of the details desirable, but the general outline should be followed as closely as is physically and financially possible. In working up such a plan it is necessary to anticipate the capital requirements of the enterprise and the probable receipts.

In any livestock enterprise requiring annual replacement of a considerable portion of the stock, equalization of the income over the entire year is advisable. Thus as the young stock is growing and maturing there will be sufficient income for their needs, for the feed bill of the layers and breeders, and for other expenses incidental to production. Unfortunately many have their income concentrated at some particular period of the year, putting the entire business out of balance and making the financing of feed bills difficult at some seasons.

Some practices which aid in providing a balanced year 'round income include the following:

TABLE 76—FLOCK REDUCTION*

PULLET AND HENS—SIXTEEN-YEAR SUMMARY

Year	Birds left from 1,000 end of year	Birds Culled	Mortality—Prolapsus & Disease	% Mortality Pullets	% Mortality Hens
1932-33	392	456	152	15.8	13.2
1933-34	428	432	140	15.0	11.0
1934-35	389	489	122	12.6	11.1
1935-36	427	451	122	12.6	11.9
1936-37	364	525	111	11.7	8.4
1937-38	360	542	98	10.3	7.1
1938-39	402	490	108	11.1	8.2
1939-40	395	499	106	11.0	8.2
1940-41	334	556	110	11.6	6.4
1941-42	323	572	105	11.0	6.7
1942-43	318	577	105	11.2	5.9
1943-44	250	646	104	11.2	4.9
1944-45	263	665	72	7.6	4.6
1945-46	217	707	76	7.7	6.9
1946-47	351	558	91	9.5	6.2
1947-48	233	619	88	9.2	6.0

1. Maintaining two or more groups of pullets of different ages to obtain better distribution of egg production.

2. Maintain highest possible production throughout the entire laying year.

3. Retaining a larger percentage of the layers for summer and early-fall production to furnish more income during months otherwise short in this respect.

4. Diversifying the business.

Diversification—This latter method of balancing the income has a stabilizing influence on the business. The adage, "Don't put all your eggs in one basket," applies even better to the poultry business than to others. The advantages are principally that it provides more money-making phases, better utilization of labor, and more regular cash returns.

Diversification is generally viewed as a combination of various types of enterprises. While this is of course advantageous in many cases, the poultry business is well adapted to diversification within itself. This may be accomplished by a combination of any or all of the following phases:

1. Production of market eggs—This in general constitutes the largest item of return on the average poultry farm.

*Home Egg-Laying Contest Bulletin, Connecticut State University.

TABLE 77

MONTHLY COMPARISON OF PULLETS VS. HENS AS TO CULLING AND MORTALITY.
Culling and Mortality—Pullet Flocks—1940-1941

Month	No. Birds Involved	As Reported by Months: Percent Culled	As Reported by Months: Prolapsus & Cannib.	As Reported by Months: Other Causes	Total	Decrease Basis 1000 Bird Flock: Flock Count	Decrease Basis 1000 Bird Flock: Culled	Decrease Basis 1000 Bird Flock: Prolapsus & Cannib.	Decrease Basis 1000 Bird Flock: Other Causes
Oct. 1940	74,491	4.00	.15	.79	.94	1000	40	2	8
Nov. "	88,432	1.95	.29	.90	1.19	950	19	3	9
Dec. "	97,987	3.17	.22	.99	1.21	919	29	2	9
Jan. 1941	101,726	4.53	.19	1.19	1.38	879	40	2	10
Feb. "	90,019	4.16	.18	.94	1.12	827	34	1	8
Mar. "	77,398	11.00	.32	1.11	1.43	784	86	3	9
April "	70,598	7.72	.31	1.21	1.52	686	53	2	8
May "	63,278	5.22	.25	1.14	1.39	623	33	2	7
June "	56,304	6.72	.31	1.17	1.48	581	39	2	7
July "	53,349	9.69	.17	1.30	1.47	533	52	1	7
Aug. "	55,383	11.37	.38	1.27	1.65	473	54	2	6
Sept. "	60,431	10.44	.24	1.15	1.39	411	43	1	5
Totals						362	522	23	93

TABLE 78

MONTHLY COMPARISON OF PULLETS VS. HENS AS TO CULLING AND MORTALITY.*

Culling and Mortality—Hen Flocks—1940-1941

Month		As Reported by Months				Decrease Basis 1000 Bird Flock			
	No. Birds Involved	Percent Culled	Prolapsus & Cannib.	Other Causes	Total	Flock Count	Culled	Prolapsus & Cannib.	Other Causes
Oct. 1940	24,237	25.52	.06	.74	.80	1000	255	1	7
Nov. "	16,837	19.85	.10	.70	.80	737	146	1	5
Dec. "	14,696	9.89	.16	1.08	1.24	585	58	1	6
Jan. 1941	13,834	16.01	.19	1.46	1.65	520	83	1	8
Feb. "	12.410	10.28	.08	.98	1.06	428	44	0	4
Mar. "	9,398	12.65	.27	1.28	1.55	380	48	1	5
April "	8,279	8.18	.39	1.03	1.42	326	27	1	3
May "	5,738	3.75	.45	1.45	1.90	295	11	1	4
June "	4,418	13.38	.23	1.38	1.61	279	37	1	4
July "	7,245	10.43	.25	1.49	1.74	237	25	1	4
Aug. "	5,108	16.74	.29	1.00	1.29	207	35	1	2
Sept. "	5,045	15.24	.14	1.19	1.33	169	26	0	2
Totals						141	795	10	54

*Courtesy R. E. Jones, Connecticut State University Home Egg-Laying Contest 1941.

2. Sale of hatching eggs—This offers a possibility for the disposal of a considerable quantity of eggs at a premium price.

3. Marketing of baby chicks—Because of the transformation of eggs into a product usually sold on a higher price level, this allows for increasing the income.

4. Custom hatching—Where practiced it is advisable to custom hatch only eggs from flocks of the same health status as the home flock.

5. Production of started chicks—Chicks a week or more in age up to pullets twelve weeks old may be profitably handled. This phase of the business can undoubtedly be developed to a very appreciable extent.

6. Sale of mature laying pullets—This can supplement the income during fall months.

TABLE 79

DIVERSIFICATION ON 23 NEW HAMPSHIRE POULTRY FARMS, NEW HAMPSHIRE AGRICULTURAL EXPERIMENT STATION BULLETIN NO. 265

RECEIPTS:

	Number		Value	Per Cent
Market Eggs	9955.3	doz.	$4392.18	43.53
Hatching Eggs	1613.7	"	1124.09	11.14
Live Fowl	634.	No.	827.62	8.20
Ducks and Geese			1.78	.02
Dressed Fowl	167.8	"	228.08	2.86
Pullets Sold	111.7	"	185.99	1.84
Cock Birds—Live	49.7	"	56.84	.56
Capons	2.4	"	7.23	.07
Cock Birds—Dressed	5.4	"	6.78	.07
Cockerels	1.2	"	3.10	.03
Roasters—Live	182.3	"	225.89	2.24
Roasters—Dressed	17.7	"	24.91	.25
Broilers—Live	1657.3	"	1051.64	10.42
Broilers—Dressed	16.1	"	20.28	.20
Started Chicks	11.1	"	6.52	.06
D. O. Chicks	3339.	"	673.60	6.70
Misc. Poultry Receipts			48.26	.48
Grain Sold			314.69	3.12
Trucking			35.95	.36
Commission on Grain Sold			9.91	.10
Custom Hatching	2701.1	"	80.46	.80
Equipment Sold			23.15	.23
Misc. Farm Receipts			13.42	.13
Fruit			33.77	.34
Wood			4.43	.04
Crops			88.66	.88
Milk			45.82	.45
Other Stock			19.61	.19
Rent of House to Hired Man			18.70	.19
Increase in Value Inventory			453.53	4.50
Total Receipts			$10,088.90	

7. Handling of breeding birds—Where a reputation has been established for quality stock, this may represent an important item, particularly in the case of men interested in carrying on a breeding program.

8. Poultry meat production—This is usually incidental to the regular business, but may be expanded where a special market is available. Fowls, broilers, roasters, capons, etc., may be sold either live, dressed, or dressed and drawn. By far the greater portion on the average farm is sold alive.

9. Sale of other poultry—Ducks, geese, turkeys, etc., may be sold as a specialty or side-line proposition, where the market warrants such diversification.

10. Distribution of supplies—Many farms engage in the selling of equipment, feed, and similar products, which may yield an appreciable portion of the annual income.

11. Local buying and selling—In some communities poultrymen with advantageous market outlets are able to act as dealers in the purchase of both eggs and poultry for resale. Some poultrymen have found this practice so profitable that they have given practically their entire time to it.

12. Custom slaughtering and dressing—This practice has developed because of the advent of picking machines. Hazards include the introduction of disease onto the owner's premises.

Cash or Credit Buying

In the purchase of feeding stuffs cash buying will save money. Credit by retail grain stores increases the cost of feed distribution and consequently adds to the cost of feed stuffs to the poultryman. It is well to establish credit with both the grain store and the bank for use in an emergency. Otherwise it is best to operate chiefly on a cash basis. The costs of credit buying are well summarized by the following statements by E. H. Rinear, New Hampshire Agricultural Experiment Station, Bulletin 251:

"Other important costs resulting from supplying credit are interest, credit accounting, office supplies and collections costs.

"Usually, the greater part of the accounts was in book and not note form, so that no direct income was received for the extension of credit. This means that a charge for credit had to be added at the time of sale, or else absorbed as part of the general overhead.

"A study of grain-store credit in New York State showed an annual cost of 13.35 per cent—more than twice the 6 per cent rate charged by banks. On this basis, the annual cost to the 89 New Hampshire grain stores extended $870,520.50 of credit would be $116,214.48, or $63,983.25 more than it would have cost if obtained from banks.

"When nothing was allowed for cash, the average per cent of cash sales for 15 stores was 41 per cent; when a discount of 5 cents a bag was made, 47 stores had an average of 53 per cent cash sales; and likewise for 10 and 15 cents a bag the average respective cash sales were 57 and 60 per cent.

"When the stores are arranged according to the percentage of annual cash sales along with their respective gross margins, it is seen that store credit costs money and that the trade is paying for it. The customers obtaining store credit are not paying in full for the service rendered because the cash men are also contributing toward the cost of credit and are paying for a service which they do not obtain."

TABLE 80

MONTHLY ANALYSIS OF EGG PRODUCTION
Ohio State University, Bulletin 60

Months—	160 Egg Standard	543 Flocks	High Flock	Low Flock
November	5	4.6	5.9	
December	8	7.1	12.8	1.0
January	10	8.0	13.4	3.2
February	13	10.0	16.4	3.2
March	17	15.3	20.3	10.0
April	21	17.9	23.5	15.5
May	20	16.7	25.4	12.6
June	18	16.8	23.7	9.7
July	16	15.3	23.6	9.3
August	14	14.2	23.8	8.6
September	11	12.0	20.4	6.5
October	7	6.8	16.7	3.4
Total	160	144.7	225.9	81.5

Annual and Monthly Egg Production

Since the greater percentage of income on the average poultry farm is derived from egg sales, a high annual egg production is essential. As previously stated a good distribution table is interesting from this view and may be used in anticipating income based on the production capacity of a flock.

Labor Requirements

Essentially the poultry business is one of attention to detail and of repetition of routine work. Consequently the labor requirements present an important problem. Prof. J. E. Rice, during his tenure at Cornell University, stated that the cost of producing eggs with commercial flocks in New York state over a 14 year peroid, 19.7 per cent was due to human labor, 1.2 per cent to horse labor, and 1 per cent to the use of automobiles and trucks.

TABLE 81

LABOR AND AUTOMOBILE REQUIREMENTS
From "Economic Studies of Poultry Farming in New York," I and II Cornell, Ithaca, New York

	New York—Up-State		New York—Long Island	
	Hours per 100 layers	Rate	Hours per 100 layers	Rate
Human labor	198	$0.423	212	$0.514
Horse labor	16	0.163	3	0.172
Use of automobile	57	0.083	108	0.074

An operator's labor accounts for 50 to 65 per cent of the total labor requirements; 20 to 25 per cent is hired labor, and the balance is unpaid labor done by the operator's family. Additional information on labor requirements is illustrated in the accompanying tables.

TABLE 82

MAN-LABOR REQUIREMENTS OF POULTRY
(By months—1912-1922)
Missouri Experiment Station Bulletin 219

Month—	Hours per Hen	Cost	Per Cent of Tot. Labor Cost
January	0.150	$0.0286	8.77
February	0.140	0.0267	8.18
March	0.169	0.0323	9.90
April	0.212	0.0405	12.41
May	0.179	0.0342	10.48
June	0.144	0.0275	8.54
July	0.112	0.0213	6.53
August	0.109	0.0208	6.37
September	0.107	0.0204	6.25
October	0.103	0.0196	6.01
November	0.143	0.0273	8.37
December	0.142	0.0271	8.30
Total	1.710	$0.9263	100.00

A study of the information presented will indicate that the labor requirements are very definite but that there is considerable variation under different conditions. The requirements may change, due to varying types of enterprises, the location and layout of the plant, the application of labor-saving devices, the size of the enterprise, and the amount of unproductive labor engaged in such work as beautification of grounds.

Cost of Production

The poultryman in business for a livelihood is vitally interested in the cost of his operations and of production. Figures of this nature are available from an economic survey of 23 poultry farms in New Hampshire. Table No. 83 represents the average expense and distribution costs entailed in their operation. The major items are feed, which represents 65 per cent of the total cost, and labor, 10.9 per cent. The balance is distributed over a wide range of items.

TABLE 83

AVERAGE EXPENSES ON 23 NEW HAMPSHIRE POULTRY FARMS, NEW HAMPSHIRE AGRICULTURAL EXPERIMENT STATION BULLETIN NO. 265

	Amount	Value	Per Cent
Real Estate		$ 348.43	4.34
Equipment—Poultry		232.72	2.90
Equipment—General		71.40	.89
Poultry Feed		5,212.42	65.00
Poultry Supplies		76.40	.95
Shavings	124.3 bales + 1.2 loads	53.76	.67
Litter			
Straw	25.6 bales	18.32	.23
Peat	8.3 bales	18.24	.23
Coal		129.99	1.62
Day Old Chicks	679.5	136.61	1.70
Stock Purchased		24.85	.31
Hatching Eggs	2347.3 eggs	169.74	2.12
Hired Labor		873.80	10.90
Misc. Poultry		181.61	2.26
Market Eggs Purchased	38.9 doz.	16.92	.21
Custom Hatching	1163.6 eggs	42.53	.53
Testing and Certification		64.83	.81
Taxes		145.43	1.81
Insurance		62.45	.78
Misc. Farm		49.48	.62
Other Feed		34.97	.44
Truck		28.60	.36
Stock Purchased		21.68	.27
Apples		3.16	.04
Total Expenses		$8,018.38	

Cost of Producing Eggs:

Egg-production costs are dependent on a number of factors. Feed, labor, and depreciation of the flock are the three main ones, and they account for approximately 85 per cent of the total. The following figures from "Economic Studies of Poultry Farming in New York," Cornell University, show the distribution of

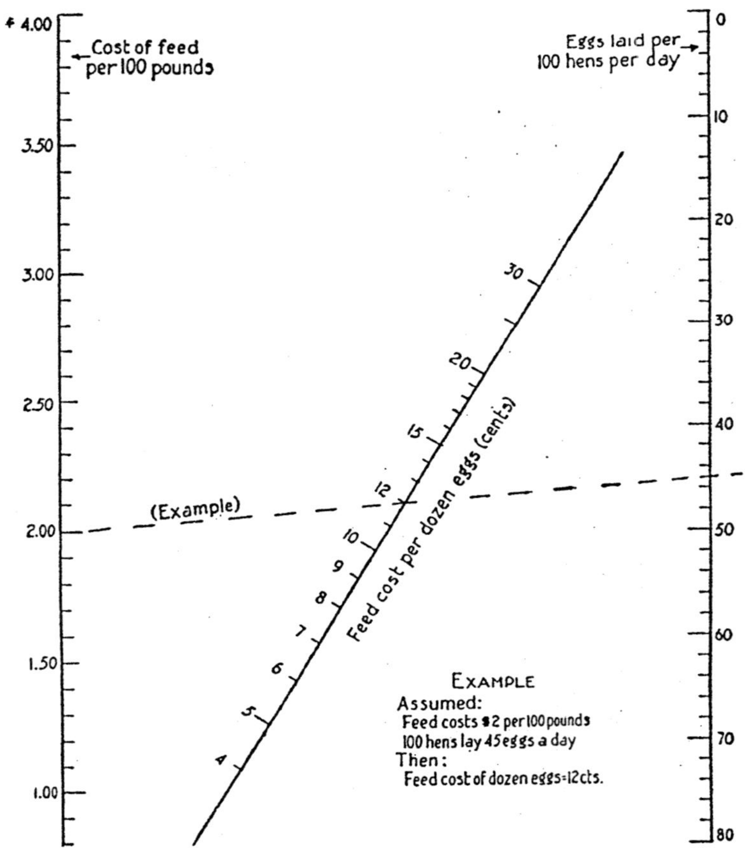

Fig. 192—A Quick Method for Calculating the Cost of Eggs.

To find the feed cost of a dozen eggs when the cost of feed per 100 pounds and the number of eggs laid per 100 hens per day are known, lay a straight-edge cross the chart so that it connects the proper points on the two outside scales (**feed cost and egg production**). The feed cost of a dozen eggs is then read directly from the inclined scale where it is crossed by the straight-edge. In constructing the chart it has been assumed that 100 hens will use 22.5 pounds of total feed a day.

[Courtesy Card & Henderson.]

TABLE 84

COST OF PRODUCING EGGS, AND PULLETS[5]

1. Cost of producing eggs:

Inputs		Products
17 lbs. feed 0.6 hours labor 4 cents supplies 13.7 cents overhead	produce	0.4 lbs. broiler 0.7 lbs. fowl 0.3 day old chick 1.0 dozen eggs

2. Cost of producing pullets:

Inputs		Products
\$ 47.40 estimated value of chicks 100.94 cost of feed 31.00 cost of labor 18.83 overhead costs 10.67 cost of supplies	produce	100 pullets and 210 lbs. broilers

\$208.84 equals 100 pullets and 210 pounds of broilers.

costs in cents and percentages, and also the distribution of returns. These cost-account records do not show the effect of annual production per bird on the cost of a dozen eggs.

Cost of Producing Mature Birds:

The annual replacement of the laying flock commands considerable time, attention, effort and finances. It is obviously a costly phase of the routine procedure in the operation of poultry

TABLE 85

COST OF PRODUCING EGGS IN NEW YORK

	Farms on Long Island Per dozen eggs Produced	Farms on Long Island Per cent of Total	Farms not on Long Island Per dozen eggs Produced	Farms not on Long Island Per cent of Total
Cost factors—				
Feed	\$0.184	44.5	\$0.183	46.2
Litter	0.006	1.5	0.005	1.3
Human labor	0.093	22.5	0.079	19.8
Horse labor	----	0.1	0.002	0.6
Use of auto	0.007	1.7	0.004	1.1
Use of buildings	0.021	5.0	0.025	6.6
Use of equipment	0.006	1.5	0.002	0.5
Use of yards and ranges	0.001	0.3	0.002	0.5
Depreciation on flock	0.073	17.6	0.072	18.2
Interest	0.007	1.8	0.008	2.0
Miscellaneous	0.015	3.5	0.013	3.2
Total	\$0.413	100.0	\$0.396	100.0
Returns—				
Eggs sold	\$0.444	91.1	\$0.381	88.8
Eggs eaten	0.004	0.8	0.006	1.3
Eggs incubated	0.033	6.8	0.034	7.9
Manure	0.005	1.1	0.007	1.7
Miscellaneous	0.001	0.2	0.001	0.3
Total	\$0.487	100.0	\$0.429	100.0

[5]N. H. Agr. Exp. Sta. Bull. 265.

farms. For this reason, if for no other, study of production costs is advisable. It should be noted, in the accompanying table, that there are certain definite costs entering into the reproduction of the flock, many of which cannot be materially reduced. Skillful rearing may, however, materially affect the quality of the pullets produced, so that careful attention should be paid to every detail in order to be able to place in the laying houses high quality pullets that will produce a satisfactory profit.

TABLE 86

ITEMIZED COST OF PRODUCING PULLETS (TO 6 MONTHS OF AGE)

Averages for 22 flocks, producing 11,785 pullets, 1927 and 1928. Oregon Agricultural Experiment Station Bulletin 287

Expense item	Cost per farm	Cost per pullet	Percentage of total cost
			Per cent
Operator's labor	$132.54	$0.25	15.7
Unpaid family labor	.73	.....	.1
Total labor	$133.27	.25	15.8
Purchased feed	432.22	.81	51.3
Farm-produced feed	5.50	.01	.6
Total feed	$437.72	.82	51.9
Baby chicks purchased	172.22	.31	20.4
Brooding fuel	18 13	.03	2.1
Use of automobile	12.09	.02	1.4
Hatching eggs (farm)	4.81	.01	6
Hatching eggs (purchased)	2.45	.01	.3
Litter (purchased)	5.14	01	.6
Taxes	1.69	.....	2
Spray and medicine	4.05	01	5
Seed for range	1.37	01	.2
Miscellaneous	6.64	01	8
Total miscellaneous	228.59	.42	27.1
Depreciation on brooders and equipment	18.14	.04	2 2
Depreciation on incubators	.09	.....	.....
Depreciation on range houses	.73	.....	1
Depreciation on other equipment	1.91	.....	2
Total depreciation	20 87	.04	2.5
Interest on brooders and equipment	$13.77	.03	1.7
Interest on land in range	7.37	01	.9
Interest on range houses	.32	.....	....
Interest on incubators and equipment	05	.....	.....
Interest on other equipment	1.04	.....	.1
Total interest at 5 per cent	$22.55	.04	2.7
Total gross cost	$843.00	$1.57	100.0
Receipts from broilers	126.59	.25	15.0
Receipts from cull pullets	2.28	.01	.3
Receipts from eggs to 6 months	34.22	.06	4.0
Total credits	$163.09	.32	19.3
Total net cost	$679.91	$1.25	80.7

The cash cost is 78.4 per cent of the total cost or 98 cents net per pullet.

"The effect on the quality of the pullet should always be given careful consideration before using a new management method designed to lower costs. There was nothing, however, which indicated that low-cost pullets were inferior to those costing more. On the contrary, such factors as prompt separation of the broilers from the pullets, low death loss, efficient feeding, efficient care, and rigid and prompt culling of inferior pullets seem to be of major importance in lower pullet production cost."

Pedigree Stock—In the production of pedigree stock, Hays and Locke, "Poultry Science," Vol. X, No. 4, state: "The total cost of producing a pedigreed chick 70 days of age is $1.73, and the total cost of producing a pullet or cockerel 162 days of age, including interest on working capital, is $2.87."

Additional information in regard to the cost of producing high quality pedigree cockerels, has been compiled by L. C. Beall,

TABLE 87

SHOWING COST OF TRAPNESTING, ALSO VALUE OF BREEDERS
(L. C. Beall, Jr.)

To 1,500 Extra Superior Chicks valued at $1.00 ea.	$1,500.00	
1,000 Double A Chicks valued at $3.25 ea.	3,250.00	
(50 per cent of which 2,500 chicks will be cockerels).		
Cost of rearing the pullets to laying age	2,000.00	
Result, 1,000 pullets at laying age	$6,750.00	$6.750.00
(Allows for culling 250 female chicks, including mortality).		
Cost of trapnesting 1,000 pullets:		
Labor	$1,200.00	
Office	200.00	
Supplies	50.00	
Interest and repairs	200.00	
Total	$1,650.00	$1,650.00
		$8,400.00

On hand second fall—

600 pullets with records 200 to 300 eggs. Further loss of 25 per cent for selection for type, size, Standard qualities, egg size and quality.
150 and failure to molt properly

450 trap-record hens suitable for use as breeders.

Estimated cost value as follows, to offset or equal the actual cost:

150 with records 200-224, value $10.00 each	$1,500.00
150 with records 225-249, value $15.00 each	2,250.00
100 with records 250-275, value $24.00 each	2,400.00
50 with records 275-300, value $45.00 each	2,250.00
Total	$8,400.00

Jr., of Vashon, Washington, and was included in his 10th Annual Catalogue.

Mr. Beall states, "that on a successful breeding plant for highest egg production, in combination with standard type and other qualifications the rigid culling starts with the eggs to be incubated, is carefully applied to the chicks in the brooding quarters and is kept up all the time the growing stock is on range. Then, later, in the case of pullets, the trapnest does its part and of course, 'plays no favorites'. In the case of cockerels it is much worse. Cockerels of the vigorous 'scrappy' kind do a large amount of 'eliminating' on their own account. We save for our own use as select breeders and to sell for that purpose, *only one cockerel in five.*"

The accompanying (Tables 87 and 88) tabulation gives Mr. Beall's estimate of the actual cost of producing high grade pedigree cockerels.

TABLE 88

SHOWING ESTIMATED VALUE OF COCKERELS

(L. C. Beall, Jr.)

Number taken as basis:	
2500 cockerels at 8 weeks of age, pedigreed @ $2.00	$5,000.00
2500 cost of feed and labor to 3 months of age	2,000.00
1000 culled	
1500 cost of feed and labor to 4 months of age	750.00
100 culled	
1400 cost of feed and labor to 5 months of age	750.00
200 culled	
1200 cost of feed and labor to 6 months of age	750.00
200 culled	
1000 cost of feed and labor to 7 months of age	750.00
200 culled	
800 cost of feed and labor to 8 months of age	600.00
300 culled	
500 cost of feed and labor to 9 months of age	400.00
Total	$11,000.00

Must be valued as follows, to offset or get actual cost of the 500 suitable to use for progressive breeding results:

50 @ $ 10.00 each	$ 500.00
200 @ $ 15.00 each	3,000.00
150 @ $ 20.00 each	3,000.00
50 @ $ 30.00 each	1,500.00
25 @ $ 40.00 each	1,000.00
20 @ $ 75.00 each	1,500.00
5 @ $100.00 each	500.00
500 cockerels	$11,000.00

TABLE 89

COST OF RAISING PULLETS*
(New York State)

Cost Factors:	Long Island		Up-State Farms	
	Per Pullet	% of Total	Per Pullet	% of Total
Feed	$0.840	60.1	$0.782	64.3
Litter	0.011	0.8	0.008	0.7
Human labor	0.259	18.5	0.212	17.4
Horse labor	0.001	---	0.006	0.5
Use of auto	0.004	0.3	0.005	0.4
Use of buildings	0.059	4.2	0.078	6.5
Use of equipment	0.018	1.3	0.021	1.7
Use of yards and ranges	0.120	8.6	0.034	2.8
Interest	0.023	1.7	0.022	1.8
Miscellaneous	0.063	4.5	0.047	3.9
Total	$1.398	100.0	$1.215	100.0

* Does not include credit for sale of broilers or pullets.

Poultry Accounts

Included in this chapter will be found a complete poultry account record devised by Dr. M. F. Abell, New Hampshire Agricultural Experiment Station.

Keeping the Farm Record:

To determine the profits from farming, it is necessary to have a record of the whole farm business for the year. Four things are essential in this: (1) inventory of the livestock, machinery, and feed and supplies on hand at the beginning of the year; (2) farm receipts for the year; (3) farm expenses for the year, and (4) inventory of the livestock, machinery, and feed and supplies on hand at the end of the year.

This record is designed to assist in keeping an account of these things, and is concerned with the profitableness or the nonprofitableness of the farm business as a whole, and not with the different enterprises within it.

One may start the farm record at any time, but the most convenient is at housing, since the work of taking the inventories is partly done by the count of pullets housed. On the other hand, a record system maintained on a calendar year basis greatly facilitates the filing of income tax returns. In taking the inventories at the beginning and end of the year, place fair values on everything. It is unfair to the farm business to use either extremely low or extremely high values.

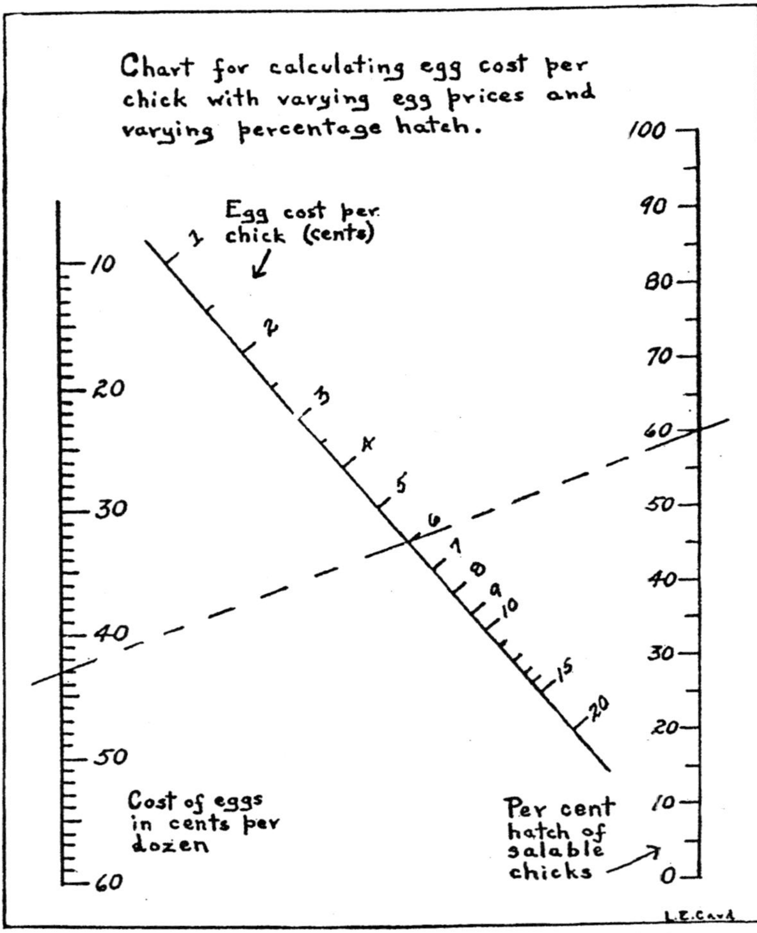

Fig. 193—Chart for Calculating Egg Cost per Chick with Varying Egg Prices and Varying Percentage Hatch.

The method of using the chart is indicated by the broken line drawn in for illustration. One simply lays a ruler across the chart, connecting the proper points on the two verticle scales, representing egg cost and percentage hatch, and the result is read on the inclined scale at the point where the rules crosses it.

The illustration shows that when eggs are costing 43 cents a dozen and giving a 60 per cent hatch, the egg cost per chick is six cents. Similarly, with 31 cent eggs and a 65 per cent hatch, the egg cost per chick is four cents.

[Courtesy Dr. L. E. Card, Univ. of Ill., and Hatchery Tribune.]

This record is concerned primarily with the farm business. In making entries of receipts and expenses, include only those which belong to the farm business. The personal and household expenses of the farm family may be included if desired, but should be entered only in the column so designated and not included under total farm expense.

Entries should be made at the time of each transaction, regardless of whether cash is paid or received at the time. Money borrowed or paid on debts should not be entered. Money borrowed is charged against the farm business when spent.

At the end of the year, do not fail to summarize the farm business to see how profitable it has been. Subtract the total expenses from the total receipts. This leaves the income from the capital invested and from the operator's labor. Deduct five per cent for the use of the capital invested at the beginning of the year. The amount left is what the operator received for his year's work—his labor income. In addition to his labor income, the operator also has free house rent and part of his family living on the farm.

When the farm record is completed, it should be carefully studied to see how the farm business may be made more profitable. By working out the efficiency factors the operator may compare his farm business with that on successful poultry farms, to see which are his strong and weak points.

Many farm businesses are unprofitable because of serious leaks. By stopping these leaks, the operator may materially increase his labor income. Full directions for figuring the efficiency factors are also given.

TABLE 90

PULLET REARING COST TO 24 WEEKS.

Regular chicks (straight run):

2.36 chicks+34 lbs. mash+14.5 lbs. scratch+21 cents cash cost —(minus)—3.25 lbs. broiler=Cost of pullet.

Sexed pullet chicks:

1.18 chicks+24 lbs. mash+12 lbs. scratch+14 cents cash cost=cost of pullet.

Assumptions:

1. Feed consumption and weights, N. H. Agr. Exp. Cir. 52.
2. Males sold as broilers @ 3 lbs., at 12 weeks.
3. Mortality—8% for first 12 weeks.
   7% for second 12 weeks.
4. Labor and overhead costs not included.

(F. D. Reed, Poultry Tribune.)

TABLE 91

SUMMARY OF EGG PRODUCTION BY MONTHS

Months	Number of birds beginning of month	Number of birds end of month	Average number per month	Number of eggs	Per cent production	Eggs per bird	Mortality	Culls
January								
February								
March								
April								
May								
June								
July								
August								
September								
October								
November								
December								
Totals								

NOTES REGARDING POULTRY

Date of hatch	Number hatched	Number eggs set
Totals		

TABLE 92

RECEIPTS*

Date	ITEMS	Total farm receipts	Egg Market	Eggs Hatching	Broilers live and dressed	Fowl live and dressed	Roasters Etc. live and dressed	Day-Old chix	Miscel-laneous poultry receipts		Other farm receipts	

TABLE 93

EXPENSES*

Date	ITEMS	Total farm expenses	labor	Feed layers	Feed growing stock	Day-old chix hatching eggs poultry stock	Poultry supplies	Poultry equip-ment	Poultry miscel-laneous	Real estate	Other farm expense	To be used if desired for personal and household expense

* Each of the above account records occupies a full length page in the account book.

TABLE 94—MONTHLY SUMMARY OF EXPENSES*

Date	ITEMS	Total farm expenses	labor	Feed layers	Feed growing stock	Day-old chix hatching eggs poultry stock	Poultry supplies	Poultry equipment	Poultry miscellaneous	Real estate	Other farm expense	To be used if desired for personal and household expense

TABLE 95—MONTHLY SUMMARY OF RECEIPTS*

Date	ITEMS	Total farm receipts	Egg market	Egg hatching	Broilers live and dressed	Fowl live and dressed	Roasters etc. live and dressed	Day old chix	Miscellaneous poultry receipts		Other farm receipts
	Total receipts for: January										
	February										
	March										
	April										
	May										
	June										
	July										
	August										
	September										
	October										
	November										
	December										
Total for year											

* Each of the above account records occupies a full length page in the account book.

TABLE 96

INVENTORY OF REAL ESTATE

	A.	$	
Total acreage and value of the farm			
Land:–			
Tillable			
Range			
Woodland included in range			
Woodland not in range			
Farmstead, lands, waste land, etc.			
Buildings:			
Dwelling house			
Remodeled barns for hens			
Other buildings			
Laying Houses			
Brooder houses			
Other poultry buildings			
Total			

(To be used in connection with credit statement.)

	Beginning of year	End of year
Cash on hand		
Cash in bank		
Stocks and bonds at present cash value		
Cash surrender value of life insurance–		
Total		
Notes and accounts receivable–		
Total		
Mortgages and accounts payable		
Store accounts		
Notes		
Mortgages		
Total		
Insurance data	Amount	Premium
Fire–		
Auto		
Personal		

TABLE 97

INVENTORY OF FEED AND SUPPLIES

	Quantity	Value beginning of year	Quantity	Value end of year
General farm		$		$
Hay				
Home grown grain				
Other supplies				
Poultry				
Grain				
Mash				
Grits and minerals				
Other poultry feeds				
Litter				
Fuel				
Miscellaneous supplies				
	Total		Total	

Increase in total $..................

Decrease in total $..................

Notes:

TABLE 98—INVENTORY OF LIVESTOCK

	No.	Value beginning of year	No.	Value end of year
Poultry				
Hens				
Pullets				
(Age in months)				
(Age in months)				
(Age in months)				
Cockerels				
Cocks				
Miscellaneous poultry stock				
(Broilers, capons)				
Total inventory of poultry				
Other livestock				
Cows				
Horses and mules				
Sheep				
Swine				
Bees				
Total other livestock				
Total all livestock				

Notes:

TABLE 99—INVENTORY OF EQUIPMENT

	Value beginning of year	Value end of year
Poultry		
Incubators		
Brooders		
Shelters		
Miscellaneous poultry equipment		
(hoppers, fountains)		
(egg cases, coops, etc.)		
Small tools		
Total inventory of equipment		
Other farm equipment		
Farm machinery–		
Total other farm equipment		
Total all equipment		

Increase in total value $......................
Decrease in total value $......................

Note: On most farms the value of the old machinery will be 10 to 15 per cent less at the end of the year to cover depreciation.

TABLE 100—SUMMARY OF THE YEAR'S BUSINESS
(Not to be filled out until end of year)

Capital		
Real estate—Value at beinning of year		
Livestock—Value at beginning of year		
Machinery—Value at beginning of year		
Feed and supplies—Value at beginning of year		
Total capital		
Receipts		
Sale of market eggs		
Sale of hatching eggs		
Sale of day-old chicks		
Sale of broilers		
Sale of other live poultry		
Miscellaneous poultry receipts		
Other farm receipts		
Increase in value of livestock		
Increase in value of machinery		
Increase in value of feed and supplies		
Total receipts		
Expenses		
Hired labor		
Purchased feed for all poultry		
Hatching eggs		
Day-old chicks		
Poultry stock		
Poultry supplies		
Poultry equipment and repairs		
Miscellaneous poultry expenses		
Real estate (taxes and insurance)		
Real estate (improvements and repairs, etc.)		
Other farm expenses		
Decrease in value of livestock		
Decrease in value of machinery		
Decrease in value of feed and supplies		
Estimated depreciation of buildings (% value)		
Total expenses		
Farm income (receipts minus expenses)		
Interest on total capital at 5 per cent		
Labor income*		
Value of food, fuel and house rent		

* Labor income is what the operator of the farm received for his year's work after all expenses are paid and five per cent interest on the capital invested is deducted. In addition he also has a house in which to live and part of his living from the farm. Labor income as above determined includes unpaid family labor.

Directions for Summarizing the Year's Business

When the monthly receipts and expenses have all been summarized, and the inventory has been taken at the close of the year, fill out the summary to see how successful the year's business has been.

Enter the value of the farm (land and buildings) at the beginning of year and transfer the totals for the value of the livestock, machinery, and feed and supplies at beginning of year. Find total capital.

Transfer the sum of the monthly totals for receipts. Include under receipts any increases in the value of livestock, ma-

CREDIT STATEMENT

☞ Please answer all questions and fill in all blanks.

Statement of ....

Business .... Address ....

To .... Bank of ....

I make the following statement of all of my assets and liabilities at the close of business on the date named below and give other material information for the purpose of obtaining from you advances on notes and bills bearing my signature or endorsement and for obtaining credit generally upon present and future applications.

☞ CONDITION AT CLOSE OF BUSINESS ON THE .... DAY OF ...., 19......

ASSETS:	VALUED AT	
Farm (No. acres ....) including buildings thereon.		
Assessed for $..................	$..........	
(Location ....)		
Horses (No. ....)		
Cattle (No. ....)		
Sheep (No. ....)		
Hogs (No. ....)		
Poultry (No. ....)		
Feeds and crops on hand		
Miscellaneous supplies		
Equipment		
Fall plowing		
Other growing crops		
Growing timber (Included in value of farm $..........)		
Cash in banks		
Cash on hand		
Other personal property		
Notes receivable		
Accounts receivable		
Total		$..........
LIABILITIES:		
Mortgage on real estate (Mortgagee ....)	$..........	
Notes		
Owing for fertilizer		
Other debts		
Store accounts		
Chattel mortgages		
Total		$..........
Net worth		$..........

Fire insurance on buildings $..........

Fire insurance on other property ....

The legal and equitable right to all the above real estate is in my name solely, except as follows: ....

Automobile insurance: Liability .... Property damage.......... Fire and theft.......... Collision ....

Life insurance .... Who is the beneficiary ....

....

I am also liable:

On indorsed notes ....

I have pledged future income as follows:

Milk checks ....

Field crops ....

Taxes $..........

My age is .... Married or single ....

Wife's name ....

I hereby certify that the figures, information and statements contained on this sheet all of which have been read by me before signing, are true and give a correct showing of my financial condition on the date stated.

Signed this .... day of ...., 19...... Name ....

chinery, and feed and supplies during the year. Find the total receipts.

Transfer the sum of the monthly totals for expenses. Include under expenses any decreases in the value of livestock, machinery, and feed and supplies during the year. (There cannot be both an increase and a decrease in the separate inventories during the same year.) Even with ordinary repairs, buildings gradually wear out; so include under expenses a charge for depreciation on buildings. Depreciation on ordinary farm buildings varies from one quarter per cent to two per cent, depending upon the construction and previous care. Portable poultry houses will depreciate six to ten per cent. The depreciation on machinery is considered in taking the inventory at end of year. Find total expenses.

The expense for new buildings, fences, drains, and other permanent improvements should not be included under expenses. Since it is offset by an increase in the value of the farm, it is unnecessary to make both entries.

Subtract total expenses from total receipts and deduct five per cent interest on total capital to find the labor income of the operator. The size of the labor income measures the success of the year's farm business.

A successful farm is one which, without decrease of fertility, pays all running expenses and five per cent interest on the capital, and returns the operator good wages for his year's work. Figuring the efficiency factors on the next page may indicate ways in which the farm business may be made more successful.

As a rule it is not the farms that are exceptionally good in one or two respects that are the most successful, but rather it is those that are fairly good in all of the more important efficiency factors. The farm business should be well balanced.

Factors that Determine Profits in Poultry Farming

The most important factors that influence profits in farming are: Size of business, as indicated by the capital invested, the number of acres of crops grown, the number of hens kept, the number of men employed, etc.; diversity of business, as indicated by the number of important sources of income, the per cent of receipts from crops, etc.; quality of livestock, as measured by the receipts per hen, or per animal unit,* etc.

*Does not include work animals. One animal unit is 1 cow, bull, steer, 2 head of young cattle, 1 horse, 2 colts, 7 sheep, 14 lambs, 5 hogs, 10 pigs, or 100 head of poultry.

TABLE 101
DIRECTIONS FOR FIGURING EFFICIENCY

Efficiency Factors	This farm	
	Last year	This year
Labor income		
Size of business–		
Total capital invested		
Total area––		
Acres tillable		
Acres other land suitable for range		
Acres used for range this year		
Number of men (12 month basis)		
Average number layers		
Number of chicks brooded in season		
Total receipts		
Total expenses		
Diversity of business		
Important sources of income		
Market eggs		
Hatching eggs		
Day-old chicks		
Broiler sales		
Roasters and chickens		
Fowls		
Poultry stocks		
Miscellaneous poultry sales –		
Sales other than poultry		
Stock production		
Receipts per bird from eggs		
Receipts per 100 hens		
Eggs per bird		
Cost of feed per hen		
Labor efficiency and management–		
Number of layers per man		
Number of chicks brooded per man		
Chicks per pullet housed		
Mortality on layers		
Mortality on chicks		
Value Poultry buildings per 100 hens–		
Value equipment per 100 hens–		

Total capital invested: From summary page.

Total area: The total number of acres farmed by the operator, whether owned or rented.

Acres tillable: Includes all land suitable for cropping other than pasture.

Acres other land suitable for range: All pasture land used for range.

Number of men: Divide number months hired labor plus 12 months for operator, by 12.

Average number of layers: The total of the monthly averages divided by the number of months.

Total receipts: From summary page, minus expense for purchased stock and increase in inventory of stock, if any.

Total expense: From summary page, minus expense for purchased stock and decrease in inventory of stock, if any.

Market eggs: Total sales of market eggs. Hatching eggs, day old chix, broiler sales, roasters and chickens, fowls, poultry stock, miscellaneous poultry sales are determined in the same way.

Sales other than poultry: Include all other sales not already listed.

Receipts per bird from eggs: Divide receipts from eggs by average number of layers.

Receipts per hen: From the sum of the inventory of all the poultry at end of year, the sales of poultry, and the sales of poultry products, subtract the sum of the inventory of all poultry at beginning of year and the purchases of poultry. Divide by number of layers above.

Cost of feed per hen: Add to value of feed on hand at the beginning of year the value of the feed raised and the feed purchased for all poultry. From this sum subtract feed sold and the value of feed on hand at end of year. Divide by the average number of layers.

Number of layers per man: Divide number of layers by average number of men.

Number of chix brooded per man: Divide number of chix brooded by average number of men.

Chix per pullets housed: Divide number of chix hatched by number of pullets housed.

Mortality on layers: Include all birds dead or killed having no meat value.

Mortality on chix: Same as for layers.

Value of poultry buildings per 100 hens: Divide value of poultry buildings by number of 100 hens.

Value of equipment per 100 hens: Divide value of equipment by number 100 hens.

Questions on Chapter XVIII

1. Does location of farm and the type of poultry enterprise conducted have any relation to the amount of capital invested?
2. List important factors that influence profits in the poultry business.
3. Is there any relationship between the number of birds per man employed on the farm and the labor income?
4. Explain what is meant by "the chore-route factor".
5. Does the level of egg production have any effect on the efficiency of feed consumption?
6. In general, is there any relationship between the number of years of experience that a man possesses and his labor income?
7. How may diversification be practiced within the poultry business?
8. Will the purchase of feed supplies on a cash basis reduce expenditures?
9. What are the three main factors determining the cost of producing eggs?
10. What are the important factors in the cost of raising pullets?
11. What are the four essential records necessary to determine the profitableness of the year's business?
12. What is the comparison of culling and mortality between pullets and hens?
13. Why is it highly desirable to have as large a part as possible of capital invested in live stock (pullets and hens)?

CHAPTER XIX

POULTRY DISEASES

Subject Matter

Infectious Laryngo Tracheitis
Coryza
Bronchitis
Respiratory Form of Cholera
Newcastle Disease
Chronic Respiratory Disease
Avitaminosis A
Pressure Necrosis of the Mandible
Coccidiosis
Crazy Chicks
Pullorum Disease
Omphalitis (Navel Disease)
Slipped Tendon Disease
Epidemic Tremors
Rickets
Blackhead
Sour Crop
Fowl Pox
Botulism
Round Worms
Capillaria Worms
Tapeworms
Leucosis
Cholera
Blue-Comb
Typhoid
Tuberculosis
Internal Layers
Prolapse
Mites (Common Red Mites)
Northern Fowl Mite
Lice
Post-Mortem Examination and Diagnosis

Poultry diseases are constantly confronting the poultryman. They must be recognized and controlled if he is to stay in business. The adage "an ounce of prevention is worth a pound of cure" certainly applies to them.

The industry has recognized this advice by adopting various sanitation programs, with the goal of producing healthy stock relatively free from disease. It has been proven practically and experimentally that such stock is capable of producing more profitably.

Diseases take an enormous toll from the industry annually, representing many millions of dollars. It is a well known fact that as intensification of livestock industries develops, livestock disease also intensify. Centralization and commercialization in the poultry industry have concentrated large numbers of chickens into smaller areas, thus increasing the frequency of disease outbreaks among them and making control measures more important.

One of the major problems of the poultryman today is to reduce the relatively high mortality in adult birds. Possibly the adoption of intensive methods of rearing coupled with high egg production contribute to this mortality.

Some diseases are sectional in nature and not common to all localities. The producer must recognize major ones in his locality and study control measures. Unless birds are extremely valuable, treatment of individual ailments are not to be justified. The old custom of maintaining a hospital for sick birds is not ordinarily desirable.

To successfully combat diseases one must develop his powers of observation so that he can detect typical symptoms of any specific condition in its early state before it becomes acute or chronic. This of course entails a knowledge of diseases and their symptoms. Skill in analyzing early symptoms makes possible the inauguration of control measures at a time when they will be much more effective than if applied after the disease is well established.

Since some symptoms are common to many diseases, it is not always possible to identify a particular trouble without post mortem and bacteriological examination. This calls for coöperation and the use of established state or private disease laboratories.

Disease outbreaks or epidemics may be caused by one or a combination of many factors. In a general way, they can be traced chiefly to the following causes:

1. Specific infection by the causative organism. This can happen only when the causative organism is present. Since the germ theory of disease was promulgated, the causative agents of many diseases have been isolated, and their relation to these diseases definitely proven. Many are still to be identified before definite control measures and treatment can be developed. Introduction of stock from outside sources may be responsible for the entrance of contagious or infectious diseases or parasitic infesta-

tion. New stock can best be added to the flock through the purchase of hatching eggs from stock free from disease.

2. Lack of sanitation coupled with poor management. This is responsible for outbreaks of many diseases, particularly those parasitic in nature. Mismanagement lowers the resistance of birds, making them more susceptible to disease.

3. Lack of knowledge. Specific information on poultry diseases is invaluable in the prevention and control of outbreaks. The accumulation of current information has largely resulted

Fig. 194—Pellagra in chicks. Note the scabs at the corners of the mouth and the eye, and on the feet.
[Courtesy Cornell University. Cornell Ext. Bull. No. 348.]

from research work of comparatively recent years, but much yet remains to be accomplished. It is thus wise to keep in contact with new discoveries along this line to be able to apply the latest methods of control as they are developed.

However, consult your local state extension service on the adaptability of all new developments.

4. Faulty nutrition. Until recent years the nutritive value of a ration has been based on its nutritive ratio—the proportion of protein to carbohydrates and fats. This ratio has been used as a guide in formulating rations. Research workers have demonstrated, however, that certain other requirements must be met

and that failure to meet any or all of them will result in deficiency diseases. Avitaminosis diseases are examples.

In this chapter no attempt is made to deal with all the diseases that poultry fall heir to. That is the responsibility of books devoted solely to the subject. Therefore, only the relatively more important are reported herein.

Infectious Laryngo-Tracheitis:

Cause—Infectious laryngo-tracheitis is due to a filterable virus.

Symptoms—One of the first symptoms is stretching of the neck and gasping for breath due to difficulty in breathing. Many of the birds make a rattling sound when breathing, which is quite typical of the disease. Death, usually sudden, is caused by stoppage of respiratory tubes effecting strangulation. Post-mortem examination may show severe inflammation and hemorrhage of the trachea and a bloody exudate.

Control—Where outbreaks of laryngo-tracheitis have occurred in the past, it is well to immunize the developing birds each year by vaccination, and in a commercial establishment where outbreaks have not previously occurred, as soon as the outbreak is recognized, vaccination of all unaffected pens on the premises should be inaugurated. Due to the close similarity of many of the respiratory diseases and the overlapping of general symptoms, it is advisable to make absolutely sure that the disease outbreak is laryngo-tracheitis before vaccination. In general, vaccinating laying birds against laryngo-tracheitis does not drastically curtail the rate of egg production.

Coryza:

Cause—The organism Hemophilus gallinarum is responsible for at least one form of this disease as demonstrated by the Rhode Island Agricultural Experiment Station.

Symptoms—The eyes show profuse watering and the sinuses and nasal passages become plugged with exudate material. Swelling of the wattles and the face is common. The face may take on a greenish appearance. The incubation period is from 24 to 48 hours, indicating the rapidity at which the disease spreads.

Control—Care should be exercised in introducing stock onto the premises that may be carriers of this disease since observation would indicate that this has been the basis for the introduction of coryza in many flocks. The Rhode Island Station has shown that sulfathiazole has definite value in controlling outbreaks. It should be used at 1/4 to 1/2 per cent of the total ration. Other sulfa drugs may prove feasible in control.

Bronchitis:

Cause—This disease is classified as a virus disease.

Symptoms—This disease is characterized by a rattling in breathing. The lungs, trachea, and the upper respiratory tract with accompanying nasal discharge may all be involved. The result is that it may be readily confused with other respiratory diseases, particularly Newcastle disease. Mortality, unless in

Fig. 195—Nutritional paralysis of chicks. Note position of hocks, with toes curled inward.
[Courtesy Cornell University. Cornell Ext. Bull. No. 348.]

very young stock, does not normally run to great levels and recovered birds appear to be immune to re-infection but some remain carriers of the disease. In some cases the spread is rapid in a flock of birds and in such cases the duration of the disease is short. Egg production in layers is seriously curtailed for a period of approximately 8 weeks. After recovery shell texture usually remains poor for many months.

Control—While no treatment is known, the disease may be controlled by segregation rearing of young stock entirely free from the disease and the complete disposal of all old stock well in advance of the housing of any pullets. This method breaks the cycle of infection. Some poultrymen prefer to give the disease to the young growing stock so as to preclude an outbreak during the laying year. This was accomplished in the past by exposing young stock to carried birds. A much more satisfactory procedure is to inoculate a few birds for seed stock purposes on the range with egg propagated virus.

Respiratory Form of Cholera:

Cause—The organism Pasteurella Avicida is responsible.

Symptoms—Individual birds may show swollen and distended sinuses, nasal discharge, swollen wattles and involvement of lower respiratory passages.

This condition is frequently referred to as roup. The nasal discharge possesses obnoxious odor. Birds affected exhibit a depressed appetite which is frequently manifested by a loss of flesh attended by some mortality.

The bottoms of the feet may be swollen and give the appearance of bumble-foot. In adult birds, egg production is depressed for an extended period.

Control—Sulfaquinoxaline used at the rate of 1 pound to 3,000 pounds of feed may be used to prevent spread of unaffected birds in the flock. Affected birds should be removed. Eradication from the premises can be affected by rearing young stock entirely out of contact from old birds which should be disposed of before young birds are housed. This disease is spread by contact between affected and susceptible birds.

Newcastle Disease:

Cause—A filterable virus.

Symptoms—Mortality may vary from a relatively small percentage to extensive losses. Birds generally show respiratory distress which may be accompanied by nervous manifestations, paralysis and spasms of the head, neck, wing and leg muscles. A rapid decrease in egg production may be expected when laying birds contract the disease.

Control—Vaccination has proven practical in prevention efforts. Two types of vaccines are available, (1) killed virus and (2) live virus. The use of the killed virus vaccine evidently gives only temporary immunity. Live virus vaccine however, gives apparently lasting immunity and therefore is a much more valuable tool for control.

Breeders which have recovered from a natural outbreak are immune and transmit a degree of immunity to their chicks.

This is also true where the breeders have been vaccinated with live virus vaccine. Such gengenital immunity last for approximately four weeks. Then it would be necessary to follow up this chickhood immunity with vaccination, otherwise they would be susceptible to a natural outbreak.

The stick method, utilizing the web of the wing as the site for vaccination, is employed.

Chronic Respiratory Disease:

Cause—A virus.

Symptoms—The respiratory symptoms displayed by affected birds are indistinguishable from those caused by infectious bronchitis or Newcastle infections. However, it does spread slowly through a flock and may persist for weeks and even months. The decline in egg production is generally gradual while with infectious bronchitis or Newcastle the decline is percipitious.

Under laboratory conditions turkeys have been found to be as susceptible or more susceptible than chickens. In turkeys inoculated in the sinuses, the disease manifests itself by swollen sinuses, and the lower respiratory involvement are indistinguishable from what may be described as turkey sinusitis. Turkeys have displayed nervous symptoms which have not been observed in chickens.

Control—At the present time no medicinal agents or vaccines are known to have any value. Therefore, opportunities for control will probably lie in breaking the cycle of infection. All replacement stock should be grown completely out of contact with the old birds and no replacement pullets should be housed in laying quarters until all the old stock have been disposed of and the buildings cleaned and disinfected.

Avitaminosis A:

Cause—This disease, commonly called nutritional roup, is caused by a deficiency of vitamin A in the ration.

Symptoms—The individuals exhibit eye symptoms similar to ordinary roup, but the cheesy exudate is white instead of yellow. The eyes become watery and inflamed, and finally fill up with a cheese-like material. Growth slows, finally stops and loss in weight follows. Post-mortem reveals severe kidney injury, with that organ presenting a corrugated appearance, and swollen and inflamed.

Control—Treatment consists of supplying vitamin A in the ration with such feed materials as cod-liver oil, yellow corn, and green foods.

Pressure Necrosis of the Mandible:

Cause—The use of finely ground feed mixtures is the apparent cause. This tends to pack within the front of the mandibles. Pressure causes inflammation and necrosis of the tissue. Either mandible may be affected and the ultimate result is that it "rots" off.

Symptoms—The mandible presents an abnormal enlarge ment. Part or all of either or both mandibles may gradually slough off.

Control—Control consists of feeding hard grains or grit which will tend to keep the mandibles free from adhering food material. The impaction may be scraped off by handling the birds individually, after which iodine or some other equally good antiseptic should be applied to the inflamed area.

Coccidiosis:

Cause—The organism, coccidium, is the direct cause of this disturbance. It is a protozoan, microscopic in size. Serious consequences arise when the number of coccidia in the intestinal tract is large enough to interfere materially with the digestive processes.

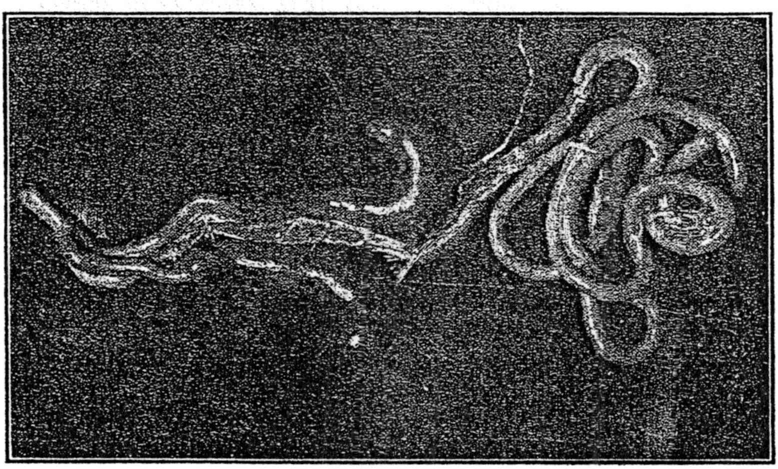

Fig. 196—Cecal Coccidiosis.

[Courtesy N. H. Agr. Exp. Sta.]

Symptoms—The bird becomes dopey, the wings droop, feathers around the vent may become soiled. The bird presents a general unthrifty appearance and usually loses weight rapidly, tends to pale in pigmentation and exhibits a depraved appetite. In many cases coccidiosis in the acute form in young chickens will manifest itself in the evacuation of bloody fecal material.

Post mortem will show, depending upon the causative type of coccidia, a thickened intestinal wall with hemorrhage areas, or the ceca may be distended and filled with bloody, cheesy or semi-liquid material. For definite diagnosis it is necessary to make a microscopic exmination of the contents of the intestinal tract, to identify the causative organism.

Control—Methods formerly used in the control of cecal coccidiosis have been largely supplanted by sulfa drugs having considerable preventive and treatment value. Sulfaguanidine, used at the rate of one pound to 100 pounds of feed fed intermittently has been used. Generally, the medicated feed is fed two days, followed by four days with regular feed and this treatment repeated if necessary. Continuous use of such medicated feed would result in toxic effects. Sulfamethizaine, at the rate of one pound to 250 pounds of feed using the same intermittent feeding schedule has been more effective.

Delaplane and Higgins, based on extensive research findings since 1946, recommend sulfaquinoxaline. If used on an intermittent feeding schedule such as above, it can be used at the rate of one pound to 2000 pounds of feed. It is effective in preventing cecal and intestinal coccidiosis when given continuously at the rate of one pound to 8000 pounds of feed. This procedure does not prevent the birds acquiring desirable immunity. Used continuously from three weeks of age until the birds are transferred to a clean range is an effective prevention program. This has been one of the most important contributions in coccidiosis control.

Research workers whose interest in coccidiosis control has been stimulated by these recent findings are making preliminary reports on many promising materials.

Crazy Chicks:

Cause—The cause of this disturbance is nutritional in nature. In many cases, it will not constitute a very serious problem although in individual cases the mortality may rise to serious levels.

Symptoms—Frequently birds, in attempting to walk or run, lose their balance and may lay on the side or back with the legs twitching. In many instances the birds show some trembling characteristics of the head. In post mortem examination, the most notable symptom appears to be small hemorrhages in the brain tissue. In general, many of the fastest growing birds are among those affected, and autopsies of the affected chicks often indicate a predominance of males. The disease is most likely to occur between three and six weeks of age, although chicks affected earlier than this minimum age and prolonging their symptoms beyond the maximum age are not entirely infrequent.

Control—Experimental work has indicated that the nutritional preventative factor is to be found in many of the vegetable oils, and for this reason soybean oil or corn oil has been of practical value in field outbreaks.

Pullorum Disease:

Cause—The organism, Salmonella Pullorum, is the causative agent. The disease is transmitted from adult to chick through the egg. Other means of transmission are through direct contact with infected chicks or mature birds, or infected feed, eggs, water, or premises.

Symptoms—This disease is particularly fatal with high mortality to be expected within the first 12 days. Under average brooding conditions, the degree of mortality is likely to be in direct proportion to the amount of infection in the group of chickens. Chicks do not necessarily show evidence of diarrhea; so their condition in this respect cannot be used as a safe guide in identifying the trouble. The birds do appear dumpy and less active, however, and will exhibit in later stages a tendency toward excessive wing development. Positive identification is obtained in a pathological laboratory by the isolation of the causative organism from the chicks.

Fig. 197—Chicks Affected with Pullorum Disease (B.W.D.)
[Courtesy C. E. Sawyer.]

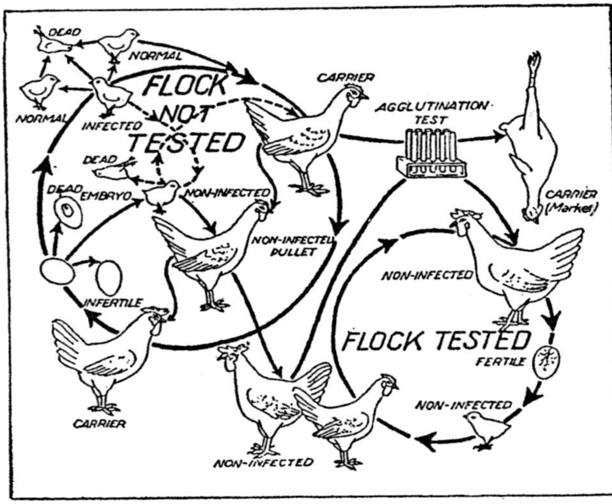

Fig. 198—Cycles Showing Method of Eliminating Pullorum Disease (B. W. D.) by the Agglutination Test.

Control—The satisfactory method of control is to purchase chicks or hatching eggs from stock tested and found free from infection. Prevention is the only control measure practical. Eggs from clean sources should not be incubated along with eggs from infested or questionable flocks, if spread through incubation process is to be prevented. Neither should clean chicks be brooded with infected or questionable chicks.

In the disinfection of the incubator as a control measure, fumigation with formaldehyde gas in advisable. See Chapter I for further information.

Omphalitis (Navel Disease):

Cause—Bacterial infections are associated with incidence of omphalitis in chicks. Bacteria of the clostridium type are associated with the so-called wet omphalitis or, more commonly, mushy chick disease. In the so-called dry form of omphalitis, non-specific bacteria are apparently responsible.

Observations on some cases in Rhode Island would indicate that omphalitis is more likely to occur where insufficient evaporation of eggs has taken place during the incubation procedure. Extremely high humidities, insufficient ventilation, or a combination of both, are believed to be conducive to poor navel healing and subsequent infection.

Symptoms—Chicks show results of poor healing of the navel and hemorrhagic areas in the navel region. In one type of the disease, a watery-like condition of the abdomen is evident. The absorbed yolk of the chick is frequently attached to the inner abdominal wall and exudate is present between the intestines and the liver. The liver normally is enlarged and yellowish in appearance and the gall bladder, in addition, shows enlargement. Normally, substantial losses are incurred.

Control—Fumigation of the incubator for disinfection purposes with formaldehyde is generally recommended. Normally a heavier application is used between hatches than is recommended to be employed in the control of pullorum disease. Brandly recommends two fluid ounces of formalin evaporated from cheese cloth suspended in the incubator per hundred cubic feet incubator space with the wet bulb humidity reading of 90° Fahrenheit.

Regulating the incubator to permit normal evaporation of eggs may assist in preventing subsequent outbreaks of this disease in future hatches.

Slipped-Tendon Disease (Perosis):

Cause—This is a nutritional disease in which an unbalancing of the mineral intake is involved. An unbalancing of the calcium-phosphorus ratio and a deficiency of manganese, choline and biotin are suspected factors.

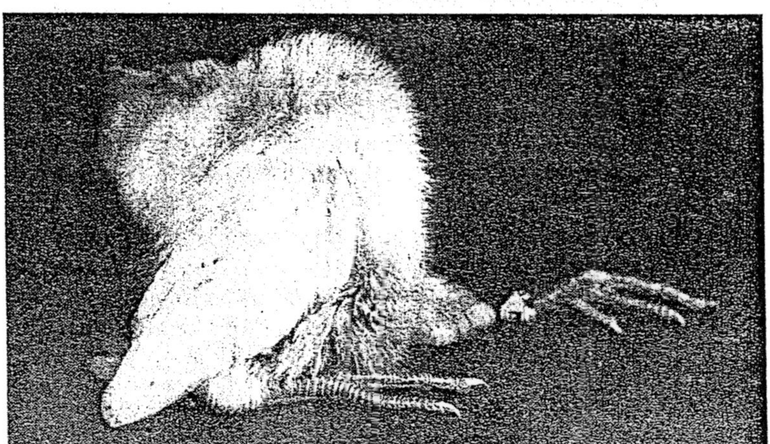

Fig. 199—Vitamin B deficiency (Polyneuritis) in chicks.
[Courtesy Cornell University. Cornell Ext. Bull. No. 348.]

Symptoms—The chick becomes unsteady in gait with a tendency to wabble at the hock joint. As the condition progresses the tendon slips from its normal position, causing the shank to be suspended at a peculiar angle from the body. The joint is noticeably enlarged and the chick ceases to place its weight on the foot,

Fig. 200—Cockerel Showing "Slipped Tendon" Condition.
[Courtesy J. E. Hunter and H. C. Knandel.]

squatting on the shank or hock instead. Due to this condition, the bird soon becomes emaciated. It may linger for a considerable period of time before finally succumbing to the condition. As soon as the condition is noticeable the bird should be marketed, if possible.

Control—Control lies in the proper provision for mineral supplies. This may involve a reduction in the phosphorous content of the ration, the addition of 4 ounces of manganese sulphate to a ton of mash mixture and the addition of small quantities of choline and biotin.

Epidemic Tremors:

Cause—This is a virus disease.

Symptoms—Rapid tremor of the head, neck, and sometimes the tail. The trembling may occur intermittently or constantly when awake. Excitement causes the trembling to become more acute. In many cases an ataxia is present. Normally only one or two hatches in a season is affected. The mortality may range anywhere from practically nothing to fifty per cent or more. Typical lesions occur in the brain, spinal cord and viscera.

Control—No suitable control measures are known.

Rickets:

Cause—Where minerals and Vitamin D are inadequate for the needs of the chicks, rickets will develop.

Symptoms—This disease is characterized by a large band of uncalcified area between the head and shaft of the tibea bones, and results in nutritional leg weakness. The chicks become weak on their legs and have a tendency to sit down in a resting position. The joint appears somewhat enlarged and the chicks gradually lose the entire use of their legs. In conjunction the ribs evidence a beaded condition. This can be determined by running the fingers along midway of the ribs.

The cranium bones as well as the beak are soft and somewhat flexible.

The shank bones are not brittle. In extreme cases the bones can be bent very decidedly before breaking.

In the majority of cases chicks are affected at three to eight weeks of age.

Control—The control of this abnormality lies in furnishing adequate minerals and Vitamin D. Vitamin D may be provided by incorporating such ingredients as cod-liver oil or feeding oils in the ration, or allowing the chicks access to direct sunlight.

Blackhead:

Cause—This disease which primarily affects turkeys is caused by a protozoa called Histomonas meleagridis. A number of cases of blackhead infection in young chickens have come under our observation.

Symptoms—Chickens affected with Histomonas meleagridis normally show a general listlessness with a loss of appetite and diarrheal discharge yellowish in color. Birds are weak and may show a very marked emaciation. Frequently internal examination will reveal the typical liver lesions normally associated with blackhead disease in turkeys, while almost invariably cecal lesions will be observed.

Control—Although essentially a disease of turkeys, chickens are susceptible to blackhead disease and for this reason should be grown on clean range or on platforms. The administration of phenothiazine, ordinarily a worming treatment, has shown promise.

Sour Crop:

Cause—Caused by the spoilage of feed in the crop, this condition may be the secondary effect of some other condition that has resulted in a digestional disturbance. Retention of feed for a prolonged period in the crop and the nature of material and feed ingested may also be classified as specific causes.

Symptoms—Chicks show general enfeeblement and exhibit a watery gaseous consistency of the crop contents. Where sour crop has resulted from retention of food through stoppage, a large bulbous crop is observed. Post-mortem examination will reveal that the entire digestive tract is affected, and its contents, like the crop, will emit a foul odor.

Control—Not much is to be gained by individual treatment. The affected birds should be disposed of.

Fig. 201—Arrow Points to Fowl Pox Lesions on Comb.
[Courtesy N. H. Agr. Exp. Sta.]

Fowl Pox:

Cause—Fowl pox is caused by a filterable virus.

Symptoms—Wart-like swellings appear on the unfeathered parts of the head of the bird. These swellings develop and break, forming a scab over the injury to the epidermus. The infection

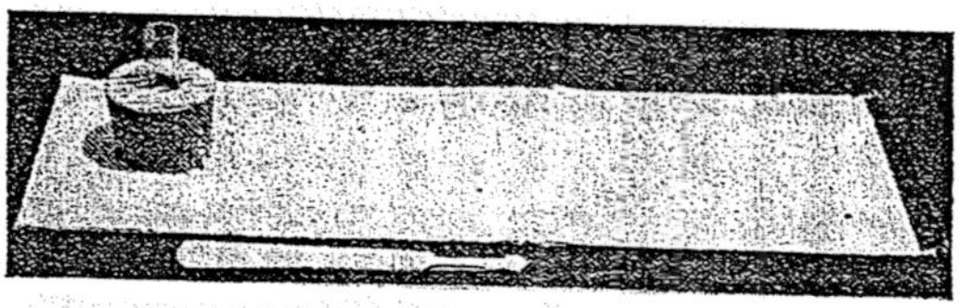

Fig. 202—Fowl Pox Vaccination Equipment. Note Vial is Set in Cork to Prevent Tipping.

[Courtesy N. H. Agr. Exp. Sta.]

may spread quite rapidly over the head and its adjuncts, and in severe cases the lesions may entirely cover the head and close the eyes. Lesions may also appear within the mouth.

Control—The control of fowl pox lies in vaccination to produce the desired immunity. One fowl pox vaccination is sufficient for the duration of the economic life of the bird. Pigeon pox virus vaccination does not confer lasting immunity. It is advisable, where practical, that birds be vaccinated previous to their entry into laying quarters. The birds may be vaccinated at the time they are transferred to range shelters or at any time that fits into the management scheme but preferably before maturity. If a natural outbreak of pox occurs in any birds on the premises and progresses fairly uniformly throughout the flock before dis-

Fig. 203—Dipping the Knife. Attendant Holding Bird.

[Courtesy N. H. Agr. Exp. Sta.]

covery, it would be inadvisable to vaccinate such birds. Only vigorous birds should be treated.

There are two practical methods of vaccination: follicle and stab. The follicle method consists of removing three to five feathers and painting the vaccine over the open feather follicles. The

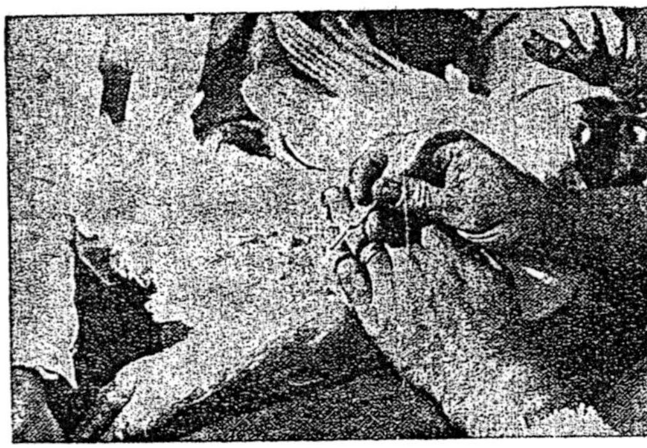

Fig. 204—Vaccinating (stab method.)
[Courtesy N. H. Agr. Exp. Sta.]

feathers are plucked from the flesh portion of the leg about three inches above the hock joint. If more feather follicles are treated the more severe will be the vaccination effects.

In the stab method, a finely pointed knife is prepared for the operation by winding adhesive tape or thread around the blade, allowing only the tip to protrude. Such preparation aids in regulating depth of incision and dosage. Vaccination consists of dipping the end of the knife into vaccine and then "stabbing" the

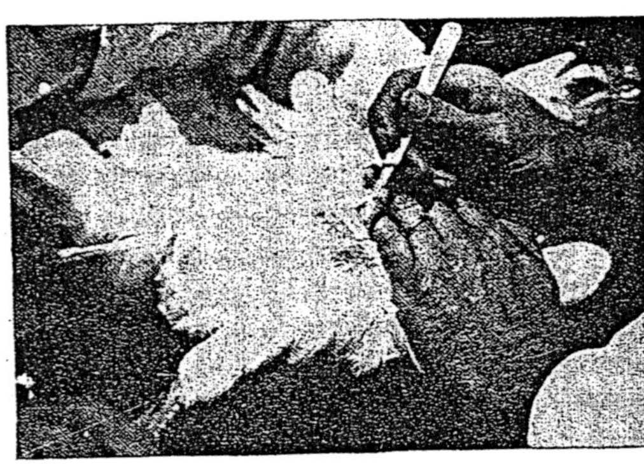

Fig. 205—Examining a Bird for "Take" at a Later Date. Pointing Out the Scab.
[Courtesy N. H. Agr. Exp. Sta.]

point into the leg of the bird at about the same location as specified for the follicle treatment. A skilled technician is capable of vaccinating from 600 to 800 birds per hour, provided sufficient help is available to keep him constantly supplied with birds.

The stab or stick method may also be applied by employing sewing machine needles. The needles may be inserted in a cork stopper so that the stopper acts as a handle for manipulation. Held in this position, the needle or needles as the case may be are dipped into the vaccine solution and then applied normally to the wing tissue. In piercing the skin, the needles carry along with them a more or less measured quantity of vaccine which has been held in the eyes of the needles.

In heavy poultry populated areas, vaccination is practically essential. In less populous areas, and where pox outbreaks has not been experienced on a farm, the poultryman will have to weigh the prospects. Vaccination, however, is cheap insurance.

Outbreaks of pox in turkeys can be very severe. Vaccination immunity in turkeys is not lasting. Therefore, revaccination of birds held over for breeders' should be carried out well in advance of their productive cycle.

Botulism

Cause—The organism, clostridium botulinus, is associated with cases of botulism. The condition ordinarily develops as a result of the consumption of putrid meat and spoiled food.

Symptoms—The bird may evidence a nervous derangement of the head, frequently holding the head in an unnatural posi-

Fig. 206—Lining of Esophagus Showing Characteristic Nodular Lesions From Avitaminosis A.

[Courtesy L. D. Bushnell.]

Fig. 207—Loss of Sight of Eye, Due to Vitamin A Deficiency (Authors).

tion. The bird will exhibit an unsteady gait and convulsive symptoms quite often.

Control—Find and remove the causative agent. Investigate the food and water supply and the premises. It is desirable to give a purgative, such as Epsom salts, in the drinking water at the rate of 1 pound of salts to 500 pounds of live weight of birds.

Fig. 208—Infectious Laryngo Tracheitis of Chickens.

[Courtesy of J. R. Beach.]

Fig. 209—Pressure Necrosis, Showing Injury of Mandible (Authors).

Worms

Worm infestations are rather widespread throughout the country. Certain types have but minor effect on the health of the birds. Other, more serious types may have only minor effects if the extent of infestation is light. It behooves the poultryman to weigh carefully the relative merits of worming a flock when a diagnosis indicates the presence of worms. This is particularly true of birds in production. Careful consideration should be given to the level of egg production and the value of that production as compared to the mortality and weight loss being sustained. The economics of the situation will frequently determine the best course of procedure.

Round Worms:

Cause—The ingestion of embryonated eggs of round worms is responsible for worm infestations. Many types of roundworms affect poultry. Some of the most common are the large intestinal worm, the small macroscopic roundworm of the intestine, and the cecal worm. The cecal worm evidently has little effect on the health of the bird other than when it plays host to the blackhead organism.

Symptoms—The birds show emaciation and a depraved appetite. They are droopy and unthrifty and in extreme cases pass a diarrheal discharge. Definite diagnosis can be made only by post-mortem examination to locate roundworms in the intestinal tract or by microscopic examination of fecal material.

Control—Control lies mainly in the maintenance of proper sanitation so that the numbers of embryonated eggs will not be

Fig. 210—Roundworms from the Small Intestine of a Chicken. [Courtesy of J. E. Ackert.]

large enough to be disastrous. Frequent cleaning of the dropping boards or wiring them to prevent birds from consuming droppings is effective in preventing ingestion of worm eggs.

Many satisfactory worm remedies are on the market as specific for the treatment of round-worm infestation. To be satisfactory most treatments must be individually administered.

Providing clean range is a valuable procedure in prevention.

Capillaria Worms (Macroscopic Round Worms):

Cause—Infestation is the result of the ingestion of embryonated eggs.

Symptoms—Similar to those of birds affected with other forms of round worms. This similarity, therefore, makes a microscopic post-mortem examination necessary for identification.

Control—Tetrachlorethylene 1 c.c. capsules, have seemingly given good results. Because of the imbedded condition of the worms in the intestinal wall many preparations are inefficient. The inauguration of sanitation measures is strongly advised. It is frequently desirable to flush the birds with a laxative previous to administering the worming treatment in order to aid the efficiency of the application.

Tapeworms:

Cause—Unlike a case of roundworms, infestation of tapeworms can never be attributed to consumption of worm eggs that have passed from the intestine in the manure. Tapeworm eggs must pass through an intermediate host before they can infect the primary host. House flies, stable flies and snails are such intermediaries, and no doubt many others are yet to be identified. Consumption of these hosts harboring the parasite in its intermediary stage result in infestation of the birds.

Symptoms—Symptoms are very similar to those produced by round worms. Feathers become ruffled; the birds become thin and possess an abnormal appetite for food as long as they are physically able to get to the hoppers. Definite diagnosis depends on identification of the worms by post-mortem examination, or by investigating the fecal material for mature segments containing tapeworm eggs, or for "tapes" that have been passed along with the feces.

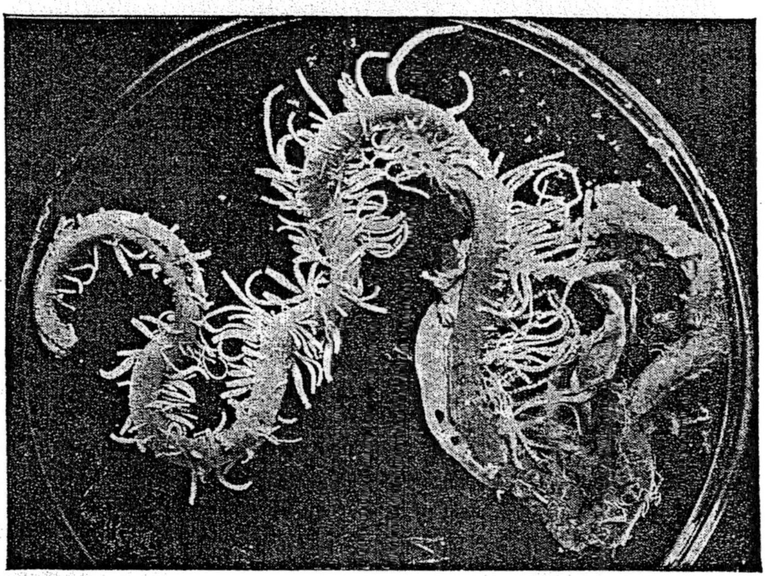

Fig. 211—Portion of Everted Chicken Intestine Showing Heavy Infestation of Tapeworms.

[Courtesy of J. E. Ackert.]

Control—This largely means preventing the development of intermediate hosts. Fly screening of the poultry houses and manure pits is a valuable preventive measure. Of equal importance in fly control are many of othe new insecticides, many of which require additional research work to determine their limitations and proper use.

There is no satisfactory treatment for the removal of all species of tapeworms.

Leucosis:

Cause—A filterable virus.

Symptoms—Birds tend to lose the coordinating power of the muscles. Any or all of its appendages may be paralyzed. Many birds will show blindness in varying degrees, in one or both eyes. Post mortem examination may show paleness of the organs, an enlarged and nodular liver, and an enlarged spleen. The disease may be characterized by an increase in the number of white blood cells and a decrease in the red ones. Hemorrhage of the liver may be the immediate cause of death.

Fig. 212—Incubation of inoculated eggs, in vaccine production.
[Courtesy C. A. Bottorff, Lederle Laboratories.]

Fig. 213—Drilling eggs for inoculation of virus.
[Courtesy C. A. Bottorff, Lederle Laboratories.]

Birds show emaciation, droopiness and general debility. Frequently tumors are associated with this disease.

Some or all of these symptoms may be found in an infected flock. The leucosis complex has many manifestations. For example, it may be broken down as follows:

1. Lymphomatosis
   a. Ocular (Blindness)
   b. Neural (Paralysis)
   c. Visceral (Tumors)
   d. Osteopetroyic (Enlarged bones)
2. Erythroblastosis (Paleness, the blood form)
3. Granuloblastosis
4. Myelocytomatosis (Large white tumors)

Control—Much research work still needs to be done but on the basis of present knowledge the following needs to be considered: (1) Watch the source of stock and secure chick or hatching eggs from flocks that have had a low incidence of the disease or from breeding stock that is resistent to the disease, (2) Raise young stock on areas that are free of the disease, (3) Follow a good sanitary program, (4) Avoid outside contact and bringing in the disease through the introduction of stock carrying the virus, and (5) Raise young stock in complete isolation from older stock at least for the first few weeks.

The regional laboratory at East Lansing, Michigan has demonstrated the possibilities in breeding resistant lines.

Cholera:

Cause—A specific organism Pasteurella avisepticus causes fowl cholera.

Symptoms—Quite often this disease is characterized by a sudden loss of birds, which usually marks the beginning of the acute form. Although commonly lasting but a few days, it may persist in a chronic form for some time. The birds' combs become purplish and their plumage ruffled. They exhibit excessive thirst and are drowsy and inactive. Yellowish diarrhea often accompanies an acute outbreak, with increase in body temperature and general emaciation. Frequently one of the most noticeable symptoms may be the swelling of the wattles and the adjacent tissues of the throat.

Hemorrhagic conditions of the tissues and the internal organs, particularly the heart and duodenum, are found on post-mortem examination. Positive identification of this disease is based on bacteriological examination.

Control—Isolation of affected birds, quarantining of premises, and having one man responsible for managing affected pens only is desirable. Some of the sulfa drugs offer promise in combatting the disease.

Blue-Comb Disease:

Cause—A virus is responsible.

Symptoms—The comb may or may not be bluish in color. It, as well as the wattles, will be shrunken and of "dead" red color. The shanks are also withered. Post-mortem examination may reveal an extremely inflamed intestine, swollen and pale kidneys, ureters swollen and distended and the liver may be swollen and covered by yellowish-gray lesions. The pancreas may be affected. Hemorrhages are frequently found on the heart and in other muscular tissue.

Mortality in both young stock and mature pullets may be extensive, occurring principally during the summer and fall months.

Control—Research work by E. F. Waller has demonstrated the value of subcutaneous vaccination procedure as a preventive treatment.

Blackstrap molasses added to the drinking water is frequently employed.

Typhoid:

Cause—This disease is caused by a specific organism, Salmonella gallinarum.

Symptoms—The symptoms are somewhat similar to those listed for cholera. However, the death rate is not so rapid and the birds do not lose flesh quite so rapidly. Greenish diarrhea, enlargement of the liver, paleness, mottled spleen, lumpy condition of the heart, and enlarged and friable kidneys are all symptoms associated with fowl typhoid.

Control—Eradication programs should be instituted conforming to existing state regulations.

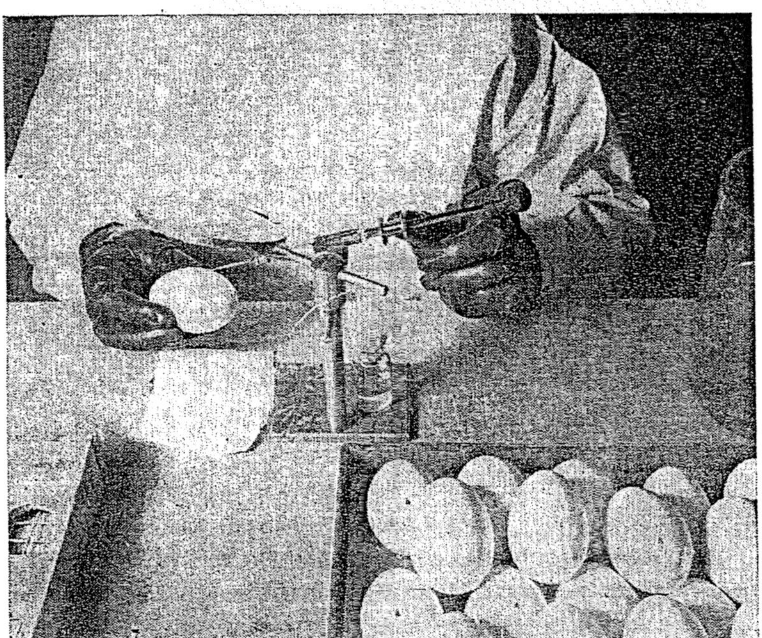

Fig. 214—Inoculation of egg with virus.
[Courtesy C. A. Bottorff, Lederle Laboratories.]

Tuberculosis:

Cause—A specific organism, Mycrobacterium tuberculosis, is responsible for avian tuberculosis.

Symptoms—This disease is most prevalent in birds over a year of age. At first the disease is not likely to be suspected because of intermittent mortality. General symptoms may not be

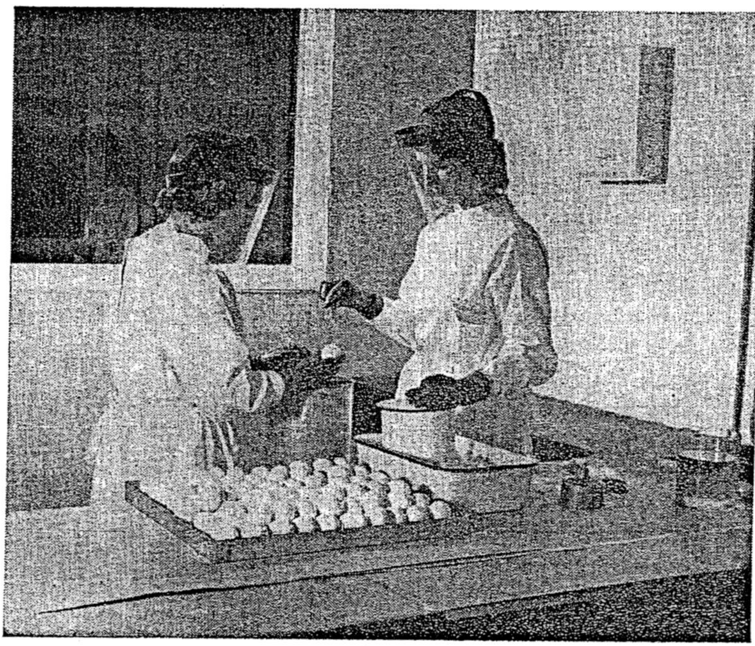

Fig. 215—Harvesting embryos, in vaccine production.
[Courtesy C. A. Bottorff, Lederle Laboratories.]

noticeable. As the disease progresses birds become emaciated; show a marked paleness of the comb, varying degrees of lameness and in some cases swelling of the joints; and act listless. Many of these symptoms are so common with other diseases that definite diagnosis must depend on post-mortem examination.

The lesions encountered on post-mortem are in the form of nodules, commonly termed tubercles, on the surface of the liver, the spleen and the mesentery along the intestinal tract. These nodules may be white or yellowish in color and of cheesy consistency, and when cut open present a softer central portion.

Control—Control lies in avoiding contact with infected stock or carriers of the disease.

Tuberculosis is widespread in domesticated poultry in some sections. It is most prevalent in sections of the country where a considerable percentage of old birds are kept over from year to year. This suggests one possible means of control; the annual replacement of the entire laying flock by a new crop of pullets.

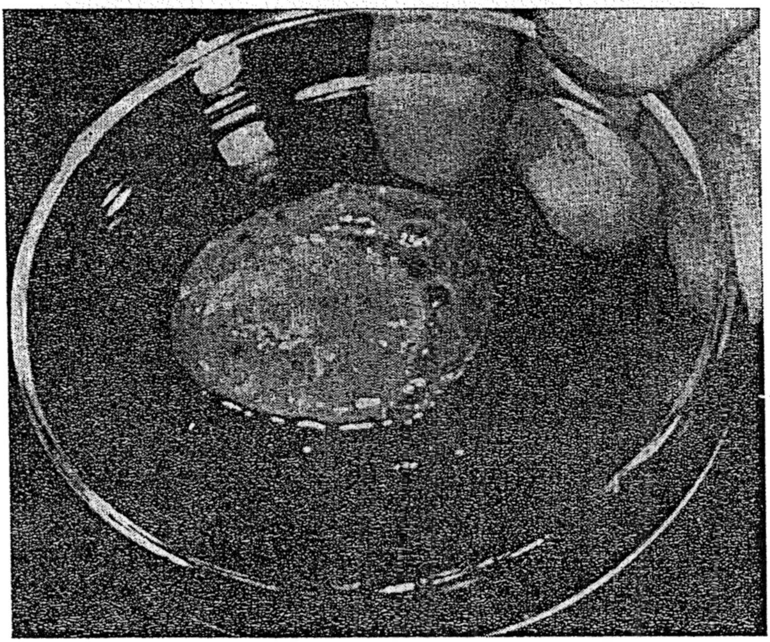

Fig. 216—Virus growth—Laryngotracheitis.
[Courtesy C. A. Bottorff, Lederle Laboratories.]

Internal Layers:

Cause—This condition is the result of the inability of the funnel portion of the oviduct to receive the yolks as they are ovulated, thus permitting them to be accumulated in the body cavity. Rupture of the oviduct is an additional cause. It is one of the most common causes of adult mortality, but because of the chronic nature of losses, it is seldom recognized as such.

Symptoms—A common symptom is that of a bagging abdomen. By manipulation the hard mass of yolk material in the body cavity can be discerned. Post-mortem will show an irregular lump of yolk material somewhat darkened on its surface. When cut into this will present the odor and appearance of a hard-boiled egg. Sometimes a number of completely formed eggs are found.

In addition, the irritation of the peritoneum may have resulted in a severe peritonitis.

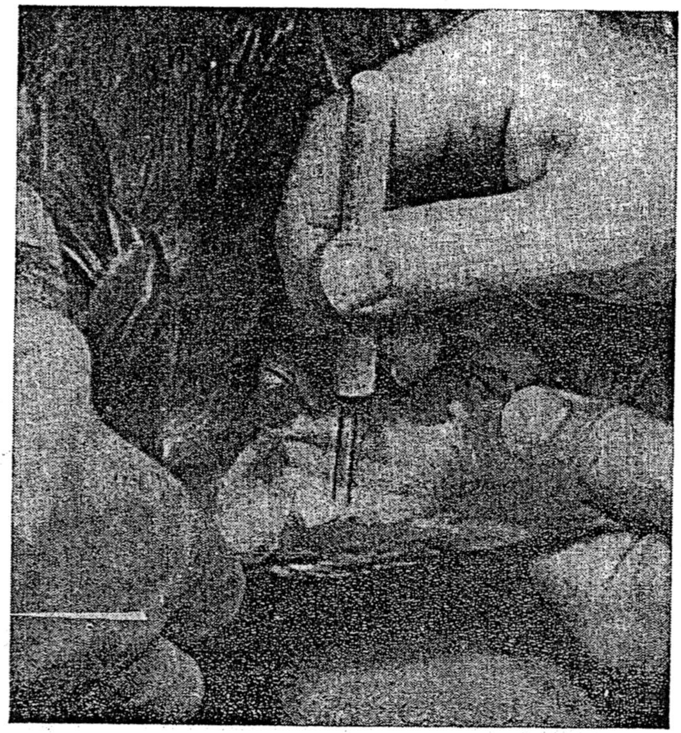

Fig. 217—Fowl Pox Vaccination—Stick Method.
[Courtesy C. A. Bottorff, Lederle Laboratories.]

Control—Trouble from this source usually affects only certain individuals in the flock, and treatment is not to be recommended. If, on the other hand, the condition is due to rough handling of the birds, resulting in physical injury, prevention is to avoid frightening, rough handling, or other disturbances of the flock. Losses may be due to inherited weakness.

Any disease that tend to materially weaken the birds will result in an increase of internal laying. This is frequently noted following bronchitis and newcastle disease outbreaks.

Prolapse:

Cause—Prolapse may be caused by an egg-bound condition, due to excessive fat which makes the passage of an egg difficult if not impossible and injury due to picking by other birds immediately following laying.

Symptoms—The main symptom of prolapse is the protrusion of the oviduct through the anus. In extremely severe cases a portion of the cloaca may also appear. Cannibalism,—involving the picking of the vent and the posterior portion of the intestinal tract, may sometimes be an indication that prolapse has occurred.

Control—Prevention of the cause is the only effective control. If due to an overfat condition reduction of the supply of fat is necessary, either by maintenance of production or by a temporary readjustment of the ration. If due to picking protected nests should assist in reducing losses.

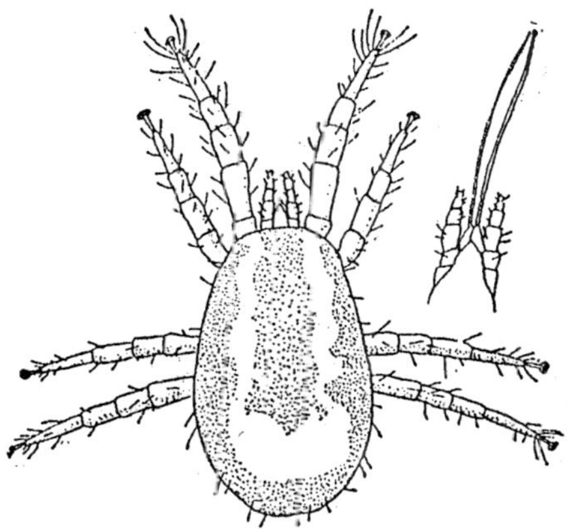

Fig. 218—The Common Chicken Mite.
[From Ext. Cir. 115, Cornell University.]

Mites (Common Red Mites):

Cause—Trouble from this source is due to the harboring of mites and mite eggs in the poultry houses. The mite is grayish in color, but appears reddish when it has sucked blood.

Symptoms—Heavy infestation will result in considerable irritation to the affected fowl and in emaciation, general weakness and sometimes death. Since mites suck blood extensively at night their presence is not usually recognized, until the infestation is heavy. Definite diagnosis depends on examination of cracks and crevices, particularly around the roosting and nesting quarters.

Control—Definite cleaning and disinfecting of the poultry house and its equipment are the only effective means of control. Removable equipment materially aids in the cleaning operation. Coal-tar disinfectants are advisable. Carbolineum has given very satisfactory results on many practical poultry farms.

For all types of external parasites it is well to make periodical observations and examinations, then act accordingly.

Northern Fowl Mite Infestation:

Cause—Occasional infestation in northern states of this parasite is to be found particularly during the fall and winter months. Apparently birds are infested at the time they are brought in from summer range shelters to permanent laying houses. It would appear, perhaps co-incidentally, that many an infestation has occurred where birds have been reared on a woodlot range.

Symptoms—This is apparently primarily an infection of mature or nearly mature fowl as infestations do not appear to be found on growing chicks. Comparatively little is known about the life cycle of this mite as they do not appear to be found on birds during the summer months. Unlike common red mite infestations, the mites appear to remain on the body of the bird at all times and, peculiarly, infestations on male birds are likely to be heavier than on females. They may readily be observed on birds infested and symptoms of such infested birds are similar to those of birds infested with the common red mite.

Control—An effective procedure in many cases is the use of nicotine sulphate sprinkled lightly into the plumage of the bird. Commercial poultrymen sometimes paint the floor of an ordinary shipping coop with nicotine sulphate and place a number of birds therein overnight.

Lice:

Cause—Infestation is the result of the introduction of this parasite into the poultry flock. Unlike common red mites, lice spend their entire life cycle on the host.

Symptoms—Irritation is very evident as the bird appears to be "preening" and picking its feathers much of the time.

The lice are most readily observable under the wings and in the region of the vent. Examination will reveal their presence.

Control—There are many methods of treating birds so affected. Many of these have become antiquated because of their limitations. Dipping can only be practical in warm weather.

Dusting and the blue-ointment treatments are reasonably effective, but like dipping require considerable time for application.

The nicotine-sulphate treatment of more recent usage is effective and requires but a minimum of time. It is painted on or applied by means of an oil can to the top of the roosting perches about 15 or 20 minutes before the birds go to roost for the night. Be sure that all birds roosting elsewhere are placed on the perches so that the entire flock receives the treatment. The vaporization of this material allows it to come in contact with the base of the feathers on the body of the fowls, killing the lice. It may be advisable to repeat the treatment in ten days in order to destroy any lice which have hatched since the previous treatment.

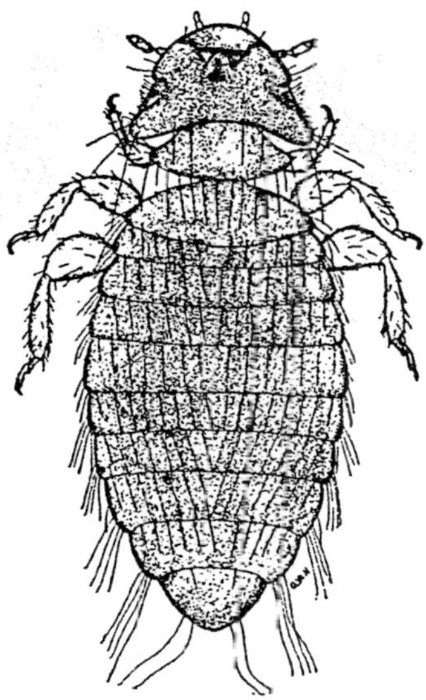

Fig. 219—Common Hen Louse.
[From Bulletin 359, Cornell University.]

Post-Mortem Examination and Diagnosis

Post-mortem examination is necessary with many diseases to accurately differentiate and determine the cause and nature of the trouble.

Recognition of external parasites is not dependent on post-mortem examination. As above indicated, however, a careful study of the internal organs of the bird is necessary where the disease is not external. In many cases a bacteriological examination is needed.

The practice of regularly dissecting and examining birds that die is a recommended procedure as it will indicate the chief sources of mortality and make possible the inauguration of preventive measures and flock treatment before an outbreak becomes acute. Identification of a disease is not always possible by post-mortem examination, due to a lack of laboratory facilities and of sufficient knowledge of all poultry diseases.

In the routine "posting" a dead bird, a convenient method is to place the bird back downward on a flat board or table. The legs may be broken at the hip joints and spread outward from the body. The wings likewise should be spread outward to facilitate operations. Make an incision in the abdomen and cut down each

Fig. 220—Incinerator for Burning Dead Birds.
[Courtesy New England Poultryman.]

side of the breast through the ribs to the wing joints. This portion may then be bent forward and the internal organs exposed for examination. It is necessary for the operator to acquaint himself with their normal appearance so that he can readily recognize the abnormal.

The heart and liver can be observed before the internal organs are removed from the body cavity. Then remove the gizzard, liver and intestines so that the kidneys are exposed to view. Observe the condition of the lungs in regard to congestion or nodular appearance. The gizzard may then be opened and examined. Slit the intestine its entire length with a small knife or scissors so that the contents may be easily seen. In examination of the intestine the use of a microscope is often necessary to identify and observe microscopic worms, worm eggs, and coccidia.

The next procedure in "posting" a bird is to examine the respiratory apparatus and the contents of the crop. Cut by shears through the corners of the mouth and tear the skin downward so that the larynx opening is exposed. Slit the trachea its entire length, and observe. Then remove the crop and open it so that the contents are revealed.

Where impossible or impractical for the poultryman to make a post-mortem examination and diagnosis, or where severe outbreaks or disease epidemics occur, reliance should be placed on an examination by a recognized private or state pathological laboratory. In most cases, particularly those of flock nature, it is advisable to confer with a trained veterinarian relative to the identification of the disease and the best procedure for control.

Questions on Chapter XIX

1. In general, disease outbreaks of epidemics may be traced to what four causes?
2. Describe and give control measures for infectious laryngotracheitis, coryza, and bronchitis.
3. How would you control the respiratory form of cholera, newcastle, and the chronic respiratory disease?
4. Describe pressure necrosis of the mandible.
5. Give cause, symptoms and control of coccidiosis?
6. Give cause, symptoms and control of pullorum disease?
7. Discuss omphalitis?
8. Is slipped-tendon disease due to a nutritional or non-nutritional cause?
9. To what is the development of rickets due?
10. Is blackhead a disease that primarily affects turkeys or chickens?
11. What would be your procedure in vaccinating a flock of young stock against fowl pox?
12. Discuss control measures in worm infestations?
13. Discuss control measures for leucosis?
14. What are the symptoms of cholera?
15. What are the symptoms of blue-comb disease?
16. Does the replacement of the entire laying flock each year with a new crop of pullets have any value in the control of tuberculosis?
17. Discuss internal laying?
18. Cite control measures for the common red mite, northern fowl mite, and lice infestations?
19. How would you proceed to autopsy a bird?

CHAPTER XX

EGG FORMATION

Subject Matter

The Ovary
The Oviduct
 Funnel
 Albumen-Secreting Portion
 Isthmus
 Uterus
 Vagina
 Cloaca and Vent
Passage of Egg Through Oviduct

Production, in so far as the hen is concerned, is essentially a matter of reproduction. If the newly laid egg has been fertilized, it is in fact a new individual which, when continued through the incubation and brooding process, will itself be capable of reproduction. For this reason nature has supplied within the egg the food requirements of the developing organism until it is capable of taking care of itself. Since it contains all the essentials to the production of life, the egg has become an important item of human diet.

The components of the egg are produced by various sections of the reproduction system of the female. There are two main divisions to this system; namely, ovary and oviduct.

The Ovary

The ovary is composed of numerous ova joined together and attached to the spinal column, adjacent to the anterior lobes of the kidneys. If a laying bird is killed and examined, the ovary will be found to contain many ova (yolks) of varying sizes. The number of well-developed ova will determine the rate of production. Each ovum is contained within a membranous sac or follicle which ruptures when the ovum becomes mature and allows the yolk to be picked up by the funnel portion of the oviduct. The line of rupture of the follicle is referred to as "suture line."

The Oviduct

The oviduct consists of a tube-like duct, the various sections of which fulfill special functions in the formation of the completed egg.

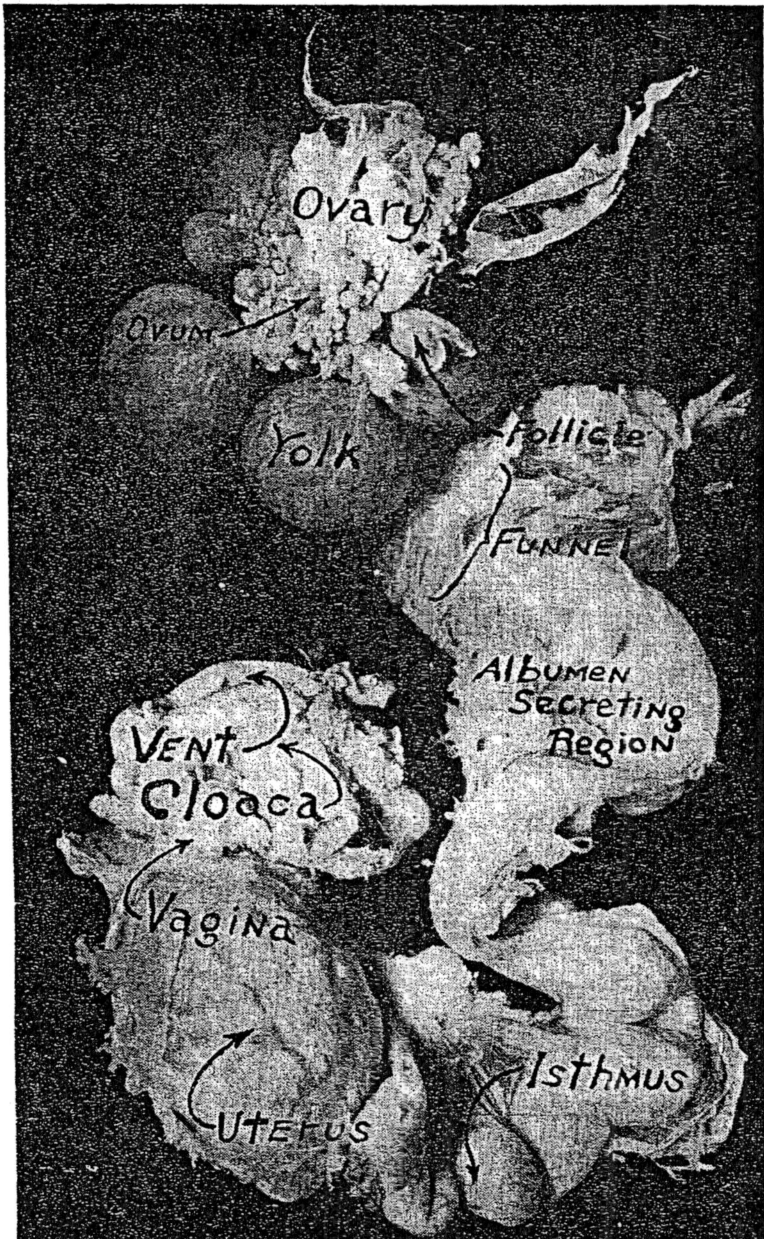

Fig. 221—Ovary and Oviduct After Being Removed from the Body Cavity of hen. Essential parts indicated as to Relative Location.

[Photo used by special permission of Rice & Botsford, Authors of Practical Poultry Management, and John Wiley & Sons, Publishers.]

1. Funnel

This is the part where the yolk is first received after ovulation and where fertilization takes place providing mating has occurred.

2. Albumen-Secreting Portion

As the yolk passes down the oviduct the dense layer of albumen and the chalazae are deposited around it.

3. The Isthmus

This is the next special-function area through which the developing egg passes. Here is added additional albumen, and the two shell membranes are formed.

4. Uterus

In this part of the oviduct the thin albumen of watery consistency is included by osmosis through the shell membranes. Here is also added the shell material and here the egg assumes its typical shape. It is very probable that considerable of the shell pigment is likewise provided in this section.

5. Vagina

The balance of the shell pigment and the bloom are supplied here.

6. Cloaca and Vent

The tissues of the cloaca protrude through the opening of the vent in the expulsion of the completed egg from the vagina through the vent.

Passage of Egg Through Oviduct

Time required for passage of egg through oviduct—opinion differs as to the time consumed in the passage of the egg through the oviduct, but in general 18 to 27 hours are considered necessary for completion of the process. Warren and Scott have reported as follows: Mean time spent in funnel—18 minutes; albumen secreting section—2 hours and 54 minutes; isthmus—1 hour and 14 minutes; uterus and vagina—20 hours and 51 minutes. This is probably influenced by rate of production. It is possible that a heavy layer may have two eggs passing through the different sections of the oviduct simultaneously.

The final egg is composed of approximately 60 per cent albumen, 30 per cent yolk and 10 per cent shell.

Fig. 222—Body cavity of bird in heavy production, showing major portions of ovary and oviduct. Note many egg yolks mature or in process of growth, and suture line along which follicle usually ruptures to release egg yolk, which then is enveloped by the funnel portion of the oviduct. Complete egg also can be recognized in the uterus.

[Photo used by special permission of Rice & Botsford, Authors of Practical Poultry Management, and John Wiley & Sons, Publishers.]

Questions on Chapter XX

1. What are the two main divisions of a hen's reproductive system?
2. What is meant by the term "funnel portion of the oviduct"?
3. List the various sections of the oviduct.
4. Approximately, how long a period is required for the development and passage of the egg through the oviduct?
5. What is the approximate composition of the egg in yolk, albumen, and shell?

CHAPTER XXI

CLASSES, BREEDS, VARIETIES, AND STRAINS

Subject Matter

Class
Breed
Variety
Strain
Standard Classification
Vigor
Beauty and Utility
Common Defects and Disqualifications
Standard Weights

Much misunderstanding exists concerning the correct use and application of these terms. Frequently people will refer to the White Plymouth Rock, for example, as a breed of poultry; in reality it is a variety of the Plymouth Rock breed. Similarly people will consistently assign all of the economical characteristics of a strain to the variety or breed without recognizing that strains within an individual variety vary considerably in their commercial poultry value. This is well illustrated by the old tendency to refer to certain breeds as "egg breeds" meaning to imply that they excelled all others in productive ability. Yet many of those classified as dual purpose breeds, meaning those that were reasonably proficient in the production of both eggs and meat, contained individual strains that outdistanced some strains in

Fig. 223—Type of an Ideal Male and Female New Hampshire (as interpreted by the New Hampshire Club of America.

[Courtesy Poultry Tribune.]

the so-called "egg breeds". There is little wonder, therefore, that confusion exists. For a thorough understanding, one would do well to study a copy of the American Standard of Perfection published by the American Poultry Association. It is our goal here to highlight some of the important points only.

Class

In general breeds are classified according to their place of origin, although similarity of general characteristics may be the basis of classification. Thus we find breeds such as the Plymouth Rock, Wyandotte, Rhode Island Red, New Hampshires, classified

Fig. 224—Male Heads Showing Different Types of Combs.
[Courtesy U. S. D. A.]

as American. Breeds such as the Brahma, Cochin and Langshan as Asiatic in view of their origination in Asia. Those that were developed along the Mediterranean coast, such as the Leghorn and Minorca, are listed under the Mediterranean class. Such basis for classification does not apply, however, for some classes as illustrated by the Ornamental Bantam and Miscellaneous classes.

Breed

The term breed implies a group of birds that possess similar physical and observable characteristics which will reproduce these characteristics in at least fifty per cent of their progeny. They are invariably groups of birds of similar shapes. This is further testified to by the statement of fanciers, "shape makes the breed, color or comb type the variety".

Variety

Individual breeds are further subdivided into individual varieties. These individual varieties are distinguishable by varia-

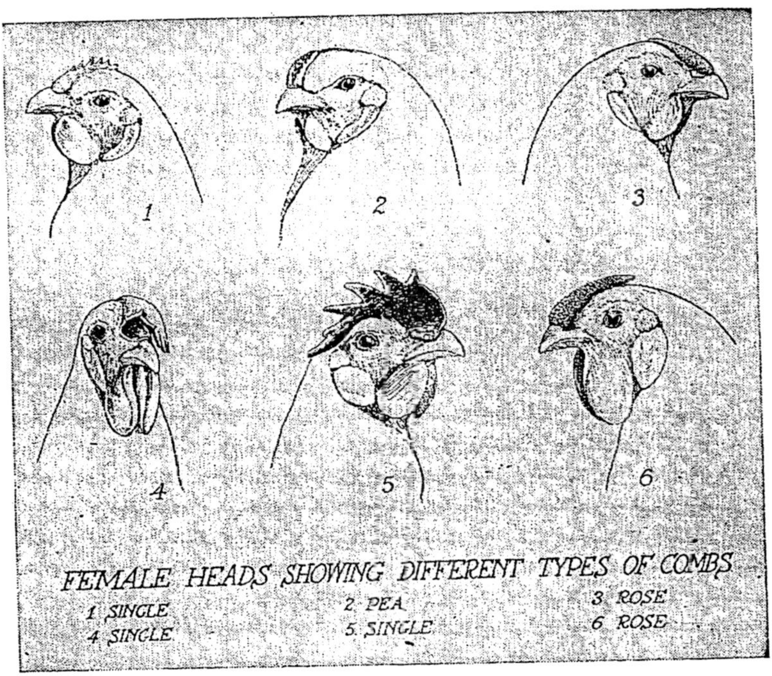

Fig. 225—Female Heads Showing Different Types of Combs.
[Courtesy U. S. D. A.]

tions in either comb type or plumage color. Single and Rose Comb White Leghorns, for example, are distinguished by the difference of comb types. White and Barred Plymouth Rocks possess varietal differences of plumage coloration. Such are the distinguishing features of varieties.

Strain

All of the preceding bases for the classification of poultry have as their foundation solely the observable external characteristics. Some slight differences of external features may also characterize individual strains of a given breed or variety. For example, variation in size of bird frequently is attributable to strain differences. On the other hand, heritable internal differences contribute to the establishment of individual strains. For instance, one strain may excell in egg production or in egg size, or in non-broodiness, or in the rate of feathering of the young stock, or in livability, or in any combination of these and in still other factors. Such are the demarcations of strains.

No one breed, variety or strain can be considered best for all the purposes for which poultry is raised. One should select a breed and variety that appeals to him and then carefully select the strain that possesses the characteristics most nearly approaching the ideals that he has in mind. Success or failure in commercial poultry production is quite often attributable to strain differences.

Other Methods of Classification

There are, of course, many other methods by which poultry may be classified. In most instances, such classifications have merit for specific purposes. The market classifies birds into groups that meet its specific needs. The fancier in speaking of generalized breeding problems may divide birds into solid and parti-colored varieties, the latter referring to those possessing more than one color in their plumage. Or one may classify them according to shank or skin color. Or on the basis of the color of egg shell, such as brown or white egg breeds. Or on the basis of adaptability, such as meat, dual purpose, and egg breeds. Or ornamental and productive breeds and varieties. These are but a few instances of the many ways in which poultry may be classified. However, the fundamental basis for all classifications are the breed and variety standards.

Fig. 226—Nomenclature of Male.[3]

[3]Courtesy U.S.D.A., Farmers' Bull. 1506.

Standard Classification

AMERICAN CLASS*

The American class includes the following breeds and varieties:

Breed	Variety
Plymouth Rock	Barred, White, Buff, Silver-Penciled, Partridge Columbian, Blue.
Wyandotte	White, Buff, Silver-Laced, Golden-Laced, Partridge, Silver-Penciled, Columbian, Black.
Wyandotte Bantam	White, Black, Buff, Partridge, Silver-Penciled, Columbian.
Rhode Island Red	Single Comb, Rose Comb.

Fig. 227—S. C. Rhode Island Reds.

[Courtesy Poultry Tribune.]

*This revision of the bulletin is by Alfred R. Lee, associate poultry husbandman, Animal Husbandry Division, Bureau of Animal Industry.
U. S. D. A. Farmers Bull. 1506.

Breed	Variety
Rhode Island White.	
New Hampshire.	
Jersey Black Giant.	
Java	Black, Mottled.
Dominique.	
Chantecler	White, Partridge.
Lamona.	

ASIATIC CLASS

The Asiatic class includes the following breeds and varieties:

Breed	Variety
Brahma	Light, Dark, Buff.
Brahma Bantam	Dark, Light
Cochin	Buff, Partridge, White, Black.
Cochin Bantam	Buff, Partridge, White, Black.
Langshan	Black, White.

Fig. 228—Light Brahmas.
[Courtesy Poultry Tribune.]

ENGLISH CLASS

The English class includes the following breeds and varieties:

Breed	Variety
Orpington	Buff, Black, White, Blue.
Cornish	Dark, White, White-Laced Red, Buff.
Cornish Bantam	Dark.
Dorking	White, Silver-Gray, Colored.
Sussex	Speckled, Red, Light.
Red Cap.	
Australorp.	

MEDITERRANEAN CLASS

The Mediterranean class includes the following standard breeds and varieties:

Breed	Varieties
Leghorn	Single-Comb White. Rose-Comb White. Single-Comb Light Brown. Single-Comb Dark Brown. Rose-Comb Light Brown. Rose-Comb Dark Brown. Single-Comb Buff. Single-Comb Black. Silver. Red. Black-Tailed Red. Columbian.
Minorca	Single-Comb Black. Rose-Comb Black. Single-Comb White. Rose-Comb White. Single-Comb Buff.
Ancona	Single-Comb. Rose-Comb.
Spanish	White-Faced Black.
Blue Andalusian.	
Buttercup.	

Fig. 229—S. C. White Leghorns.
[Courtesy Poultry Tribune.]

CONTINENTAL EUROPEAN CLASS[2]

The Continental European class contains the following breeds and varieties:

Breed	Variety	Size
Polish	White-Crested Black	Normal.
	Bearded Golden	Normal
	Bearded Silver	Normal.
	Bearded White	Normal.
	Buff-Laced	Normal
	Non-Bearded Golden	Normal.
	Non-Bearded Silver	Normal.
	Non-Bearded White	Normal.
Hamburg	Golden-Spangled	Normal.
	Silver-Spangled	Normal.
	Golden-Penciled	Normal.
	Silver-Penciled	Normal.
	White	Normal.
	Black	Normal.
Campine	Silver	Normal.
	Golden	Normal.
Houdan	Mottled	Normal.
	White	Normal
Crevecoeur	Black	Normal.
La Fleche	Black	Normal.
Faverolles	Salmon	Normal.

Fig. 230—S. C. Anconas.

[Courtesy Poultry Tribune.]

[2]U. S. D. A. Farmers' Bulletin 1507.

ORIENTAL CLASS

The Oriental class contains the following breeds and varieties:

Breed	Variety	Size
Sumatra	Black	Normal and bantam.
Malay	Black-Breasted Red	Normal.

GAME CLASS

The Game class contains the following breeds and varieties:

Breed	Variety	Size
Game	Black-Breasted Red	Normal and Bantam.
	Brown Red	Normal and Bantam.
	Golden Duckwing	Normal and Bantam
	Silver Duckwing	Normal and Bantam.
	Birchen	Normal and Bantam.
	Red Pyle	Normal and Bantam.
	White	Normal and Bantam.
	Black	Normal and Bantam.
	Old English Black-Red	Bantam.
	Old English Spangled	Bantam.

ORNAMENTAL CLASS

The Ornamental class contains the following breeds and varieties of bantams:

Breed	Variety	Size
Sebright Bantam	Golden	Bantam.
	Silver	Bantam.
Rose-Comb Bantam	White	Bantam.
	Black	Bantam.
Booted White Bantam		Bantam.
Brahma Bantam	Dark	Bantam.
	Light	Bantam.
Cochin Bantam	Buff	Bantam.
	Partridge	Bantam.
	White	Bantam.
	Black	Bantam.
Japanese Bantam	Black-Tailed	Bantam.
	White	Bantam.
	Black	Bantam.
	Gray	Bantam.
Silkie		Bantam.
Mille Fleur Booted Bantam		Bantam.
Polish	White-Crested Black	Bantam.
	Bearded Golden	Bantam.
	Bearded Silver	Bantam.
	Bearded White	Bantam.
	Buff-Laced	Bantam.
	Non-Bearded Golden	Bantam.
	Non-Bearded Silver	Bantam.
	Non-Bearded White	Bantam.

Fig. 231—Bantams Showing Types of Feathering and Combs.

(1) Rose Comb White.
(2) Rose Comb Black.
(3) Light Brahma.
(4) Black-Tailed Japanese.
(5) White-Crested Polish.

[Courtesy U. S. D. A.]

Fig. 232—Frizzled Hen.

[Courtesy U. S. D. A.]

MISCELLANEOUS CLASS

The Miscellaneous class contains the following breeds:

Breed	Size	Breed	Size
Sultan	Normal.	Frizzle	Normal.

Vigor

Our domesticated breeds of chickens are supposed to have been derived from the wild jungle fowl. With domestication, the responsibility for perpetuation and development became man's. Breeding efforts in the succeeding generations have been directed to adaptation for specific purposes. Many are of the belief that man has permitted the inherent vigor of the stock to wane and that our present breeds and varieties do not possess the vigor or vitality of their ancestors.

In any such discussion one must remember that man has demanded and secured greater and greater performance. Performance in itself may be an important measure of what we refer to as vigor as noted in the following definition.

Vigor, as here referred to, is designated as the ability of a bird to stand up under heavy production over prolonged periods of time and withstand the ravages of disease. It implies that a bird which has come through a long laying year of high production possesses superior vigor or vitality over one that has succumbed to the same environment. Wide differences exist between different strains as attested to by the records of egg laying contests. The commercial poultryman would do well to investigate the adult livability of stock that he contemplates making purchases from as well as for other considerations.

Beauty and Utility

The poultry industry owes much to the fancier for the establishment of breed and variety characteristics. Many commercial poultrymen are prone to ignore the services the fanciers have contributed. On the other hand, fanciers may ignore utility value of stock provided the bird excells in some other characteristic such as plumage coloration. There is no need for this wide discrepancy if the breeding program followed takes into consideration both beauty and utility. Birds of good standard qualities have made highly satisfactory production records under trapnests. It should be the aim of the breeder to keep in mind standard qualifications as well as economic considerations.

Defects and Disqualifications

"The most common defects and disqualifications include:[1]

Fig. 233—Side Sprig.

Side Sprig—a point of fleshy material coming out of the side, rather than the top, of a single comb.

Crooked Toes—not a disqualification, but a male whose toes are badly bent may have trouble in mating.

Crooked Beak—The upper and lower portions of the beak do not meet, thereby making it hard for the bird to eat enough food.

[1]N. H. Ext. Cir.

Crooked Breast—not a disqualification in chickens, but considered a defect. This condition may or may not be inherited but is extremely undesirable in either sex.

Split Comb—The rear part of the blade of a single comb divided perpendicularly so that the two parts overlap.

Lopped Comb—when the comb of a single-combed male falls over to one side far enough to obstruct vision. Lopped comb does not disqualify a New Hampshire female, though it does disqualify a Barred Plymouth Rock female.

Fig. 234—Split Comb.

Comb Foreign to the Breed—each breed must have the type of comb that is standard for that breed.

Split Tail—a tail showing a decided gap between the main tail feather; this gap extending all the way down to the base of the feathers. This is a defect in young specimens but a disqualification in all old specimens.

Wry Tail—a tail carried permanently to one side.

Fig. 235—Split Tail.

Squirrel Tail—a tail a portion of which extends forward beyond a line perpendicular to the back of the bird and meeting the back at the base of the tail.

Feathered Shanks—in the American breeds, it is a disqualification for a bird to have feathers on the shanks. These feathers may be so small as to be hard to see. Often they will be seen behind and just below the hock joint, and appear as small quills one-sixteenth to one-eighth inch long.

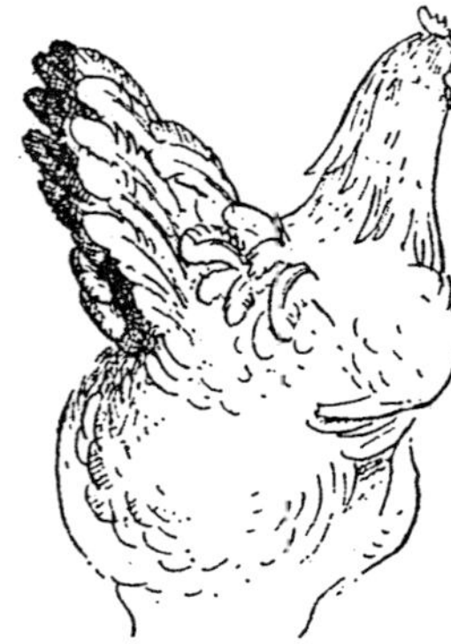

Fig. 236—Wry Tail.

Down or Stubs between the Toes—no feathers, whether they have quills (and are called stubs) or have no quills (down) are allowable between the toes of any bird of the American Class.

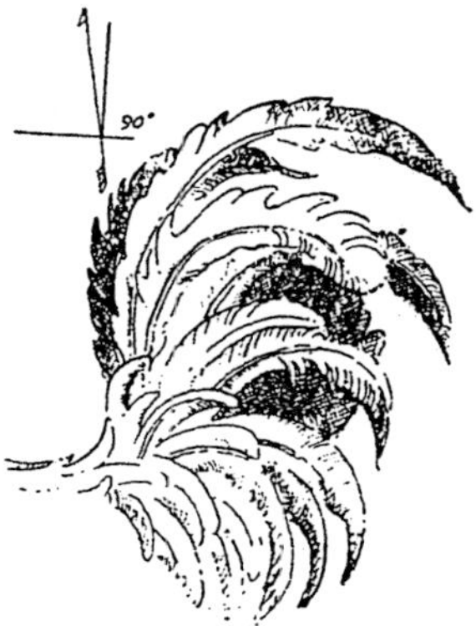

Fig. 237—Squirrel Tail.

Split Wing—a wing so formed as to show a decided gap between the primaries and the secondaries.

Slipped Wing—a wing which is not properly held up in position and therefore tends to hang down on the side of the bird.

Plumage Color—there is a wide latitude left in the matter of plumage color. To quite an extent it is a matter which the individual breeder must decide; yet he must remember that each breed has a preferred color and he should strive to reach that perfection. Each breeder should decide for himself the exact shade which he wants his birds to have, and select constantly for that shade. Only by constant selection for this ideal will he be able to eliminate color variations in his strain.

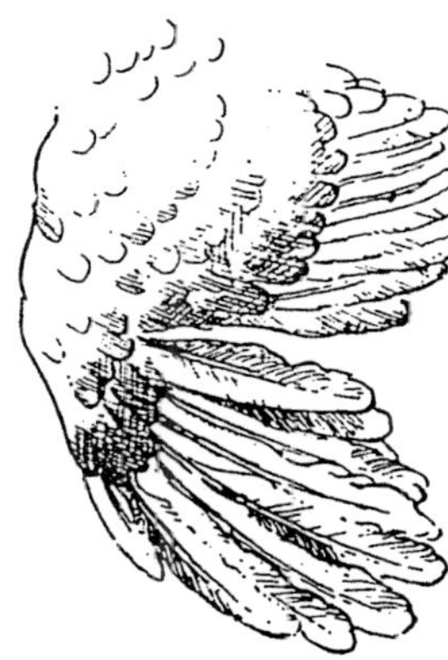

Fig. 238—Split Wing.

General Condition of Bird—It is not the purpose here to discuss selection in general. In selecting birds for the flock, remember the only bird worth keeping in a flock is one that meets reasonable requirements as far as breed characteristics and standard disqualifications are concerned and which will return a profit to the owner."

Fig. 239—Slipped Wing.

TABLE 102—STANDARD WEIGHTS*

Domestic Fowls

	Pullet	Hen	Cockerel	Cock
American Class				
Plymouth Rocks	6	7½	8	9½
Wyandottes	5½	6½	7½	8½
Javas	6½	7½	8	9½
Dominiques	4	5	6	7
Rhode Island Reds	5½	6½	7½	8½
Buckeyes	5½	6½	8	9
Chantecler	5½	6½	7½	8½
Jersey Black Giants	8	10	11	13
New Hampshires	5½	6½	7½	8½
Asiatic Class				
Light Brahma	8	9½	10	12
Dark Brahma				
Buff Brahma	7	8½	9	11
Cochin	7	8½	9	11
Langshans	6½	7½	8	9½
Mediterranean Class				
Leghorns	4	4½	5	6
S. C. Black Minorcas	6½	7½	7½	9
Other Minorcas	5½	6½	6½	8
White-Faced Black Spanish	5½	6½	6½	8
Blue Andalusians	4½	5½	6	7
Anconas	4	4½	5	6
Buttercups	4	5	5½	6½
English Class				
Dorkings				
Whites	5	6	6½	7½
Silver Grey & Colored	6	7	8	9
Redcaps	5	6	6	7½
Orpingtons	7	8	8½	10
Cornish				
Dark & White	6	7½	8	10
White-laced Red	5	6	7	8
Sussex	6	7	7½	9
Australorps	5½	6½	7½	8½
Polish Class	None			
Hamburgs	None			
French Class				
Houdans	5½	6½	6½	7½
La Fleche	6½	7½	7½	8½
Faverolles	5½	6½	7	8
Continental				
Campines	3½	4	5	6

* The above is reproduced by permission from the American Standard of Perfection (according to which all Standard-bred poultry is judged) published and copyrighted by the American Poultry Association, Fort Wayne, Indiana. (For additional standard requirements of domestic poultry you are referred to the American Standard of Perfection.)

DUCKS

	Young Duck	Adult Duck	Young Drake	Adult Drake
Pekin	7	8	8	9
Aylesbury	7	8	8	9
Rouen	7	8	8	9
Cayuga	6	7	7	8
Call	–	–	–	–
East India	–	–	–	–
Muscovy	6	7	8	10
Swedish	5½	7	6½	8
Buff	6	7	7	8
Crested	5	6	6	7
Runner	3½	4	4	4½

GEESE

	Young Goose	Adult Goose	Young Gander	Adult Gander
Toulouse	16	20	20	26
Embden	16	18	18	20
African	14	18	16	20
Chinese	8	10	10	12
Egyptian	6	8	8	10
Canada	8	10	10	12

TURKEYS

	Pullet	Hen	Cockerel	Yearling Cock	Adult Cock
Bronze	16	20	25	35	36
Narragansett	14	18	23	30	33
White Holland	14	18	23	30	33
Black	14	18	23	30	33
Slate	14	18	23	30	33
Bourbon Red	14	18	23	30	33

Questions on Chapter XXI

1. What is meant by the following terms: class, breed, variety, and strain?
2. Give examples in which strains of the same variety may vary.
3. List the standard classes of poultry.
4. To what class does the Plymouth Rock belong?
5. Are the White Wyandottes a variety or a breed?
6. Are Single Comb Rhode Island Reds a breed or a variety?
7. To what class does the Leghorn belong?
8. Can chickens be classified by other methods than through the Standard classification?
9. How may vigor be defined?
10. Is it possible for breeders to combine good standard qualifications with high utility value?

CHAPTER XXII

TURKEY PRODUCTION

Subject Matter

Breeds
Incubation
Brooding
Relation, Size Flock to Feed Consumption
Grow-Healthy Poult Program
Rearing Management
Breeding Farms
  Labor Requirements
  Egg Production
Turkey Rations
Selection of Breeding Birds
Marketing
Turkey Production Trends
Tentative U. S. Standards for Dressed Turkeys
Cost of Raising Turkeys

The turkey, a native of America, has long been considered as the festive bird for special occasions. It was carried back to European countries by early settlers and soon gained prominence on the continent. Since those early days, with wide variation in interest, a substantial number of breeds have been developed which have found adaptation to present day demands.

Some of these breeds have gained recognition because of excellent qualifications. In the United States during recent years there has been great interest in the development of breeds particularly adapted to market requirements. In a like manner much attention has been directed to the improvement of strains within the breed. Efforts in this respect involve rapidity of growth, distribution of fleshing, shorter legs and other desirable market factors.

Breeds

Well established American breeds have included the Bronze, Narragansett, White Holland, Bourbon Red, Black and Slate. In addition, a number of breeders maintain small breeding flocks of native wild turkeys. Recent additions reflecting the breeders' skill include the Broad-Breasted Bronze and Whites, Beltsville White, Nittany, Royal Palm, and the Jersey Buffs.

The producer would be well advised to study the special qualifications and his local market demands in selecting a variety

Fig. 240—Slatted wooden floors. Waterer between pens.
[Courtesy Prof. John Vandervort.]

adapted to this particular need. Wide variations exist between strains of the same variety. For this reason, careful examination of the strain should precede the inauguration of the enterprise.

Incubation

Successful incubation is dependent upon many factors including high fertility, careful handling of eggs and proper operation of incubators (Refer to Chap. I). To secure good fertility, artificial lights should be applied to the males three to four weeks before they are mated to the hens. Artificial lights have the same stimulating effect on turkeys as on domestic fowl and can be used advantageously to secure early fertility and production.

Eggs should be collected two to three times daily and stored in a suitable room providing a temperature somewhat under 60 degrees. Many authorities recommend that the eggs be turned daily if they are to be held more than a few days. Best hatching results can be secured if there is little or no delay between collection and setting. Ordinarily they should not be held much more than a week.

Artificial incubation is as equally adaptable to the hatching of turkey eggs as those of chickens. In general the same procedure is followed except that turkey eggs require 28 days to hatch and therefore transfer to hatching trays will be consummated on the 24th day. At this time all dead germs and infertile eggs should be removed and only live embryos transferred. If extensive hatching is contemplated, a specially designed incubator

Fig. 241—Welded wire fabric floors.
[Courtesy Poultry Dept., N. H. Univ.]

should be utilized. Where such a machine is used, slightly lower temperatures and humidity are recommended by Romanoff after the 24th day. In operating incubators, the manufacturer's directions should be followed strictly unless experience dictates a change.

Brooding

The principles of successful brooding of chickens apply equally well to turkeys (Chapter III). More difficulty is likely to be encountered in getting the poults to eat and drink, however. Also, turkeys should be afforded more space in the brooding operation. Brooder units which ordinarily accommodate 300 or more chicks are adequate for 150 to 200 poults. Some raisers prefer to start poults in a battery brooder for the first two or three weeks and then transfer to floor brooding units.

Regardless of whether the birds are to be reared on range or in confinement, they should be confined for the first eight to twelve weeks so that good sanitary conditions can be maintained. This also provides for the careful scrutiny by the operator during

TABLE 103—RELATION OF SIZE OF FLOCK TO FEED CONSUMPTION, LABOR REQUIREMENTS AND EGG PRODUCTION OF 61 TURKEY BREEDING FARMS IN WASHINGTON, 1942*

Item	Unit	Less than 200 hens per flock	200-399 hens per flock	400 or more hens per flock
Records	Number	25	16	17
Hens per farm	Number	76	272	747
Toms per farm	Number	8	33	81
Hens per tom	Number	10	8	9
			Average per breeder	
Feed consumed:				
Mash	Pounds	34	28	32
All grains	Pounds	45	45	39
Hay	Pounds	4	1	1
Man labor	Hours	3.9	2.8	1.9
			Average per hen	
Eggs produced	Number	35.4	39	43.3
Fertile eggs	Number	23.1	26.2	30.1
Salable poults	Number	19.5	20.8	22.4
Salable poults per 100 eggs produced	Number	55.1	53.3	51.7

*Courtesy Turkey World.

this critical period and the inauguration of corrective management.

Rearing Management

Many producers prefer to rear birds in strict confinement to market age. This involves, generally, the use of wire or slatted-floor platforms which permit the passage of fecal material through to the ground and out of reach of the birds. Where used, the platforms should be high enough to facilitate the re-

Fig. 242—Turkeys on Landino clover range.
[Courtesy A. Mackie.]

moval of the manure. Ordinarily, six to eight square feet of platform space should be provided for each bird. Feeders and water containers may be fastened to the sides to facilitate chores and provide full use of platform area for the birds.

Roosting perches will need to be furnished. These should provide a flat area of at least three to four inches so that a larger proportion of the keel bone may carry the weight of the bird when roosting. This will assist in preventing indentations of the keel. Ten or twelve inches of linear perch room should be available for each bird.

Size of group is an important consideration in successful rearing. It is the belief of many that not more than 200 poults should be reared as a group on the platform.

If it is the intention to range rear the birds, a suitable range rotation program will need to be followed. This should provide a large uncontaminated area of land each year. Substantially this will involve a three-year rotation system. Not over 75 to 100 birds should be ranged per acre. This is assuming that the range area is well sodded and will provide a reasonable forage. It has been demonstrated that a good, succulent sod will effect a saving in feed costs.

Adequate roosts should be provided, either sheltered or unsheltered. In any case, they should be a sufficient height off the ground to reduce losses from predatory animals. Locate them far enough away from other buildings so that if birds are scared, stampede, and take off on flight at night they may land without extensive injury. Some folks use night lights on range to forestall stampeding and to deter thievery.

Whether raised in confinement or on range birds should be vaccinated during the rearing season if fowl pox has been a problem on the premises or in the community.

Feeding and general management procedures are similar regardless of the type of rearing. Starting and growing rations generally contain from 22 to 26 per cent protein. The turkey appears to utilize these high protein levels to better advantage than chickens. In some cases, the mixture is reduced to 20 per cent protein during the last third of the growing period. The ration might be finely ground as a mash mixture or be pelleted for feed conservation. From six to eight weeks before marketing the feeding of grain should be inaugurated in order to properly finish the birds. However, the birds will not consume a sufficient quantity of grain unless the weather is at least cool. Best finish is obtained in cold weather.

In feeding practice it is well to provide wire in the vicinity of the hoppers to facilitate beak cleaning.

New Hampshire Six-Point Grow-Healthy Poult Program

To reduce the mortality in rearing poults, to grow turkeys that are healthy, thrifty, and free from disease, and that will make a good profit for the owner, the following practices are recommended:

Clean Brooder Houses

1. Clean, scrape and scrub. (If desired, hot water and lye, one 13-ounce can to 15 gallons of water, may be used.)
2. Use reliable disinfectants.
3. Clean thoroughly all equipment used for brooding.
4. Do all cleaning of brooder houses before moving to clean range.

5. The use of wire or slatted wooden floored sun porches is recommended.

Clean Litter

1. Keep litter as dry and clean as possible at all times.
2. Use a dry, clean, inexpensive litter.
3. If a wire platform is used inside or outside of house, it should be thoroughly cleaned as needed.

Clean Feed and Water

1. Use wire or slat platform and keep area dry beneath water fountains.
2. Place all feed and water containers on wire platforms inside house.
3. On range, move all hoppers to clean spot each day.
4. Place feed on egg cup flats for first two days, and feed often.
5. Feed in hoppers which poults cannot contaminate, having feed continuously before poults.
6. Hard grain should be hopper-fed.

Clean Poults

1. Healthy, disease-free poults.
2. From stock which has demonstrated ability to produce thrifty, healthy, vigorous birds.
3. Hatched from clean eggs and in clean incubators.
4. Purchased from a reliable breeder.

Clean Ground

1. Land not contaminated by drainage water or poultry manure.
2. Land not used by poultry for three years or more.
3. A sodded range desirable.

Clean Management

1. Maintain a quarantine between old and young stock, also from other poultry.
2. Avoid crowding, chilling, and overheating.
3. Teach poults to range by moving feed containers regularly.
4. Encourage early roosting.
5. Move to clean range when disease threatens.

6. Allow one square foot of floor space per poult. Not over 150 to be started under one hover.
7. Number of poults should be reduced when their growth or limited floor space causes crowding.
8. Confine poults to house for four weeks. Allow range on wire or wooden slatted platforms in front of the brooder house after this time.
9. Avoid wet area around the water fountains.
10. Provide some shade on range. A few rows of corn or sunflowers is ideal if natural shade is not available. Avoid a dense shade.

Fig. 243—Confinement rearing of turkeys. Note easy to fill feeders. [Courtesy Prof. John Vandervort.]

Selection of Breeding Birds

The selection of breeders should precede marketing so that nly the birds approaching the breeders' ideal will find their way ato breeding units. Unless done before marketing, some of the hoice birds will find their way to the consumers' table and less esirable birds will be available for perpetuation.

Select birds that are well fleshed over the breast and thighs; hat have a straight and long keel with good fleshing throughout s length; that have relatively short rugged legs, set well apart;

TABLE 104—NEW ENGLAND COLLEGE CONFERENCE TURKEY RATIONS

Mash Ingredient	Starter	Grower[1]	Breeder[1]
Ground Yellow Corn	940	1100	800
Standard Wheat Middlings	100	100	300
Ground Oats	—	—	100
Fish Meal (60%)	100	50	50
Meat Scraps (50%)	250	200	150
Soybean Meal (41%)	450	400	350
Alfalfa Meal (17%)	60	60	100
Corn Gluten Meal	40	40	—
Butyl Fermentation Product (2)	40	40	40
Bone Meal	—	—	40
Salt	10	10	10
Limestone	—	—	40
Unextracted Liver Meal (3)	20	—	20
Nicotinic Acid	45.4gm.	45.4 gm.	45.4gm
Choline Chloride (100%)	0.7	0.7	0.7
Fortified A & D Oil (400D - 2000A)	6.0	6.0	10.0
Dry Vitamin D	.6	.6	—
Manganese Sulfate	.50	.50	.50
Total	2017.8	2007.8	2011.2

(1) To be fed with hard grain.
(2) Use 2% of a butyl fermentation or 3% corn distillers dried solubles made from a base containing 90% corn and 10% barley malt. If the corn distillers dried solubles are used they must be fortified with riboflavin so that the finished ration contains not less than 2.0 mg. of riboflavin per pound for the starter and grower and 1.8 mg. for the breeder.
(3) Three per cent Fish Solubles (50% solids) may be used to replace the liver.

TABLE 105—CALCULATED ANALYSIS OF TURKEY RATIONS

		Starter	Grower	Breeder
Protein	%	26.22	22.54	21.43
Fat	%	5.46	5.01	3.45
Fiber	%	4.39	4.38	5.53
Calcium	%	1.73	1.30	2.49
Total Phosphorus	%	1.14	.92	1.15
Inorganic Phosphorus	%	.80	.58	.76
Managese	mgs/lb.	33.65	32.91	40.33
Vitamin A	I.U./lb.	5755	6109	312
Vitamin D	A.O.A.C. units/lb.	815	815	1008
Riboflavin	mgs./lb.	2.29	1.97	8 2.15
Pantothenic Acid	mgs./lb.	10.36	9.79	17.54

that are vigorous and active; free from deformities; and approac in type and conformation the standard for the breed. Such youn vigorous breeders are relied upon by most producers. On th other hand, old breeders are utilized to a considerable degre where pedigreeing is practiced. In pedigree matings normally or male is mated with ten to fifteen females. In flock matings th ratio will be nearer one to twenty. Pedigree breeding requires the utilization of trapnests so that identity may be established. Detail records will need to be maintained as discussed in Chapter XIII. In any breeding operation it will be found advantageous to apply saddles to all females to prevent injury by the males. Toe-nails of the toms may also be clipped.

Fig. 244—Farmers' Bulletin 1409, turkey trapnests open at top for removing female.

[Courtesy U. S. D. A.]

Improvement is to be obtained through progressive breeding practices and attention must be so directed in order that a larger percentage of birds will meet exacting requirements.

Efforts towards the development of a suitable breeding program for the benefit of the industry culminated in the adoption of the National Turkey Improvement Plan in 1943. (Refer to Chapter XIV).

Marketing

Turkeys are marketed in the same channels as live and dressed chickens (Chapter XV). Practice varies in different sections, as illustrated by the fact that in sections of dense population direct sales to the consumer of dressed and drawn birds has

Fig. 245—Killing, semi-scalding, and picking turkeys.
[Courtesy Poultry Dept., N. H. Univ.]

been popular since it provided a premium market. Regardless of market channel used, it is desirable that only finished birds of matured plumage and flesh be disposed of. Birds without substantial subcutaneous fat and with pin-feathery plumage do not present an attractive carcass or have sales approval. The industry needs to cater to the consumers in order to further develop a year 'round outlet for turkey meat.

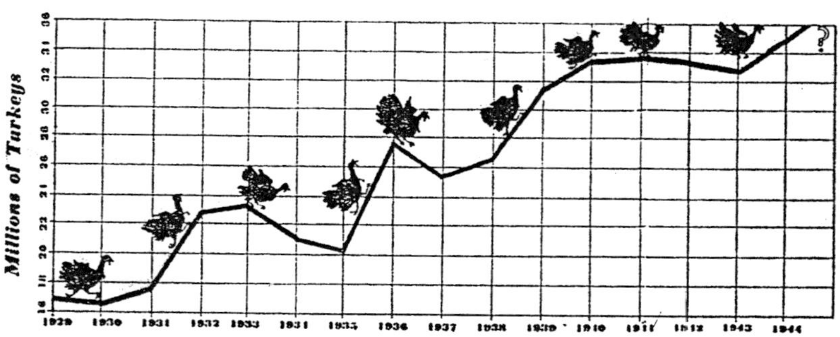

Fig. 246—Trends in turkey production—1929-1945.
[Courtesy Turkey World.]

By far the largest percentage (possibly as much as 80 per cent) of turkeys are marketed in the last three months of the calendar year. Americans associate turkey with the holiday season. With the enormous increase in production, as shown in an accompanying chart, greater effort must be directed towards cultivating a demand over a longer period of time.

Trends in Turkey Production—1929-1945

In catering to the public, consumer preferences must be given proper consideration. A survey in 1936 by the National Association of Food Chains showed that 73 per cent of the consumer demand was for turkeys ranging in weight from eight to thirteen pounds. Large sized turkeys, while not desired by the individual housewife, can be most advantageously utilized in the hotel and restaurant trade. The turkey producers will need to

TABLE 106—COSTS OF RAISING TURKEYS, 30 FARMS, 1938, AND 32 FARMS, 1943*

	1938			1943		
	Cost of raising 40,764 young turkeys			Cost of raising 58,880 young turkeys		
	Per bird	Per cent	Per cent	Per bird	Per cent	Per cent
Costs:						
Feed	$1.64	57.7	47.6	$3.24	60.6	51.3
Litter	0.01	0.5	0.4	0.04	0.7	0.6
Human labor	0.69	24.3	20.1	1.21	22.7	19.2
Horse labor	0.01	0.4	0.3	0.00	0.1	0.1
Miscellaneous	0.11	4.0	3.3	0.25	4.6	3.9
Interest on above costs	0.04	1.3	1.1	0.07	1.3	1.1
Use of tractor	0.01	0.2	0.1	0.03	0.6	0.5
Use of automobile and truck	0.13	4.7	3.9	0.16	2.9	2.4
Use of land	0.04	1.3	1.0	0.05	0.9	0.8
Use of buildings	0.10	3.3	2.8	0.19	3.5	3.0
Use of equipment	0.05	1.7	1.4	0.08	1.6	1.3
Interest on stock	0.02	0.6	0.5	0.03	0.5	0.5
Total	$2.85	100.0	82.5	$5.35	100.0	84.7
Poults	0.60	—	17.5	0.97	—	15.3
Total	$3.45		100.0	$6.32		100.0
Credits:						
Manure	$0.01			0.02		
Feathers	—			0.03		
Miscellaneous	0.02			0.00		
Total	$0.03			$0.05		
Net costs	$3.42			$6.27		
Return per bird	$4.58			$6.25*		
Return per pound of turkey produced:						
Live weight	$0.287			$0.377*		
Dressed weight	$0.326			$0.428*		
Cost of production per pound:						
Live weight	$0.214			$0.376		
Dressed weight	$0.243			$0.428		
Average number of turkeys reared per farm	1,359			1,840		
Average live weight per bird, pounds	16.0			16.7		
Average dressed weight, per bird, pounds	14.1			14.7		
Return for labor per pound of turkey produced, live weight	$0.116			$0.073*		

*Computed at O.P.A. ceiling prices.

[Courtesy E. G. Misner, N. Y. State Coll. of Agri.]

consider quality and market demand in competing against other producers and other meats.

In the practice of killing and dressing, the same general procedures are used as for domestic fowl. (Chapter XV). Shrinkage amounts to approximately 10 per cent where birds are dressed or 25 to 30 per cent where full drawn. Some variation is to be found depending on size of birds, smaller birds losing a greater percentage in full drawing.

TENTATIVE U. S. STANDARDS FOR CLASSES AND GRADES FOR DRESSED TURKEYS

Tentative U. S. Classes for Dressed Turkeys

According to Age and Sex:

Young Hens—Female birds usually less than one year old, soft meated with flexible breastbone.

Young Toms—Male birds usually less than one year old, soft meated with flexible breastbone.

Old Hens—Mature female birds more than one year old, with toughened flesh and hardened breastbone.

Old Toms—Mature male birds more than one year old, with toughened flesh and hardened breastbone.

According to Method of Plucking, Dressing Finishing, Chilling and Packing:

PLUCKING:

Scalded—Turkeys that have been immersed in scalding water before plucking, usually at a temperature of 170° to 180° Fahrenheit.

Semi-scalded—Turkeys that have been immersed in hot water before plucking, usually at a temperature of 125° to 130° Fahrenheit.

Dry-plucked—Turkeys that have been plucked dry; that is, without immersing in water.

DRESSING:

Drawn—Turkeys from which the entrails are removed.

Undrawn—Turkeys from which the entrails are not removed.

FINISHING:

Milk-fed—Turkeys with skin and flesh bleached by feeding milk in the ration and with muscle fiber softened by fatty deposits throughout the connective tissue.

SPECIFICATIONS FOR TENTATIVE U. S. STANDARDS FOR GRADES FOR DRESSED YOUNG HEN TURKEYS

U. S. Grades	Quality Specifications for Individual Birds
U. S. Grade AA	Young, fine grained, soft meated female bird, with broad full-fleshed breast extending well back to the rear of the bird, with the breast flat across the top rather than pointed. The entire carcass fully covered with fat. Must be fully bled, extra well dressed and free of pinfeathers. No flesh bruises allowed; and only very slight skin abrasions, bruises or discolorations permitted, none of which shall be on the breast. Slightly dented breastbone (not to exceed ⅛ inch in depth) permitted, but no crooked breast or other deformities allowed. No open tears or sewn skin permitted. A broken or disjointed wing, or a broken or disjointed leg not permitted. Must be dry picked or semi-scalded and must be dry packed.
U. S. Grade A	Young, soft meated female bird, with well-fleshed breast, and with entire carcass well covered with fat. Must be well bled, well dressed, and practically free of pinfeathers, especially on the breast. Only slight flesh or skin bruises, abrasions, or discolorations permitted, with breast practically free of such defects. Slightly dented breastbone (not to exceed ¼ inch) permitted. Slightly curved, but no crooked breast permitted. Must be free from deformities. Broken wings above the wing tips or broken legs not permitted. A disjointed leg or wing permitted if only slightly bruised. Bird with crop properly washed out or properly removed may be included in this grade. No torn skin permitted. No sewn skin permitted on the breast or fleshy part of the carcass, and only slight tears, sewn, permitted on the back. Must be dry picked or semi-scalded.
U. S. Grade B	Young female bird, with fairly well-fleshed breast, and with carcass fairly well covered with fat. Must be fairly well bled and dressed and may show scattered pinfeathers over the entire carcass. Slight flesh or skin bruises permitted, but not more than three such defects if on the breast. Fairly numerous skin abrasions or discolorations permitted. Abrasions or tears over 3 inches in length, on the fleshy parts of the carcass, not permitted unless properly sewn. Slight open tears less than 3 inches in length may be permitted if on the back or over the back of wings. Dented or slightly crooked breastbone or other slight deformities permitted. One broken wing or one broken leg permitted if bone does not protrude through the flesh and if not showing excessive bruise or blood clot.
U. S. Grade C	Young female bird which may be poorly fleshed and with carcass poorly covered with fat. May show evidence of poor bleeding and have numerous pinfeathers over the entire carcass. Numerous skin abrasions and discolorations permitted. Hunchback or other deformities allowed, if bird is fairly well fleshed. Open tears permitted in skin. Broken bones or bird badly bruised so as to make any appreciable part of the carcass inedible not permitted. Bird showing emaciation or external evidence of disease or other conditions which renders it unwholesome or unfit for human food not permitted.

Grain-fed—Turkeys that show no marked evidence of milk feeding.

CHILLING:

Fresh dressed—Turkeys that have not been hard chilled or frozen.

Fresh hard-chilled—Fresh dressed turkeys that have been hard-chilled or frozen, but have not developed any appearance of cold storage stock, and show no evidence of deterioration from freezing and have not been held at low temperature for more than 60 days.

Storage—Turkeys that have been held at a low temperature for more than 60 days or show evidence of deterioration from freezing, regardless of time held.

PACKING:

Dry—Turkeys that have been packed dry.

Iced—Turkeys that have been packed in ice.

Tentative U. S. Grades for Dressed Turkeys

U. S. Grade AA—Commercially perfect specimens of any class.

U. S. Grade A—The second highest grade.

U. S. Grade B—The third highest grade.

U. S. Grade C—Edible birds below the grade of U. S. Grade B, except such as are specifically excluded by the detailed specifications for U. S. Grade C.

Specifications for Tentative U. S. Standards for Grades for Dressed Young Tom, Old Hen and Old Tom Turkeys

Young Tom:

U. S. Grade AA—Young male bird with quality specifications the same as for U. S. Grade AA Young Hen with due allowance made for fleshing conditions characteristic of its sex. No staggy condition permitted.

U. S. Grade A—Young male bird with quality specifications the same as for U. S. Grade A Young Hen, with due allowance made for fleshing conditions characteristic of its sex. Slight staggy condition permitted, if bird is soft meated.

U. S. Grade B—Young male bird with quality specifications the same as for U. S. Grade B Young Hen, with due allowance made for fleshing conditions characteristic of its sex.

U. S. Grade C—Young male bird with quality specifications the same as for U. S. Grade C Young Hen.

Old Hen:

U. S. Grade AA—Mature female bird with quality specifications the same as for U. S. Grade AA Young Hen, with due allowance made for fleshing conditions characteristic of its age.

U. S. Grade A—Mature female bird with quality specifications the same as for U. S. Grade A Young Hen, with due allowance made for fleshing conditions characteristic of its age.

U. S. Grade B—Mature female bird with quality specifications the same as for U. S. Grade B Young Hen, with due allowance made for fleshing conditions characteristic of its age.

U. S. Grade C—Mature female bird with quality specifications the same as for U. S. Grade C Young Hen.

Old Tom:

U. S. Grade AA—Mature male birds with quality specifications the same as for U. S. Grade AA Young Tom, with due allowance made for fleshing conditions characteristic of its age.

U. S. Grade A—Mature male bird with quality specifications the same as for U. S. Grade A Young Tom, with due allowance made for fleshing conditions characteristic of its age.

U. S. Grade B—Mature male bird with quality specifications the same as for U. S. Grade B Young Tom, with due allowance made for fleshing conditions characteristic of its age.

U. S. Grade C—Mature male bird with quality specifications the same as for U. S. Grade C Young Tom.

Description of Terms Used in the "Quality Specifications for Individual Birds"

Full-fleshed—Bird with plump, broad breast which carries well back to the rear of bird and with breast flat across the top rather than pointed. Thighs, back, and all bones covered with a thick layer of flesh.

Well-fleshed—Bird with well developed breast; well covered with flesh over the thighs and back.

Poorly fleshed—Bird with narrow breast and with thin covering of flesh over thighs and back.

Poorly-bled—Bird with red pin marks on breast, thighs, or wing tips, or skin reddened from blood clots.

Slight deformities—Dented or notched breast, crooked back, curved breastbone, or misshapen legs or wings.

Deformities—Hunch back, crooked breast, or other serious abnormal conditions.

Slightly curved—Breastbone slightly out of line but not crooked.

Callouses—Hardened and darkened areas on the breastbone usually caused by roosting.

Skin abrasions—Reddened, bruised, or scuffed outer skin caused by rubbing or rough handling in dressing, or packing, or otherwise.

Pigmented skin—Skin of a bluish green color usually found over the tail and back, caused by secretion of feather pigment under the skin.

Skin bruise—A bruise of the skin sufficient to form a blood clot under it but not extending into the flesh.

Flesh bruise—A bruise in the flesh which shows through the skin but does not move with the skin.

Free of pinfeathers—No noticeable pinfeathers.

Slight dressing defects—Few scattered pinfeathers on breast or back; blood spots in wing tips; one slight skin abrasion less than three inches in diameter; or two broken wing tips.

Dressing defects—Pinfeathers scattered more or less over the entire carcass; incomplete bleeding showing in places other than the wing tips, a cut or tear three inches or more in length or more than one cut of any length; skin abrasion of more than three inches in diameter; broken leg, wing or other bone in edible parts of carcass; feed in crop; dirty feet, body or vent.

Extra well dressed—Commercially perfect dressing.

Well dressed—All conditions listed as "dressing defects" excluded.

Uniform pack and size—Light-colored and yellow-skinned birds packaged separately; packages and paper linings uniform in size, shape, and color not over two pounds variation in the weight of individual birds packed in the same box or package except in case of U. S. Grade C and Old Toms.

Edible bird—Bird that shows no external evidence of disease or other condition that would render it unwholesome or unfit for human food.

Slight staggy condition—Spurs somewhat developed but soft; slightly pouchy breast but still soft meated.

Staggy condition—Extended breast, pouchy or milky, with toughened flesh.

Appreciable part of carcass—An amount equal to ¼ of entire carcass.

Weight Specifications for Dressed Turkeys

With the exception of Old Toms, and U. S. Grade C birds, turkeys must be packed with only a 2-pound variation per bird per box or package beginning with 8 to 10 pounds; 10 to 12 pounds; 12 to 14 pounds; 14 to 16 pounds, etc.

U. S. Grade C turkeys must be packed with only a 3-pound variation per bird per box or package.

Old Toms must be packed with only a 5-pound variation per bird per box or package.

Tentative U. S. Wholesale Grades for Dressed Turkeys

U. S. Grade AA—A lot of dressed turkeys consisting of three or more wholesale packages shall contain not less than 90 per cent birds of the quality of U. S. Grade AA, the balance to be of the quality of U. S. Grade A and with no individual package in the lot to contain more U. S. Grade A birds than in the proportion of two to each twelve. In a lot consisting of one or two wholesale packages, the tolerance for individual packages mentioned above may apply to each package.

U. S. Grade A—A lot of dressed turkeys consisting of three or more wholesale packages shall contain not less than 90 per cent birds of the quality of U. S. Grade A or better, the balance to be of the quality of U. S. Grade C and with no individual package in the lot to contain more U. S. Grade B birds than in the proportion of two to each twelve. In a lot consisting of one or two wholesale packages, the tolerance for individual packages mentioned may apply to each package.

U. S. Grade B—A lot of dressed turkeys consisting of three or more wholesale packages shall contain not less than 90 per cent birds of the quality of U. S. Grade B or better, the balance to be of the quality of U. S. Grade C and with no individual package in the lot to contain more U. S. Grade C birds than in the proportion of two to each twelve. In a lot consisting of one or two wholesale packages, the tolerance for individual packages mentioned above may apply to each package.

U. S. Grade C—All birds shall be of the quality of U. S. Grade C or better. No birds obviously unfit for human food permitted.

General Grading Requirements

1. Any turkey that has feed in its crop will be considered as below grade. Turkeys with crop properly washed out or with crop properly removed (the entire crop removed through a short incision of the skin at the back and side of the neck), either sewn or unsewn, may be included in the grades of U. S. Grade A, U. S. Grade B and U. S. Grade C, provided they otherwise meet the requirements of those grades and provided that there is no appreciable leakage of feed materials into the crop cavity.

2. All feathers, including the neck feathers above the head wrap, garters around the knees and fan feathers on the wing tips, must be removed from the carcass or the bird will be considered as below grade, except that turkeys with fan feathers on the wing tips may be included in the grades of U. S. Grade A, U. S. Grade B and U. S. Grade C, provided they otherwise meet the requirements of those grades and provided further that the fan feathers are free of blood or other foreign material.

3. Unsewn torn skin is not permitted in Grade AA, nor in U. S. Grade A. Unsewn torn skin is permitted in U. S. Grade B only, if on the back, or back of wings and if less than three inches in length. Open tears are permitted in U. S. Grade C. Turkeys with unsewn torn skin, except as such tears are permitted in U. S. Grade B and U. S. Grade C, must be lowered one grade.

4. Turkeys with dirty feet are considered below grade.

5. Turkeys with dirty vents are considered below grade.

6. Turkeys with dirty or bloody heads or bodies are considered below grade.

7. No skin pigmentation is permitted in U. S. Grade AA. It is permitted only on the tail of a U. S. Grade A bird, and no further than over the hips on a U. S. Grade B bird.

8. Turkeys showing callouses on the breast bone are not permitted in U. S. Grade AA. If of a slight character and not more than one inch in length, they may be permitted in U. S. Grade A. If not over three inches in length and not too dark, they may be permitted in U. S. Grade B.

9. Turkeys with improper head wraps or head wrapping may be graded only when the grader is certain this condition will be corrected before the birds are packed.

Ice packed dressed turkeys without head wraps may be included in the grades of U. S. Grade A, U. S. Grade B, and U. S. Grade C, provided they otherwise meet the requirements of those grades and provided that the heads have been carefully and thoroughly cleaned of blood, so that there is no appreciable soiling of dressed carcasses with which the heads may come in contact.

10. An individual turkey may be marked with a tag showing the proper U. S. Grade only when it has been graded by a grader authorized by the U. S. Department of Agriculture.

11. Individual turkeys may be marked with U. S. Grade only when grade tags or labels approved by the U. S. Department of Agriculture are used and are attached to the birds in an approved manner.

12. When an individual turkey is tagged to indicate its U. S. Grade, the grade tag also must state the age; that is, whether the bird is "young" or "old."

13. A package in which officially graded birds are packed may be stamped with the proper U. S. Grade whether or not the individual birds in the package are labeled with a U. S. Grade.

14. All turkeys packed in the same container must be of the same class, grade, color, and within the same weight class except that birds of different classes, grades or weight classes may be packed together provided the number of birds of each class and grade is plainly marked on the outside of the package.

15. Turkeys shall be labeled with U. S. Grade tags only by the authorized grader who graded them or by persons under his direct supervision.

16. Each Government authorized grader must have and use a poultry thermometer for taking internal temperatures.

17. Turkeys shall not be graded and marked or tagged with a U. S. Grade until they have been pre-cooled to an internal temperature of at least 36° F., unless the grader is sure they will be so cooled before packing.

18. A grading memorandum should be made out at each shipping point, regardless of the size of the lot, and a grading certificate issued on each carlot or less-carlot shipment.

19. Where turkeys which have been previously graded and regarded and show discoloration of skin, darkening of flesh and development of "off" condition of skin, due to holding in storage or during transit or warehousing, these characteristics shall be

considered as factors of condition and not of grade. They shall not be considered in determining the grade upon regrading, but such condition factors must be noted on the grading certificate.

20. Turkeys that are dark or blistered or skin dried from freezing cannot be graded into the two top grades.

21. All turkeys must be free of any condition as evidenced by external appearance which would render them unwholesome for food.

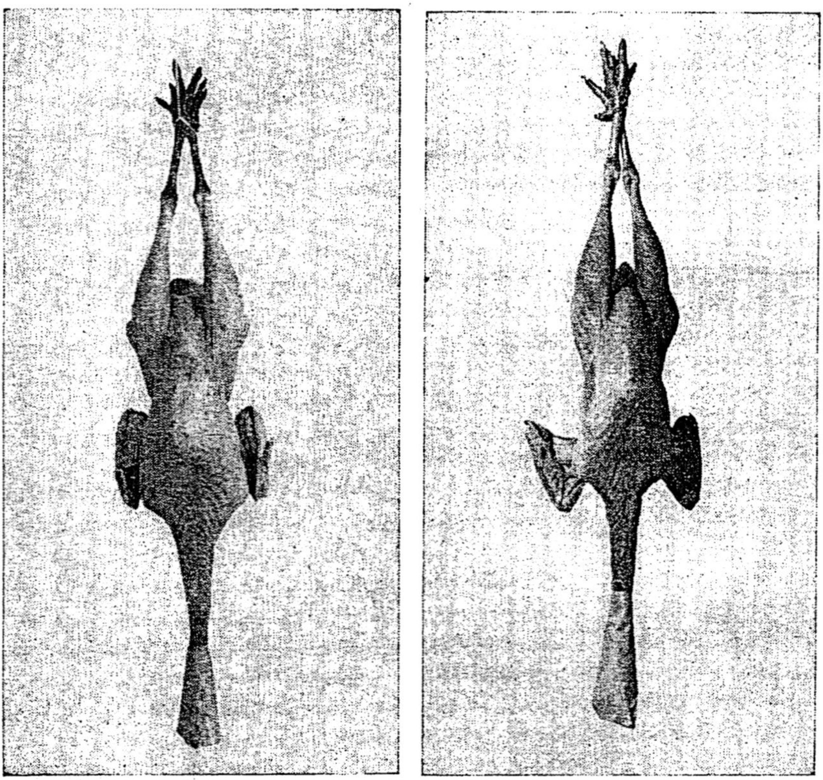

U. S. Grade AA young hen turkey. U. S. Grade A young hen turkey.

Fig. 247—(Left) U. S. Grade AA, young turkey. The breast is broad and full, the hips are well-rounded and the entire carcass is fully covered with fat. (Right) U. S. Grade A, young hen turkey. This turkey is well-fleshed but somewhat narrow over the hips. The carcass is well covered with fat.

Recommendations for Standardizing the Dressing and Packing of Dressed Turkeys

1. All turkeys should be box packed in single layer boxes.

2. Packages should be lined with good quality 30-pound waxed or parchment paper.

3. Packages should be of uniform size, color and markings.

4. Packages should be clean, odorless and of sufficient strength to carry the bird in good condition.

5. The net weight should be marked on the box, also the number of birds.

U. S. Grade B young hen turkey.

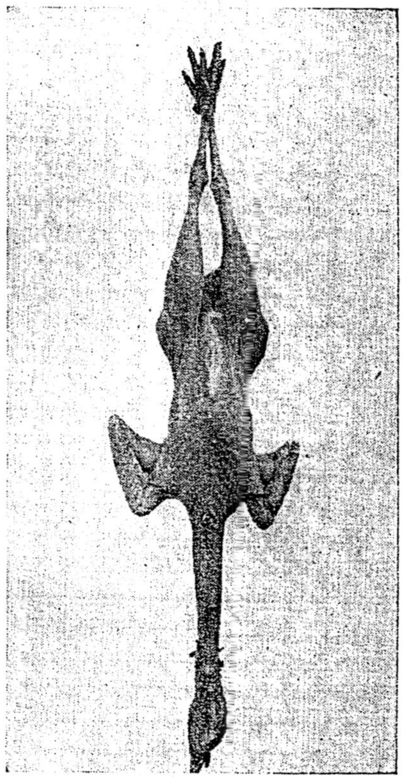

U. S. Grade C young hen turkey.

Fig. 248—(Left) U. S. Grade B, young hen turkey. The carcass is fairly well-fleshed with a slight covering of fat over the sides and back. (Right) U. S. Grade C, young hen turkey. This turkey is poorly fleshed but is edible so far as can be judged by external appearance. Scarcely any fat shows through the skin covering the carcass.
*U.S.D.A. Farmers' Bulletin 1815.

6. Barrel packed turkeys should show the number of birds, the class and the gross, tare and net weights.

7. The U. S. Grade should be stenciled plainly on the end of each box or in a prominent place on other packages of officially graded turkeys.

8. Any U. S. Grade name may be supplemented by such breed, variety or other descriptive terms as may more completely describe the lot of turkeys, as U. S. Grade A "milk fed" turkeys.

9. Private brands and class numbers may be stenciled or stamped on packages of officially graded turkeys in addition to the U. S. Grade.

10. Stencil marks on boxes should be at least ¾ inches high.

11. All heads should be wrapped in good quality 60-pound water finish fiber or brown waxed paper.

12. All turkeys should be bled through the mouth.

13. All turkeys should be packed rump lock side pack.

14. Turkeys showing pigmented skin of a bluish green character over the back should be graded according to these specifications and should be packed in separate containers showing their proper grade and condition.

Questions on Chapter XXII

1. What are the most important breeds?
2. How does turkey incubation differ from hatching chickens?
3. How many poults should be brooded as one unit?
4. What is the relation of size of flock to (a) feed consumption, (b) labor requirements, and (c) egg production?
5. What is meant by a grow-healthy-poult program?
6. Outline a method for selecting turkey breeders.
7. What is the comparative cost of producing a turkey in 1938 as compared to 1943?
8. What are the tentative classes for dressed turkeys? Grades?
9. What are the quality specifications for a dressed young hen U. S. Grade AA?

INDEX